S0-AKY-790

dictionary of the environmental sciences

compiled by
Robert W. Durrenberger
Arizona State University

NATIONAL PRESS BOOKS

Acknowledgement is made for permission to reprint the following illustrations:

Figures from pages 6, 8, 10, 12, 44, 50, 56, 80, 92, 116, 124, 172, 196, 224, 225, 238, 262, 268, 298, 304, 312, 319, 414, 430, 480, 558, 676 and 686 of *Geomorphology* by A. K. Lobeck. Copyright 1939 by McGraw-Hill, Inc. Used with permission of McGraw-Hill Book Company.

Figures from pages 8, 13, 20, 25, 26, 36, 41, 62, 63, 81, 143, 145, 149 and 165 of *Mapping* by David Greenhood, published by The University of Chicago Press. Copyright 1964 by The University of Chicago.

Figures 3-7, 4-16, 6-13 and 10-10 from the Teacher's Guide to *Investigating the Earth* by the Earth Science Curriculum Project. Copyright 1967 by the American Geological Institute and published by Houghton Mifflin Company.

First edition 1973

Library of Congress Catalog Card Number: 78-142370

International Standard Book Number: 0-87484-150-X

Manufactured in the United States of America

Design by Nancy Sears

Layout by Mary Wiley

Cover by Ireta Cooper

NATIONAL PRESS BOOKS, 850 Hansen Way, Palo Alto, California 94304

preface

Current interest in the environment has led to the development of a number of new courses and curricula in the colleges and universities in this country and abroad. Multidisciplinary teams are at work to solve problems associated with various types of pollution and with the management of our natural resources. Terminology and concepts drawn from a broad spectrum of academic disciplines are now frequently used by those concerned with the environment.

One difficulty for these individuals is in communicating their ideas to each other and to their students, for certain words mean different things to individuals trained in different academic fields. This *Dictionary* was prepared to help people concerned with the environment, whether professionals or novices, to bridge the gap between and among the various environmental disciplines so that they can convey their thoughts to one another with clarity and understanding.

The *Dictionary* was conceived and developed in its early stages by my son Daniel and my daughter Mary Ann who labored through one hot summer in Illinois to prepare an initial word list and a set of definitions. Since then, as I have encountered new terms in the daily press, in scientific journals, and in scholarly books on environmental topics, I have added to the list. Although the number of entries is extensive, it is not exhaustive: new terms constantly enter our vocabulary or old terms take on new definitions to fit new needs.

A broad range of academic disciplines will find the *Dictionary* useful, including economics, engineering, geology, geography, anthropology, architecture, botany, zoology, agriculture, and many more. Students from junior high school through college who are enrolled in courses on the environment or in earth sciences will find the book particularly helpful because of the way in which many of the concepts are illustrated. In addition, Appendix I is a Geologic Time Scale and Appendix II consists of tables of equivalents and conversions—for length, area, volume, weight, pressure and flow—that will be particularly useful to students enrolled in introductory courses.

Special recognition should go to Don Ryan who prepared the original sketches. And to all the students in my classes and to many other individuals who have assisted in the preparation of this book, I wish to express my appreciation.

A horizon The surface horizon of soil having maximum biological activity or eluviation or both. See: horizon

aa A type of basaltic lava (Hawaiian).

ABC soil A soil with a complete profile, including an A, a B, and a C horizon. See: horizon

abiotic (adj.) Characterized by a lack of living organisms.

ablation The process by which ice and snow waste away due to melting and evaporation.

abnormal (adj.) Deviating from the normal. In meteorology, applied to values of the weather elements, such as temperature, that deviate from the normal values by amounts so much greater than usual that they deserve comment.

abnormality The state of the atmosphere when abnormal values of any element occur; e.g., conditions during a January thaw in southern Canada.

aborigine The original inhabitant of an area.

Abraham's tree The popular name for a cloud form of long feathers and plumes of cirrus, which seem, by an effect of perspective, to radiate from a single point on the horizon. In popular tradition, it is a sign of rain if the base of this cloud appears to touch a sheet of water.

abrasion The removal of bedrock material by the grinding action of other material being moved by wind, water, or ice.

abrolhos A squall occurring on the coast of Brazil between May and August.

abscissa The x-coordinate, representing the distance of a point from the y-axis measured parallel to the x-axis. See: Cartesian coordinates

abscission The natural separation of parts of a plant (such as flowers, bark, fruit, leaves, or branches) by the breakdown of the absciss layer.

absciss layer In plants, a layer of cells, across the base of a branch or embedded in the bark, through which the leaf or branch or other part breaks off.

absolute (adj.) 1. In climatology, the highest and lowest values of any given meteorological element during a particular period at the place of observation. 2. The difference between the absolute extremes defined above.

absolute extremes See: absolute

absolute humidity The mass of water vapor present per unit volume of space, i.e., the density of water vapor; usually expressed in grams per cubic meter or grains per cubic foot.

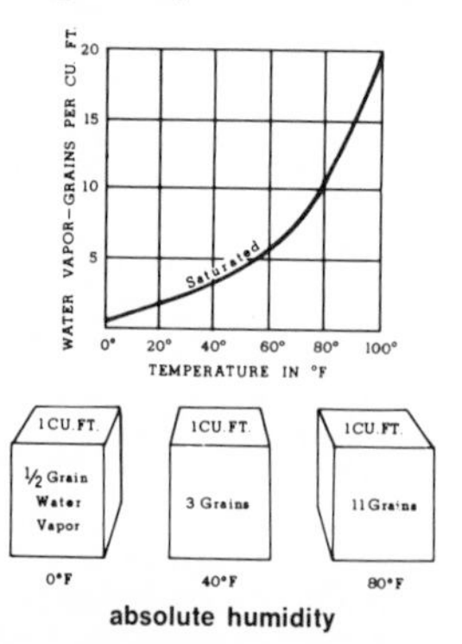

absolute humidity

absolute instability The state of an air mass with a lapse rate greater than the dry adiabatic, so that a parcel of air within it displaced vertically would continue to move vertically.

absolute maximum See: absolute

absolute minimum See: absolute

absolute stability The condition in an air mass with a lapse rate less than the moist adiabatic, which makes it tend to resist displacement.

absolute temperature Temperature measured from a zero point corresponding to the absence of molecular motion on the Kelvin scale.

absolute temperature scale See: Kelvin scale of temperature

absolute zero That point on the Kelvin scale of temperature at which the random molecular motions of an ideal gas become zero, i.e., at which the molecules are at rest. This occurs at about –273°C.

absorption 1. In meteorology, the depletion that radiant energy undergoes in traversing the earth's atmosphere (either from sun to earth or from earth to space), and also in penetrating the oceans or other bodies of water and the solid earth. 2. In physics the holding by capillary action of a liquid in the pores of a solid, as water is held by a sponge or by soil. 3. In chemistry, the holding by chemical attraction of the molecules of a gas or liquid between the molecules of a liquid or solid.

absorption bands Dark bands (aggregates of lines) in the spectrum of radiant energy resulting from the absorption of particular wave lengths of the radiation by the constituents of the medium through which it passes. The amount of absorption varies with rays of different wave length or refrangibility, and is in general highly selective, producing lines or bands according as the absorption is by atoms or molecules. See: absorption; radiation; selective absorption

absorption spectrum The spectrum that results after any radiation has passed through an absorbing substance.

absorptivity The ratio of the radiant energy absorbed during transmission through a medium to the incident amount, which ratio varies from unity for a black body to zero for a perfectly transparent body.

abyssal (adj.) Pertaining to great depths. Usually refers to those parts of the ocean more than 6000 feet below the surface; sometimes refers to great depths within the solid earth.

abyssalbenthic zone The ocean bottom of the deep sea zone.

abyssalpelagic zone The waters above the bottom layers of the deep sea zone.

acacia A woody tree or shrub of the mimosa family found throughout the tropics and subtropics; some yield valuable gums and tanning extracts.

acceleration The time rate of change of velocity, either linear or angular; it may be either positive or negative.

access road A road built into isolated stands of commercial timber so they can be reached by loggers, fire fighters, and others.

acclimate (v.) To become naturally accustomed to a different climate; not synonymous with acclimatize.

acclimation The process of becoming acclimated; not synonymous with acclimatization.

acclimatization The adaptation of animals or plants to a different climate, through a process in which nature is assisted by the intervention of man.

acclimatize To bring about the adaptation of animals or plants to a different climate, in part by the intervention of man.

accretion The process by which bodies grow larger by the addition of particles to the outside.

accretion theory A theory of the origin of the solar system through the growth of planets from a disk-shaped mass of gas and space dust.

acculturation The process by which the cultural traits of one group of people are acquired by another.

accumulated excess (or deficiency) A phrase used in the U. S. Weather Bureau to indicate the total excess or deficiency from the normal of either temperature or precipitation from any stated time, as the beginning of the month or of the calendar year to date. The accumulated excess (or deficiency) of temperature since January 1st of the current year to any date for any station is the difference between the sum of the daily means since January 1st and the sum of the daily normals. The average daily excess (or deficiency) of temperature is the result obtained by dividing the accumulated excess (or deficiency) by the number of days in the period considered.

accumulated (or cumulative) temperature An index to both the amount and the duration of an excess or deficiency of air temperature above or below an adopted standard, or base temperature; usually expressed by means of the total degree days for a given period. The standard of 42°F, adopted primarily for agricultural purposes is the temperature above which the sun's heat is said to be effectual in starting and maintaining plant growth and in completing the ripening of crops in a European climate. The summation of degree days may begin with the first of January, the first of December, or the date of sowing the seed, according to the special crop or plant.

acetone A colorless organic liquid that forms fragrant and inflammable vapor at room temperatures. It dissolves natural and synthetic resins, guncotton, and cellulose acetate, and is extensively used as a solvent in lacquers and as nail-polish remover.

acid A compound, capable of neutralizing an alkali containing hydrogen, that can be replaced by a metal or an electropositive group to form a salt, or containing an atom that can accept a pair of electrons from a base.

acidic (adj.) 1. Acid-forming. 2. Having a high hydrogen ion concentration (low pH). 3. In geology, any rock containing more than two-thirds silica is acidic.

acidity The state of being acid in reaction, which would be evident to the taste as sourness. The degree of acidity of a solution (or soil) may be expressed as pH (hydrogen-ion concentration).

acid soil Generally, a soil having a high concentration of acid throughout most or all of the parts that plant roots occupy. Commonly, only the surface-plowed layer or some other specific layer or horizon of a soil. Practically, this means a soil more acid than pH 6.6; precisely, a soil with a pH value less than 7.0.

acid wood Wood cut for use by manufacturers of charcoal, acetic acid, and methanol by destructive distillation; sometimes called distillation wood or chemical wood.

aclinic line See: magnetic equator

acoustic cloud A cloud occurring at a boundary between two air layers of sharply different densities (owing to differences in temperature or humidity, or both), so that sound waves are reflected from the interface.

acoustic pollution Sounds objectionable to man because of either their nature or their volume.

acoustics See: atmospheric acoustics

acquired characteristic A noninheritable characteristic acquired by an organism during its lifetime as a result of the use or disuse of an organ because of environmental influences.

acre Originally used to designate an enclosed or tilled field; now a measure of area equivalent to 43,560 square feet.

acre foot A unit for measuring the volume of water equal to the quantity of water required to cover one acre to a depth of one foot or to 43,560 cubic feet or to 325,851 gallons. The term is commonly used in measuring volumes of water used or stored.

actinic (adj.) 1. Applied to light capable of initiating or causing chemical changes, as in photography or the fading of colors. 2. Applied to wave lengths of light too small to affect our sense of sight, such as the ultraviolet rays.

actinomycetes A group of soil micro-organisms that produce an extensive thread-like network. They resemble soil molds in some respects but are more like bacteria in size.

activated (adj.) Rendered active, capable of reaction, or promoting reaction; descriptive of various unrelated chemical and physical conditions. Carbons and bleaching clays used for adsorbing color bodies and other substances from liquids are considered activated when they have been treated to increase their surface areas. It is said that a pound of activated carbon can have 1.5 million square feet of surface area.

activated sludge Material precipitated by removing organic material from sewage by saturating it with air and adding biologically active sludge.

active layer The layer of soil above the permafrost, which alternately thaws and freezes from summer to winter.

active storage capacity The total amount of usable reservoir capacity available for seasonal or cyclic water storage. It is gross reservoir capacity minus inactive storage capacity.

acyl The organic radical or group that remains intact when an organic acid forms an ester; more specifically, the group left by the removal of the hydroxyl group from an organic acid.

adaptation Any alteration, resulting from the processes of natural selection, by which the organism becomes better fitted to survive and multiply in its environment.

adaptive disease The physiologic changes experienced by an organism because of exposure to a new environment.

adaptive radiation The diversification of a group of organisms into subgroups because of the interaction of its genetic potentialities with the environments it encounters.

adiabat A line of constant potential temperature on an adiabatic chart or the path along which a thermodynamic change takes place in a system when there is no exchange of heat with the environment. See: adiabatic; adiabatic gradient (or lapse rate); adiabatic process; dry adiabat; wet adiabat

adiabatic (adj.) Applied to a thermodynamic process during which no heat is communicated to or withdrawn from the body or system concerned. In the atmosphere, adiabatic changes of temperature occur only in consequence of compression or expansion, accompanying an increase or decrease of atmospheric pressure. Thus, a descending body of air undergoes compression and adiabatic heating, while an ascending air parcel experiences expansion and adiabatic cooling. See: adiabatic gradient (or lapse rate); dynamic heating; dynamic cooling

adiabatic chart (or diagram) A thermodynamic diagram with pressure and temperature as co-ordinates, which shows the thermodynamic states of atmospheric air over a wide range of conditions and by which the changes

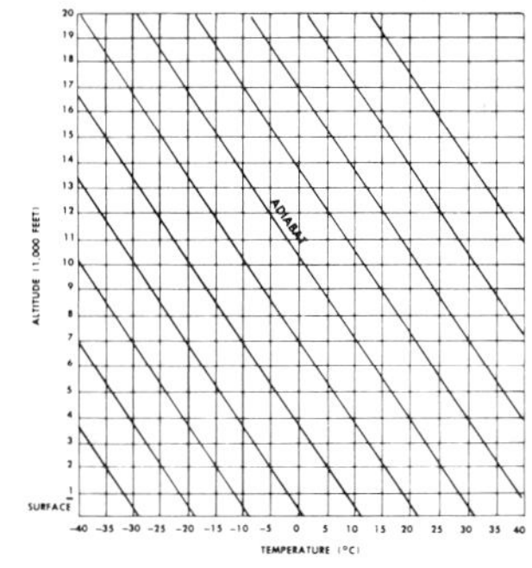

adiabatic chart

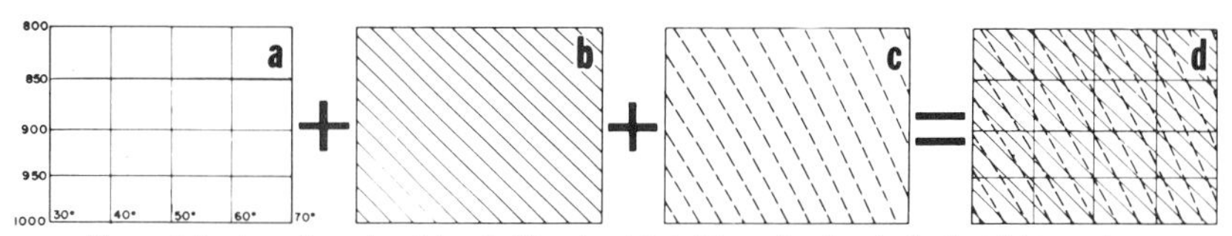

Lines relating to cooling rates of dry air (b) and moist air (c) are placed on basic chart (a) to produce adiabatic chart (d) which is used to determine degree of air's stability.

of state and the energy transformations during any prescribed process may be traced out. By plotting on this chart, the values of temperature, pressure, and specific humidity at various levels in the atmosphere may be determined and many weather phenomena may be inferred.

adiabatic compression See: adiabatic; dynamic heating

adiabatic cooling See: dynamic cooling

adiabatic expansion See: adiabatic; dynamic cooling

adiabatic gradient (or lapse rate) The rate of decrease of temperature of a sample of adiabatically ascending air. For dry (i.e., unsaturated) air it is approximately 1°C per 100 meters. After saturation is reached, the rate of cooling, now called the saturated-adiabatic lapse rate has no fixed value, but varies from 0.04°C to 1°C per 100 meters. See: adiabat; adiabatic; adiabatic process

adiabatic heating See: dynamic heating

adiabatic process A thermodynamic process in which no heat exchange occurs between the working system and its environment; closely approximated in the atmosphere by an element of air in a rapidly ascending or descending current.

adiabatic temperature change See: adiabatic process

adjuvant A material added to an insecticide to aid its action. Adjuvants may act in various ways—as emulsifying agents, wetting agents, spreaders, and stickers.

adobe Type of sun-dried clay or brick.

Adrenalin 1. A trademark name for a drug, also known as epinephrine. 2. a hormone produced in the medulla of the adrenal glands.

adreno-cortical stress The pressures on the adrenal cortex thought to be brought on by urban crowding.

adret The sunny side of a slope in the Alps also called the sonnenseite. See: ubac

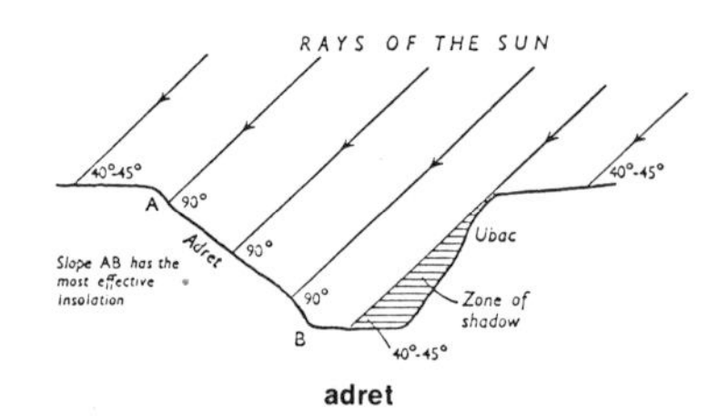

adret

adsorb (v.) To retain upon the surface through the force of adhesion some of the gas, liquid, or dissolved substance with which a solid body has been in contact.

adsorption The attachment of compounds or ionic parts of salts to a surface or another phase. Nutrients in solution (ions) carrying a positive charge become attached to (adsorbed by) negatively charged soil particles.

advection Horizontal flow of air at the surface or aloft; one of the means by which heat is transferred from one region of the earth to another.

advection fog A fog due to the transport of warm air over a cold surface,

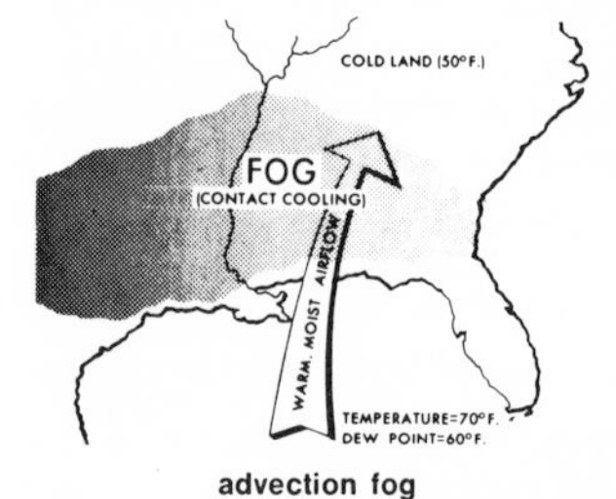

advection fog

either land or water, or to the transport of cold air over a warm-water surface. Since many fogs are caused by the joint action of two or more processes, fogs classified as of the advection type may have been formed in part in other ways.

advective (adj.) Pertaining to atmospheric phenomena or conditions in which advection is the dominating influence.

advective cooling Decrease of air temperature when cold air moves into an area previously occupied by warmer air. It may be sufficient to cause a temperature decrease on the earth's surface, even though the surface is absorbing solar radiation. Warm air advection over a colder surface will result in conductive cooling of the lower air layers.

advective thunderstorm A thunderstorm resulting from instability produced by horizontal advection of colder air at higher levels or by horizontal advection of warmer air at lower levels, or by a combination of both.

AEC The U.S. Atomic Energy Commission.

aeolian (adj.) Pertaining to the action or effect of the wind.

aeration The process whereby any substance becomes permeated with air.

aeration zone That portion of the lithosphere in which the functional interstices of permeable rock or earth are not ordinarily filled with water under hydrostatic pressure, or are filled with water that is held by capillarity.

aerobic (adj.) 1. Conditions of environment in which oxygen gas is present. 2. Living or acting only in the presence of air or free oxygen.

aerobiology The science that treats of the small organisms, both plant and animal, that are borne aloft by the air and investigates their behavior in the air and their effects on other organisms.

aerodynamic (adj.) Pertaining to the laws of motion of air or other gases.

aerodynamics A branch of physics defined broadly as the science of motion of air or an aeriform fluid. Commonly, air alone is implied in the word.

aeroembolism The formation of bubbles of nitrogen in the blood stream caused by a rapid decrease of pressure; symptoms are discomfort, pain in the joints, and loss of mental alertness.

aerography The science dealing especially with the description of the atmosphere as a whole and its various phenomena.

aerology A term formerly used synonymously with meteorology. The science of the atmosphere.

aeronomy The study of chemical and physical phenomena in the upper atmosphere.

aerosol 1. A dispersed system in which the dispersion medium is a gas. 2. An aggregation of dispersed particles suspended in the atmosphere. 3. The special type of colloid, formed by the liquid or solid particles, organic and inorganic, and the gases of the atmosphere in which these particles float.

aesthetic (adj.) Pertaining to a feeling or a sense of the beautiful.

aestivate See: estivate

afforestation The process of transforming an area into forest, usually when trees have not previously grown there.

aftercooling The cooling of a reactor after it has been shut down.

afterglow A broad high arch of radiance or glow seen occasionally in the western sky above the highest clouds in deepening twilight, caused by the scattering effect exerted upon the components of white light by very fine particles of dust suspended in the upper atmosphere.

aftershock An earthquake or earthquakes occurring soon after a larger earthquake.

aftersummer A recurrent mild period in autumn. See: Indian summer

agave A member of the amaryllis family of perennial lilylike plants with narrow, flat leaves, widely distributed in the lowlands of the tropics and subtropics. They provide soap, food, beverages, and fiber and are grown as ornamentals. The century plant, maguey, and sisal are agaves.

age Any of the great periods of time in the earth's history marked by special phases or events.

age dating See: radioactive age determination

age pyramid The graphic representing of age structure in a population; typically, the vertical dimension is graduated in groups of years, usually five, beginning with zero at the base; the horizontal axis shows the number or percentage of males to the left and females to the right of the central axis.

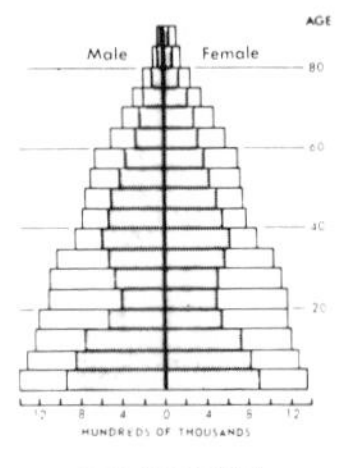

age pyramid

age ratio The ratio of the amount of daughter isotope in a given specimen to that of the parent isotope after a period of radioactive decay. Used to calculate the age of the specimen when the rate of decay of the given substance is known. See: radioactive dating; carbon 14.

age-specific birth rate Number of births for each age group in the population.

age-specific death rate Mortality rate for each age group in a population.

agglomerate 1. Rock composed of rounded or subangular volcanic fragments. 2. (v.) To gather into a cluster.

aggradation The process of building up an area by deposition.

aggrade (v.) To build up a surface by deposition.

aggregate (of soil) Many fine soil particles held in a single mass or cluster, such as a clod, crumb, block, or prism.

agonic line An imaginary line passing through points on the earth's surface at which the magnetic declination is zero, i.e., where the magnetic needle points to true north.

agricultural climatology The branch of climatology that deals with climate in its relation to agriculture.

agricultural revolution The change in agricultural production that involves: (a) farming of large-scale units of land; (b) intensive livestock production; (c) large increases in agricultural productivity due principally to larger inputs of capital in the form of equipment, fertilizer, and hybrid seeds.

agriculture The production of crops, livestock, or poultry.

agronomy The science of soil management and the production of field crops.

air The mixture of gases comprising the earth's atmosphere. See: atmosphere.

air drainage The flow of air down a slope or channel. Air pockets with greater density than surrounding air tend to flow downhill, due to gravity.

air mass A wide-spread body of air that approximates horizontal homogeneity: that is, its physical properties, level for level, are about the same over a wide area.

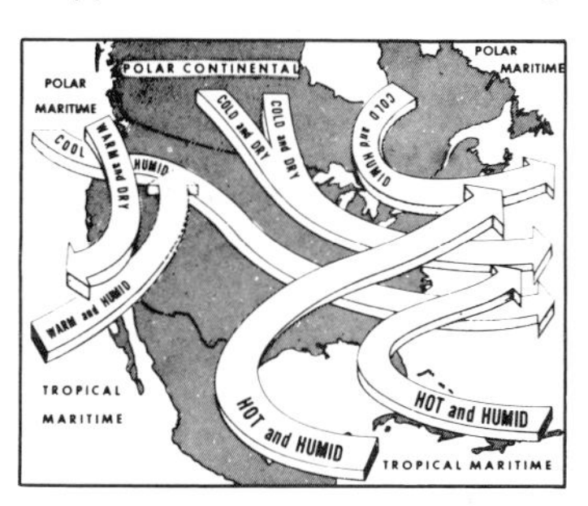

air mass

air-mass analysis The analysis of a synoptic weather map in respect to: (a) the extent and physical properties of each of the air masses over the region covered by the map, (b) the relations of the different air masses to each other, and (c) the location, structure, and movement of the fronts along which the different masses meet.

air-mass classifications The systems devised by meteorologists to characterize and identify different air masses.

air-mass climatology The study of climate from a dynamic and thermodynamic point of view, with particular reference to the air masses involved.

air-mass modifications The alterations in physical characteristics that occur in an air mass once it has left its source region.

air-mass thunderstorm A thunderstorm that occurs in an air mass whose chief characteristics are instability and high moisture content.

air pocket A downward current of air that causes an aircraft to drop abruptly in altitude.

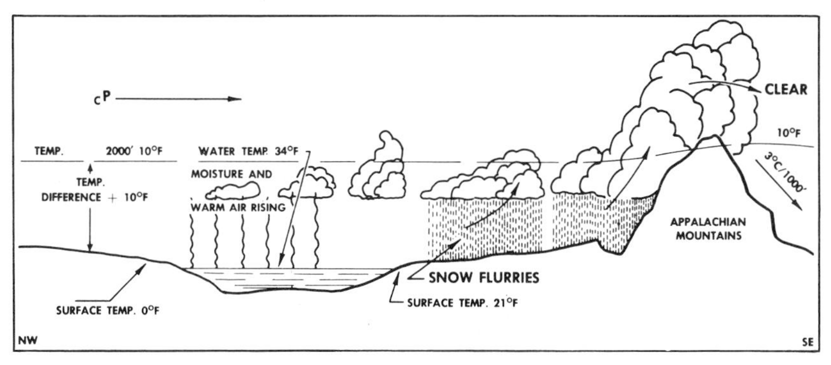

air-mass modifications

air pollution See: smog

air pollution control district Agency created by state and county legislation as early as 1947 to combat air pollution from all sources. The programs presently administered by these agencies control emissions from stationary sources only.

air pressure See: atmospheric pressure

air resistance The frictional resistance offered by air to the motion of bodies passing through it.

air shed A region with common sources and problems of air pollution; it may coincide with a drainage basin or be a part of a large urban agglomeration.

Aitken nuclei Microscopic solid and liquid particles in the atmosphere counted by an Aitken nucleus counter, that serve as condensation nuclei for droplet growth in clouds. They play a role in atmospheric electrical processes by capturing small ions to form large ions, thus bringing about a condition where the air conductivity is low.

Aitoff's projection

Aitoff's projection An equal-area map projection in which the whole of the earth's surface is represented on an ellipse.

albacore A saltwater fish belonging to the mackerel family.

albedo 1. The ratio of light reflected by a planet or satellite to that received on its whole illuminated hemisphere. 2. The reflectivity of one of the different materials forming the surface of the earth. See: reflectivity

albedo

albumin Water soluble proteins composed of nitrogen, carbon, hydrogen, oxygen, and sulfur, occurring in animal and vegetable juices and tissues.

alchemist A medieval chemist, primarily concerned with (a) the search for ways to turn one substance into another and (b) the search for medicines that would cure all diseases or prolong human life.

aldehyde A colorless, volatile liquid, CH_3ChO, obtained by the oxidation of alcohol.

aldrin A white crystalline insecticide consisting chiefly of a chlorinated derivative of naphthalene.

Aleutian low The semipermanent cyclone or low usually located near the Aleutian Islands.

alfalfa A plant grown as feed for animals.

alfisol Soil formed under boreal forests or broadleafed-deciduous forests with a high clay content and a high base saturation. See also: planosol

algae The simplest plants, including seaweeds, usually found growing in water or in damp places. (Singular: alga.)

alidade An instrument used in surveying, consisting of a telescope mounted parallel to a ruler.

aliphatic compound Organic compound having or belonging to an open-chain structure. Sometimes called fatty, since fats and oils are representative, common examples.

alkali A salt of sodium or potassium, or, more generally, any bitter-tasting salt capable of neutralizing acids.

alkali soil A soil of so high a degree of alkalinity—pH 8.5 or higher—or so high a percentage of exchangeable sodium—15 percent or higher—or both, that the growth of most crop plants is reduced.

alkaline (adj.) Having the qualities of a base.

alkaloid Any of a group of nitrogenous organic compounds (especially one of vegetable origin). They have a powerful physiological effect on animals, and constitute the active principals of the common vegetable drugs and poisons. Atropin, codeine, morphine, nicotine, quinine, and strychnine are alkaloids.

alkoxides Metallic salts of alcohols or phenols in which the hydrogen atom of the hydroxyl group is replaced by a metal.

allele A gene, usually developed through mutation, that is responsible for hereditary variation.

allelomorph One of the pair of genes that have identical positions in homologous chromosomes of two uniting gametes but have opposite effects on the same developmental element (size or color), thus producing variation among mixed offspring in accordance with Mendel's law.

allergen A foreign substance, generally a protein, which upon being introduced into the tissues of a sensitive person causes inflammation, swelling, or more serious consequences.

allocation The process by which society sets aside or uses portions of particular resources.

alloy A metal consisting of a fusion or mixture of two or more metals, or of metals with other substances.

All Saints' summer Indian summer.

alluvial (adj.) Pertaining to water-carried material.

alluvial fan A fan- or cone-shaped deposit of alluvium formed where a stream flows out onto a level area or into a slower stream; usually found along the base of desert mountain ranges.

alluvial fan

alluvial soils Soils developing from transported and relatively recently deposited material (alluvium) with little or no modification of the original materials by soil forming processes. (Soils with well-developed profiles that have formed from alluvium are grouped with other soils having the same kinds of profiles, not with the alluvial soils.)

alluvium Material, including sand, clay, gravel, and mud, deposited in riverbeds, lakes, alluvial fans, valleys, and elsewhere by modern streams.

alpaca A domestic animal of the camel family, usually raised for its wool. It is smaller than a llama and larger than a vicuña. Its fine, long, woolly hair is used to make lightweight cloth.

alpaca

alpenglow A reappearance of sunset colors on a mountain summit after the original colors have faded; also a similar phenomenon preceding the regular coloration at sunrise.

alpha particle A positively charged particle given off by the nucleus of some radioactive substances. It is identical to a helium atom that has lost its two electrons.

alpine (adj.) Pertaining to the Alps or any other high mountain range.

alpine glacier A glacier located in a mountainous area.

alpine orogeny A series of mountain-building movements beginning in the Triassic and continuing into the present.

altimeter

altimeter An aneroid barometer graduated to show elevation instead of pressure.

altithermal A period of high temperature in past geological time, especially the postglacial period.

altitude 1. Vertical distance above the surface of the earth. 2. Angular distance above the horizon of a heavenly body.

altitudinal zones See: tierra caliente; tierra fria; tierra nevada; tierra templada.

amoeba

altocumulus A form of middle cloud, composed of laminae or rather flattened globular masses, the smallest elements of the regularly arranged layer being fairly small and thin, with or without shading.

altostratus A form of middle cloud, a striated or fibrous veil, more or less gray or bluish in color.

alum A chemical substance that is gelatinous when wet, usually potassium aluminum sulfate, used in water-treatment plants for settling out small particles of foreign matter.

alumino-silicates Compounds containing aluminum, silicon, and oxygen atoms as main constituents.

aluminum A metallic element, silverish, light in weight, used for utensils, airplane parts, etc.

amber A fossilized resin from prehistoric conifers.

amino acids Nitrogen-containing organic compounds, large numbers of which link together in the formation of a protein molecule. Each amino acid molecule contains one or more amino groups and at least one carboxyl group; some contain sulfur.

ammonia A colorless gas composed of one atom of nitrogen and three atoms of hydrogen. Ammonia liquefied under pressure is used as a fertilizer.

ammonium ion The positively charged NH_4^+ ion. The form in which nitrogen occurs in many commercial fertilizers.

amoeba A simple form of life consisting of a nucleated mass of protoplasm that obtains sustenance by enveloping minute particles of plant and animal life.

amorphous (adj.) Without form. Usually applied to rocks and minerals having no definite crystalline structure.

amphibian A cold-blooded vertebrate; the young breathe through gills, the adults by lungs.

amphibole Common rock-forming minerals containing chiefly calcium, magnesium, sodium, iron and aluminum.

amplitude In hydrodynamics, a value that is one-half the wave height.

anabatic (adj.) A term applied to local winds moving up a hillside due to surface heating, and not in response to the general pressure distribution. See: katabatic

anabolism In living organisms, the synthesis of more complex substances from simpler ones.

anabranch A diverging branch of a river that re-enters the main stream.

anadromous (adj.) Of fish, ascending rivers from the ocean at certain seasons for breeding.

anaerobic (adj.) Living in the absence of air or free oxygen.

analemma A scale, shaped like a figure 8, which shows the declination of the sun and the equation of time for each day of the year.

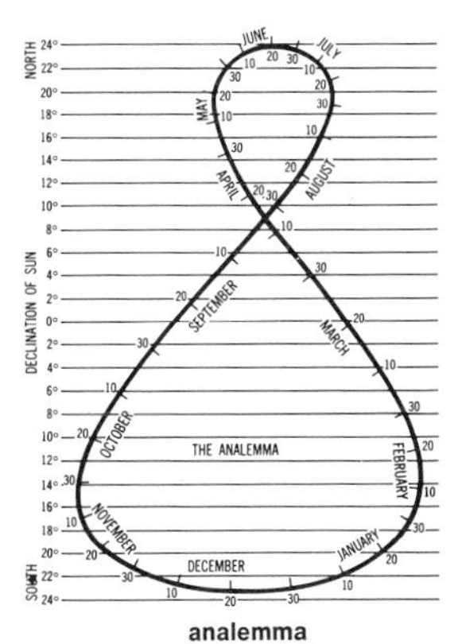

analemma

anatomy The science that deals with the structure of plants and animals.

anchor ice Ice formed on the bottom of rivers; also known as ground ice.

anchovy Any small, marine, herring-like fish.

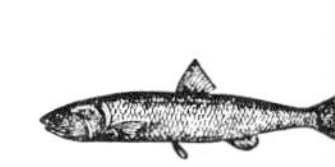
anchovy

anemometer An instrument for measuring the speed or force of the wind.

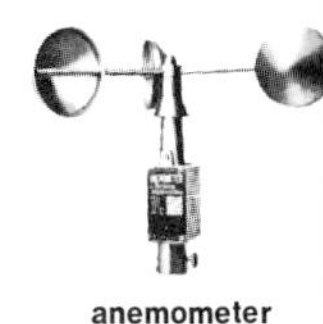
anemometer

aneroid barometer An instrument for measuring atmospheric pressure.

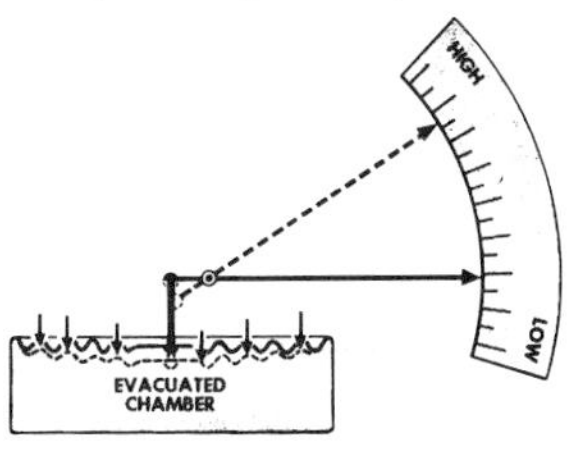

aneroid barometer

angiosperm A plant having seeds enclosed in an ovary; a flowering plant.

angle of incidence The angle at which the sun's rays strike the earth. The angle varies with latitude and time of year.

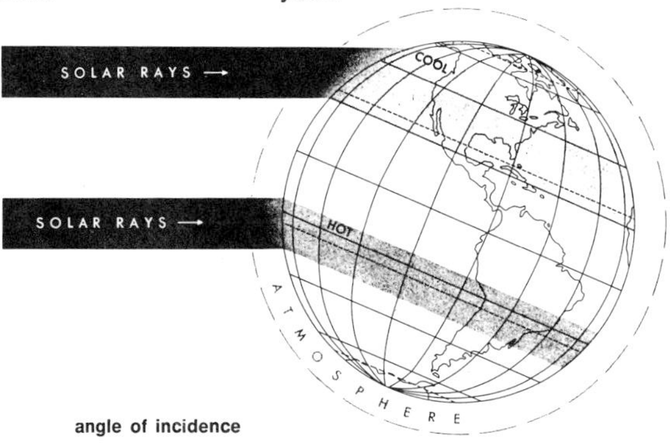

angle of incidence

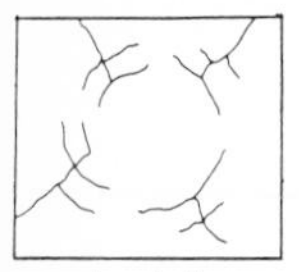

annular drainage

angstrom A unit used in the measurement of the wave lengths of light. One angstrom equals 1/100,000,000 centimeter (10^{-8} cm).

angular momentum See: conservation of angular momentum

angular unconformity In geology, an unconformity in which the older underlying rock strata dip at a different angle (usually steeper) from the younger over-lying ones. See: unconformity; trap

anhydrous (adj.) Dry, or without water. Anhydrous ammonia is water free, in contrast to the water solution of household ammonia.

anion A negatively charged ion.

annual (adj.) Occurring once during, or accumulated over, a consecutive twelve-month period of time for which the beginning date is identified.

annual march See: annual variation

annual range of temperature The difference between the highest and lowest temperature for a year: the average annual range of temperature is the difference between the averages of the yearly extremes for a suitable number of years.

annual ring

annual ring The formation of two concentric layers of new growth each year, which provide a basis for determining the age of trees.

annual variation The general progression of a particular climatological element throughout the year, obtained by plotting the normal values of the element for each month and connecting these points by a smooth curve. This procedure indicates the annual march of rainfall, temperature, or any other element. See: variation; variability

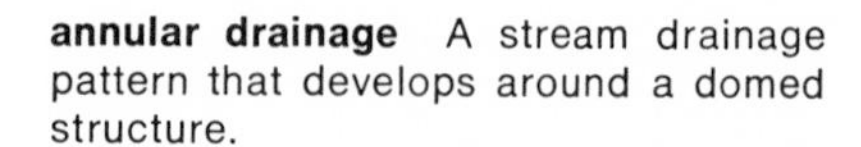

annular drainage A stream drainage pattern that develops around a domed structure.

anomalous (adj.) Exceptional, varying from the usual or mean.

anomaly 1. Any deviation from the common rule. 2. The departure of the local mean value of a meteorological element from the mean value for the latitude.

Antarctic Circle The parallel at 66½° S. which marks the northern limit of the region in the southern hemisphere that experiences a 24-hour period of sunlight at least once a year.

Antarctic Zone One of the climatic divisions of the earth, according to an old classification based on the altitude of the sun; also called the South Frigid Zone. It is that area comprised within the Antarctic Circle, which is at 66½° S. lat.

antecedent stream A river that established its course before uplift or mountain building occurred and maintained its course during and after the uplift.

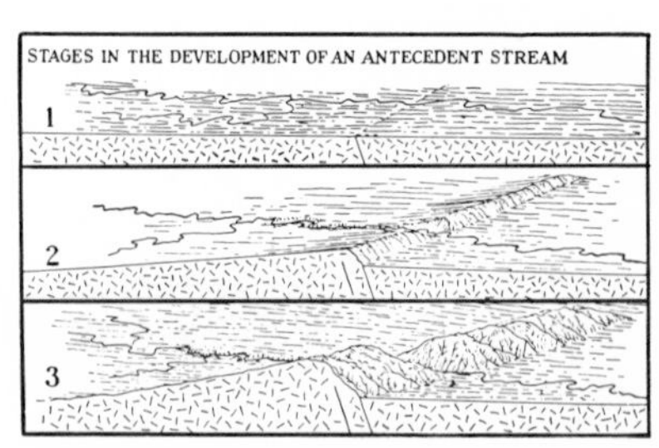

antecedent stream

antediluvian (adj.) Pertaining to the time before Noah's Flood, as described in the Old Testament.

anther The pollen-bearing part of the stamen.

anthracite Hard coal, formed from bituminous coal through the application of heat and pressure.

anthrax An infectious, usually fatal, bacterial disease of cattle and sheep that occasionally affects man.

Anthropic soil A soil produced from a natural soil or other earthy deposit by the work of man that has new characteristics that make it different from the natural soil. Examples include deep, black surface soils resulting from centuries of manuring, and naturally acid soils that have lost their distinguishing features because of many centuries of liming and use for grass.

anthropocentric (adj.) Considering man to be the central or most significant part of the universe.

anthropogeography Human geography; a term introduced by Ratzel, a German geographer, but no longer widely used.

anthropoid Any tailless ape resembling man; e.g., gorillas, chimpanzees, orangutans, gibbons.

anthropology The science of man, his origins, culture, and physical characteristics.

antibiosis An association between living organisms that suppresses the life of one of them. See: symbiosis

antibiotic Any substance that has the capacity to inhibit the growth or destroy bacteria or other microorganisms.

anticline Layers of rock folded so that they slope downward and outward from the center line of the fold. See: syncline

anticyclogenesis The sum of the processes that create and develop a new anticyclone or intensify an already existing one.

anticyclone An area of relatively high pressure with closed isobars, the pressure gradient being directed from the center so that the wind blows spirally outward in a clockwise direction in the northern hemisphere, counterclockwise in the southern.

anticylonic (adj.) Pertaining to an anticyclone.

antigen Any substance that stimulates the production of antibodies.

antimony A hard, brittle, silvery-white metal, often mixed with other metals to harden them.

antipodes Places on opposite sides of the globe.

antitrade A wind plowing over the trades in a direction opposite to them; hence, sometimes called the countertrade.

anvil cloud Popular name of a heavy cumulus or cumulonimbus having an anvil-like form, especially in its upper portions.

apatite A common mineral, calcium fluophosphate, used in the manufacture of phosphate fertilizers.

aperiodic (adj.) Phenomenon not occuring regularly, but taking place at unequal intervals of time.

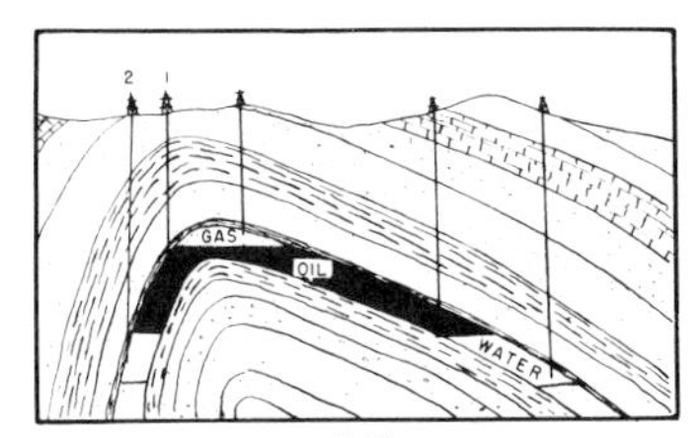

anticline

aperwind Thawing wind of the Alps.

aphelion The point on the earth's orbit that is farthest from the sun. About July 1st, but the time varies irregularly by a few days from year to year, and also has a slow secular variation. See: perihelion

apogee The point in the orbit of a satellite when it is at the greatest distance from the earth. See: perigee

apparent solar time Time based on the daily appearance of the sun and observed by using a sun dial. Because of the varying speed of the earth as it revolves about the sun, the time at which the sun appears on any meridian varies during the course of the year, at times arriving before local noon and at times after local noon. See: equation of time

arc

Appleton layer An ionized layer of the atmosphere, also called the F_1 layer. It is the highest of the recognized layers of the ionosphere; its mean elevation is about 150 miles, but it is variable and has been recorded as low as 93 miles in Australia and as high as 250 miles in England. It reflects radio waves of about 100 meters wave length.

appropriation The capture, impounding, or diversion of water from its natural course or channel and its application to some beneficial use.

appropriative water right The legal right, acquired by fulfilling certain legal requirements, to take or divert water for beneficial use.

aquaculture The production of fish, shellfish, and other types of seafood on a commercial basis.

aqueduct An artificial channel, pipe, or tunnel for carrying water over a distance.

aquiclude A rock layer that is porous but does not transmit water at a sufficient rate to supply springs or wells.

aquifer A body of earth material capable of transmitting water through its pores at a rate sufficient for water-supply purposes. See: artesian aquifer; confined aquifer

aquifuge A body of earth material that is impervious and nonabsorptive.

arable land Land that can be cultivated and made productive.

arboreal (adj.) Pertaining to or like a tree or trees.

arboriculture The science and art of growing trees, especially as ornamental or shade trees. See also: silviculture; pomology

arc That part of a circle intercepted by the rays of an angle.

Arctic Circle The parallel at 66½° N that marks the southern limit of the region in the northern hemisphere that experiences a 24-hour period of sunlight at least once a year.

Arctic Zone The North Frigid Zone, the portion of the earth's surface that lies north of the Arctic Circle.

arid (adj.) Dry.

arkosic sandstone A sandstone containing a large proportion of unweathered feldspar.

arroyo A water-carved channel or gully in an arid country, usually rather small with steep banks, dry much of the time due to infrequent rainfall and the shallowness of the cut, which does not penetrate below the level of permanent ground water.

artesian (adj.) Refers to ground water that is under sufficient pressure to flow to the surface without being pumped.

artesian aquifer An aquifer in which the water is under sufficient pressure to permit it to rise above the zone of saturation.

artesian ground water Ground water, confined under an aquiclude or an aquifuge, that rises when a nonpumping well penetrates it.

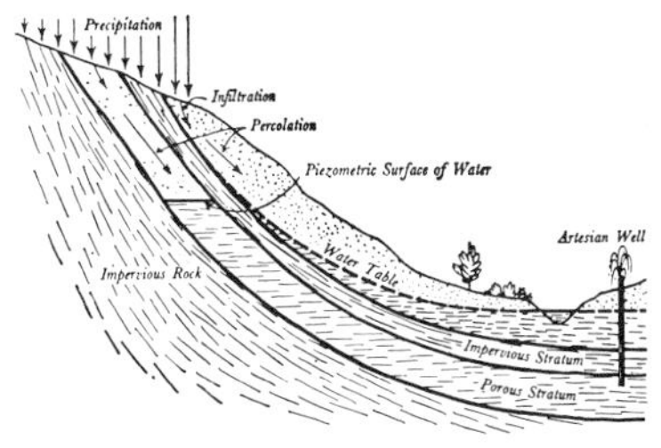

artesian ground water

artesian well A well that normally gives a continuous flow because of hydrostatic pressure, created when the outlet of the well is below the level of the water source.

artificial channel See: channel

artificial recharge The addition of water to the ground water reservoir by the activities of man, such as irrigation or induced infiltration from streams, wells, or spreading basins.

ascorbic acid Vitamin C. An essential growth factor that occurs in fruits (especially citrus) and vegetables.

Asiatic Realm One of the biotic realms identified by naturalist A. R. Wallace. See: Wallace's Line

aspect The compass direction toward which a sloping land area faces. The direction is measured downslope and normal to the contours of elevation.

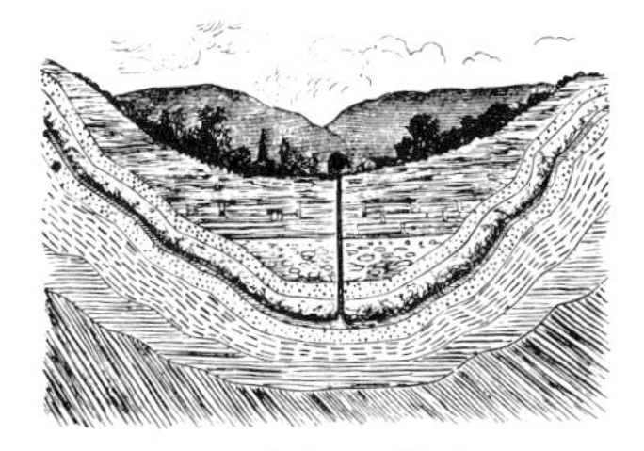
artesian well

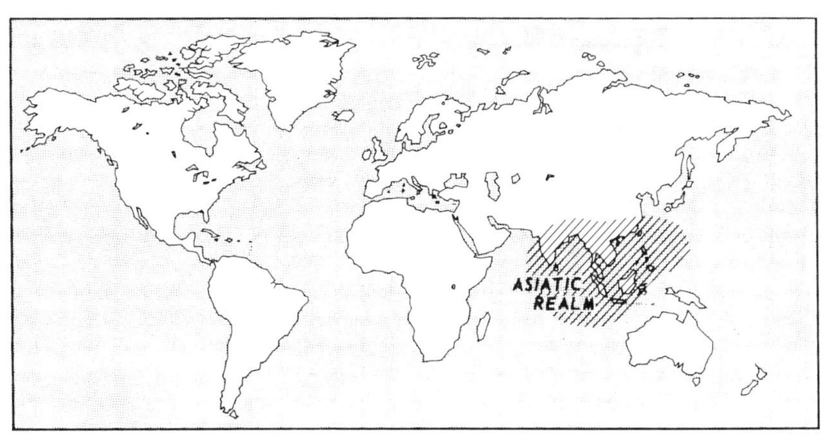

Asiatic Realm

association An assemblage of plants, usually over a wide area, that has one or more dominant species from which it derives a definite aspect.

asteroid (or planetoid) Any one of the thousands of minor planets that revolve around the sun, mostly between the orbits of Mars and Jupiter.

Astrakhan sheep A breed of sheep raised for their skins in the dry lands of central Asia.

astronomical unit A unit of measurement equivalent to the mean distance from the earth to the sun, i.e., 93,000,000 miles.

astronomy The science that studies the material universe beyond the earth's atmosphere.

atmometer An instrument for measuring the rate of evaporation; also called an atmidometer or evaporimeter.

atmosphere The gaseous envelope surrounding the earth.

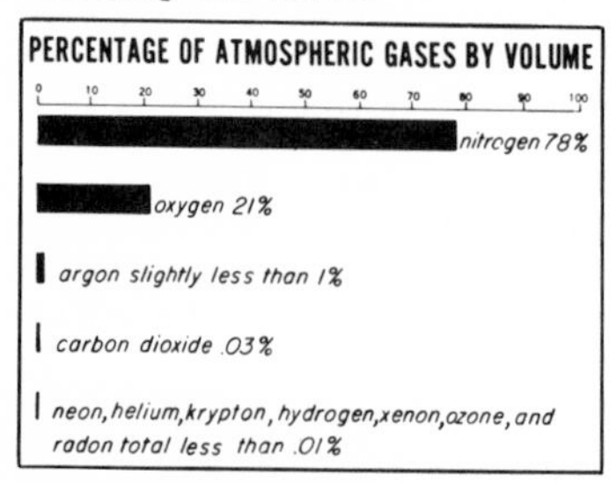

atmosphere

atmospheric acoustics A branch of the physics of the air that seeks to explain sounds having meteorological origins—the humming of wires, the murmur of the forest, thunder, etc.—and the many effects of the atmosphere on sounds in general.

atmospheric electricity The sum of the electrical manifestations of the atmosphere, among which the most familiar are lightning, the aurora, and St. Elmo's Fire.

atmospheric pressure The force per unit area exerted by the atmosphere in any part of the atmospheric envelope. Since the atmosphere is a substance, it has mass and is acted upon by gravity.

atmospheric radiation The radiation emitted by the atmosphere in two directions: upward to space and downward to the earth, and consisting mainly of the long-wave terrestrial radiation plus the small amount of short-wave solar radiation absorbed in the atmosphere.

atmospheric tides Small fluctuations in the atmosphere, created by the gravitational action of the sun and moon in the same manner that these bodies produce the tides of the sea.

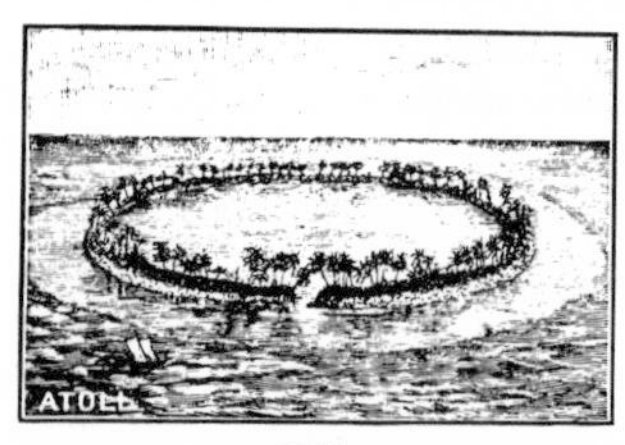

atoll

atoll A ringlike island or island group, composed of coral and surrounding or nearly surrounding a lagoon.

atom The smallest particle of a chemical element, made up of electrons moving around an inner core (nucleus) of protons and neutrons. Atoms of elements combine to form molecules and chemical compounds.

atomic energy Energy released by changes in the nucleus (protons and neutrons) of an atom, etiher by the splitting of a large nucleus or by the joining of smaller nuclei.

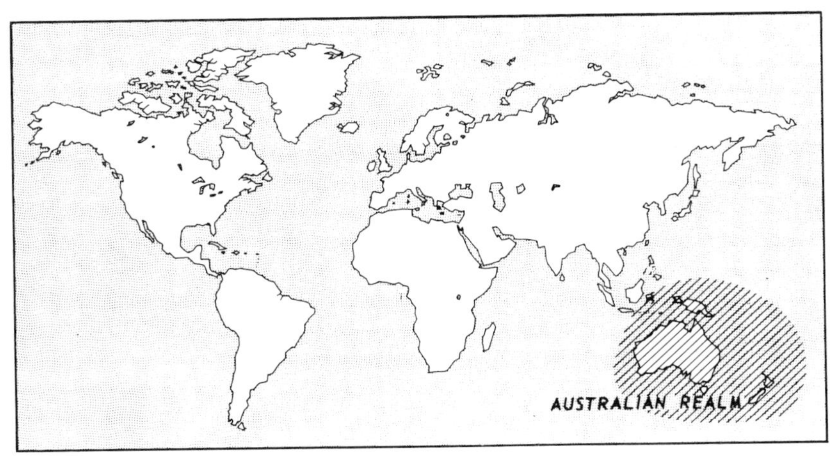

Australian Realm

atomic number The number of protons in the nucleus of an atom, and also its positive charge. Each chemical element has its characteristic atomic number, and the atomic numbers of the known elements form a complete series. See: element; isotope; Periodic Table

atomic pile A nuclear reactor, a structure of many tons of pure graphite and uranium for the production of plutonium. Used for power generation, as an essential unit for nuclear weapons, and for medical purposes.

atomic time scale See: radioactive age determination

atomic weight The weight of an atom of a given element relative to that of oxygen.

atrophy (v.) The wasting away of the body or of an organ or part.

aureomycin An antibiotically active chemical compound, with the approximate formula for the crystalline free base, $C_{22}H_{26}O_6N_2Cl$.

aurora A luminous glow sometimes seen at night in the northern and southern skies; in the northern sky it is called aurora borealis, and in the southern, aurora australis.

austausch A measure of the degree of turbulence that exists within a turbulent zone, in terms of the exchange of mass across unit horizontal surface, during unit time.

Australian Realm One of the biogeographical regions, identified by naturalist A. R. Wallace. See: Wallace's Line

autoconvection Atmospheric convection that is spontaneously initiated by a layer of air when the lapse rate of temperature is such that density increases with elevation at a sufficiently rapid rate.

autotrophic (adj.) Capable of using (oxidizing) simple chemical elements or compounds, such as iron, sulfur, or nitrates, tc obtain energy for growth.

autumn The third season of the year, the period between summer and winter; commonly called fall in America. Astronomical autumn is the period from the autmnal equinox (about September 22) to the winter solstice (about December 21).

autumnal equinox See: equinox

auxins Organic substances that cause lengthening of the stem when applied in low concentrations to shoots of growing plants.

avalanche A mass of snow, ice, rock or other material detached from its position and slipping down a mountain slope.

average An arithmetic mean, obtained by adding a number of terms and dividing the sum by the number of terms.

Avogadro's hypothesis The principle that equal volumes of all gases under the same pressure and temperature contain the same number of molecules.

axis The line about which a rotating body such as the earth turns.

azimuth

azimuth The arc of the horizon intercepted between a given point and an adopted zero point.

azimuthal projection A map projection in which a portion of the globe is projected upon a plane tangent to it.

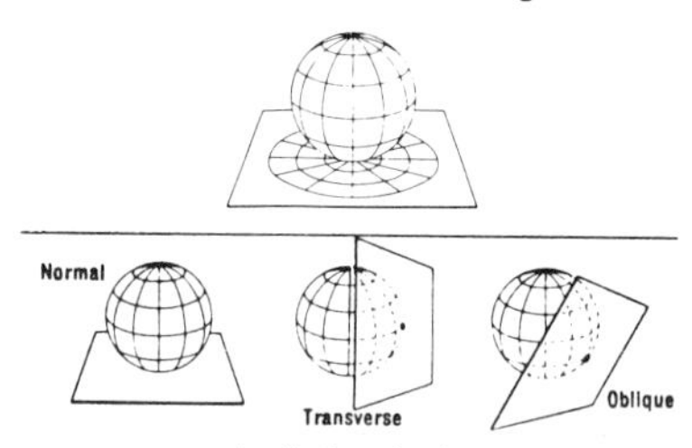

azimuthal projection

azoic (adj.) Without life.

azonal soils Soils having little or no soil profile development, mostly young. In the United States, includes alluvial soils, lithosols, and regosols.

B horizon A soil horizon, usually beneath an A horizon, or surface soil, in which clay, iron, or aluminum, with accessory organic matter, have accumulated by receiving suspended material from the A horizon above or by clay development in place. The soil has a blocky or prismatic structure. In soils with distinct profiles it is roughly equivalent to "subsoil."

back (v.) To change direction counterclockwise; applies to the wind when it so changes, as, for example, from the north to northwest, east to northeast, etc., in the northern hemisphere.

backfire A fire intentionally set along the inner edge of a control line located ahead of an advancing fire, for the purpose of facilitating control by a widening of the control line and the removal of intervening combustible materials.

background radiation The normal radioactivity shown by a counter, due principally to cosmic rays, trace amounts of radioactivity in the vicinity of the counter, and impurities in the counter.

backwater Water backed up or retarded in its normal or natural condition of flow.

bacteria One-celled, microscopic organisms, widely distributed in the air, water, soil, and animal and plant tissues, including foods. Bacteria cause many infectious diseases in plants and animals; they also exert many beneficial effects in agriculture (fixing of atmospheric nitrogen, decay of dead matter), in food industries (vinegar fermentation), and in chemical industries (acetone, butanol, and organic acid production).

bacteriology The science that studies bacteria.

badlands Any area in which soft rock material is eroded into varied forms.

bagasse The mill residues from the cane-sugar industry, consisting of the crushed stalks from which the juice has been expressed. Applied to similar residues from other plants, such as sorghum, beet, or sisal, but usually refers to sugarcane.

baguio (bagio; bagyo) Philippine term for a tropical cyclone.

bai A colored mist prevalent in China and Japan in spring and fall, when clouds of sand rise, are carried eastward, collect moisture, and fall, producing a thick coating of very fine yellow dust.

bai-u (adj.) The name of a season (April and May in Japan, May to July in China) during which copious rains occur in southern Japan and in parts of China. Important for the cultivation and transplanting of rice, the bai-u rains are also called blossom showers, plum rains, or mold rains, with reference to the season for ripening plums and to the effects of continued dampness, respectively.

bajada The nearly flat surface consisting of the pediment and the coalesced alluvial fans that surround a playa.

balance of nature A state of equilibrium in nature resulting from the interaction of all elements of the environment.

ball-hooter A slang term loggers use for a man who rolls or slides logs down a hillside.

ball lightning A form of lightning consisting of luminous ball- or pear-shaped bodies that travel or float along freely in the air or over the surface of walls, roofs, floors, or the ground. No simple explanation of the phenomenon has been found.

balsa A light wood found in the American tropics used in life preservers, rafts, toys, etc.

bamboo Any of the hollow woody tropical and semitropical grasses.

bank 1. The margins of a channel. Banks are called right or left as viewed facing in the direction of the flow. 2. An area of shallow water in the ocean, a submarine bank.

bank storage Water absorbed and stored in the voids in the soil cover in the bed or banks of a stream, lake, or reservoir, and returned in whole or in part as the level of the surface of the water body falls.

banner cloud A cloud that resembles a great white flag floating from a high mountain peak. See: cap cloud

Bantu A large group of Negro tribes living in central and southern Africa. Also, their language.

bar 1. A unit of pressure equal to 10^6 dynes/cm^2; equivalent to a mercurial barometer reading of 750.076 mm. at 0°C. (or 29.5306 inches at 32°F.), gravity being equal to 980.616 cm/sec^2. 2. A mass of alluvium (usually sand or gravel) deposited on the bottom of a stream, lake, or sea, and coming close enough to the water surface to form an obstruction to navigation.

barat Heavy northwest squall in Menando Bay on the north coast of the island of Celebes, prevalent from December to February.

barber A gale of wind with damp snow or sleet and spray that freezes upon every object, especially the beard and hair.

barber chair In loggers' slang, a stump on which is left standing a slab that splintered off the tree as it fell. Generally it indicates careless felling.

barchan A crescent-shaped sand dune with the convex side in the direction of the wind.

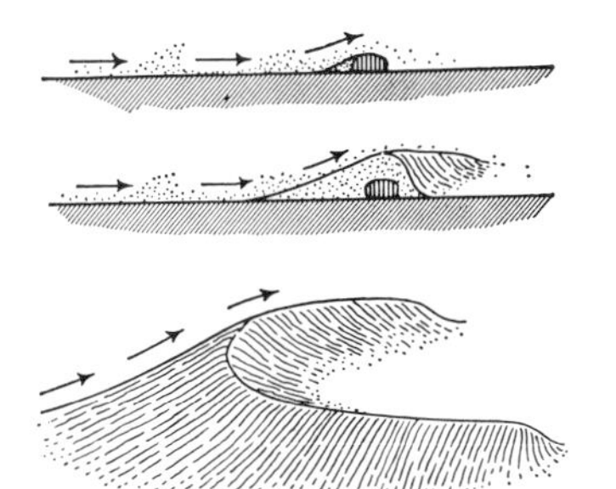
barchan

barine Westerly wind in eastern Venezuela.

barite A common mineral, barium sulfate ($BaSO_4$).

barium 140 The radioactive isotope of barium having a mass number of 140 and a half-life of 12.8 days; used principally as a tracer.

bark The external covering of the stems, branches, and roots of woody plants, distinct from the wood itself.

barley A widely distributed cereal used for both human and animal food and in the making of beer, ale, and whiskey.

barograph A barometer that makes a continuous record of barometric changes.

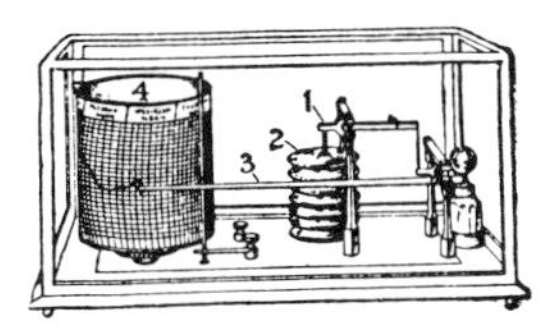

barograph

barometer An instrument for measuring atmospheric pressure. See: mercurial barometer; aneroid barometer

barometer

barometric (or pressure) gradient The rate of change of atmospheric pressure per unit horizontal distance, measured in a direction normal to the isobars. See: gradient

barometric pressure See: atmospheric pressure

barometric tendency The net change of barometric pressure within a specified time (usually three hours) before an observation.

barothermograph An instrument that automatically records temperature and pressure. It is a combination barograph and thermograph.

barracuda A member of a group of large, savage fish, resembling the freshwater pike, living in tropical seas.

barrage A large structure erected across a river in order to store water, usually for irrigation.

barrier bar A ridge of sand and gravel built parallel to a coast by current and wave action in shallow water.

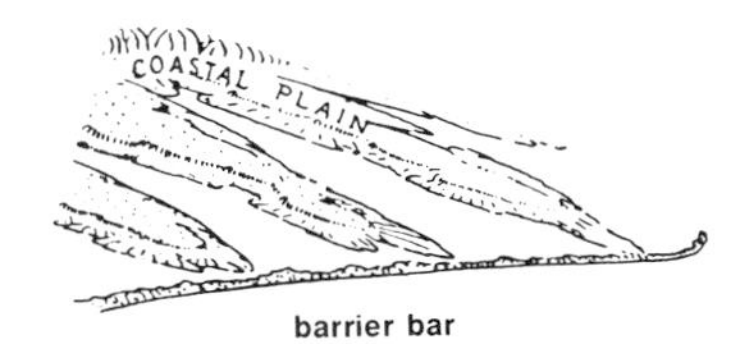

barrier bar

barrier reef A coral reef separated from the shore by a lagoon too deep for coral growth.

barrier reef

barrier theory A theory of cyclogenesis proposed by F. M. Exner, in which an outbreak of polar air into the region of the prevailing westerlies is considered to act as a barrier to the eastward motion of the warmer air and is said to cause the formation of a low-pressure center just to the east of the barrier, around which a cyclonic vortex develops and travels eastward. In a similar manner a continent, especially a mountainous continent, may act as a barrier.

basal metabolic rate The rate of oxygen intake and heat discharge of an organism.

basalt A dark-colored igneous rock.

base A compound that reacts with an acid to form a salt.

base discharge Geological Survey term for the discharge above which peak discharge data on surface-water supply are published. The base discharge at each station is selected so that an average of about three peaks a year will be presented.

base-exchange capacity See: cation-exchange capacity

base flow That portion of the stream discharge that is derived from ground water outflow or other sources outside the net rainfall that created the surface runoff.

base level The lowest level to which a stream can wear its bed. The permanent base level is the level of the ocean.

base line A line with known true bearing, length, and elevations of extremities.

base map A map having essential outlines of basic features that is used for indicating specialized data of various kinds.

basement The igneous and sedimentary rock complex that underlies sedimentary strata.

base period A period of time specified for the selection of data for analysis, sufficiently long to contain data representative of the averages and deviations from the averages that must be expected in other periods of similar and greater length.

base saturation The relative degree to which soils have metallic cations absorbed. The proportion of the cation-exchange capacity that is saturated with metallic cations.

basin (or level borders) irrigation The application of irrigation water to level areas that are surrounded by border ridges or levees. Usually irrigation water is applied at rates greater than the water intake rate of the soil. The water may stand on uncropped soils for several days until the soil is well soaked; then any excess may be used on other fields. The water may stand a few hours on fields having a growing crop.

basin listing A method of tillage that creates small basins by damming lister furrows at regular intervals of about 4 to 20 feet.

bast The phloem or sieve tissue of plants. Also, the fibers making up this tissue, also called bast fibers. Generally, any commercial fiber arising from the stem of the plant, including such important fibers as flax (linen), hemp, jute, and ramie.

batholith A very large body of igneous rock, usually granitic, that has been exposed by erosion of the overlying rocks. Batholiths form the heart of many mountain ranges, such as the Sierra Nevada in California.

bathyal (adj.) Pertaining to the deeper parts of the ocean; especially between 100 and 1000 fathoms.

bathymetry The measurement of the depths of the oceans.

bathysphere A spherical diving apparatus from which to study the ocean depths.

bauxite The principal ore of aluminum.

bay A wide indentation into the land, formed by the sea or by a lake.

bayamo A violent blast of wind, accompanied by vivid lightning, blowing from the land on the south coast of Cuba, especially near the Bight of Bayamo.

bay ice Young ice that first forms on the sea in autumn.

bayou Marshy or sluggish water, particularly a lake or small, sluggish stream in an abandoned or half-closed river channel along the Gulf Coast or on a river delta.

B-complex vitamins The term "vitamin B" was first used to distinguish water-soluble accessory food factors from those that are fat-soluble. It was gradually realized that vitamin B is a mixture, or complex, of many chemicals, each given a number; i.e. thiamine (B_1), riboflavin (B_2), pyridoxine, niacin, biotin, choline; folic acid, p-aminobenzoic acid, etc. The most recently well-characterized member of this complex, B_{12}, although pure, has not had its constitution completely worked out.

BC soil A soil with a B and a C horizon but with little or no A horizon. Most BC soils have lost their A horizons by erosion.

beach The strip of land or terrace bordering the sea that lies between high and low water marks and consists of boulders, pebbles, or sand.

beaded lightning Chain lightning seen end-on with respect to the line of discharge, so that several points, more brilliant than the rest of the path, appear as a string of incandescent beads. It is sometimes called "pearl" lightning.

bearing The horizontal angle between an object viewed by the observer and the meridian measured clockwise from true north.

Beaufort wind scale A system of estimating wind velocities, devised by British Admiral Sir Francis Beaufort in the early 19th century, originally based on the effects of various wind speeds on the canvas of a full-rigged frigate; since modified and widely used in international meteorology. See: breeze

bed A layer of sedimentary rock, more or less clearly separated from the layers above or below by differences in particle type, size, etc. (v.) To arrange the surface of fields by plowing and grading into a series of elevated beds separated by shallow ditches for drainage.

bedding plane The surface that separates one layer of sedimentary rock from another.

bed load Sand, silt, gravel, or soil and rock detritus carried by a stream on or immediately above its bed.

Bedouin A member of one of the wandering tribes of Arabs living in the Sahara Desert and the deserts of Arabia and Syria.

bedrock The solid rock beneath the soil and subsoil.

belat A strong land wind, from the north or northwest, which occasionally affects the southern coast of Arabia and is accompanied by a hazy atmosphere due to sand blown from the interior desert.

Benard (or convection) cell A type of eddy having a horizontal axis and formed when the top surface of a fluid is cooled.

bench mark A mark cut in stone that is used as a reference point in surveying to determine altitude.

bench mark

beneficial use of water The concept that in arid lands an owner of a water right must generally prove that he is using it productively to retain rights to it.

beneficiate (v.) To improve the grade of ore by milling, sintering, etc.

benthic (adj.) Of, pertaining to, or occurring on the bottom of a sea or lake.

benthos Sedentary, bottom-dwelling marine plant and animal organisms, such as seaweed or coral.

bentonite A clay, formed from the decomposition of volcanic ash, that absorbs water readily.

bentu de soli An east wind on the coast of Sardinia.

berg An iceberg.

Bergeron classification See: air mass classifications

Bergeron-Findeisen theory A theoretical explanation of the process by which precipitation particles may form within a mixed cloud containing both ice crystals and liquid water droplets. The basis of the theory is the fact that the vapor pressure of ice crystals is less than that of water droplets, drawing water to the ice crystals until they become large enough to fall out of the cloud. In falling they may pick up more water as a result of collisions with other ice and water droplets.

bergschrund The gap left on the upper rim of a glacier or a snowfield as the ice or snow moves downward.

bergschrund

berg wind A foehn wind in South Africa.

beriberi A dietary disease caused by a deficiency of vitamin B_1 and characterized by general debility and rigidity.

berm 1. A narrow shelf or ledge. 2. A nearly level part of a beach formed of material deposited by wave action. 3. A valley-wall terrace that is a remnant of an earlier valley floor.

Bermuda grass A perennial creeping grass of southern Europe used for lawns and pastures; also known as Bahama grass, devil grass, scutch grass.

Bermuda high The high pressure cell found over the Atlantic Ocean near the Bermuda Islands.

Bernoulli's principle The theorem of Daniel Bernoulli that as the speed of a fluid increases, its pressure decreases; and conversely, as the speed of a fluid decreases, the pressure increases. (Not to be confused with Jakob Bernoulli's law of statistics.)

beta particle An elementary particle emitted from a nucleus during radioactive decay, with a single electrical charge and a mass equal to 1/1837 that of a proton. A negatively charged beta particle is identical to an electron. A positively charged beta particle is called a positron.

beta ray A stream of high-speed electrons given off by the nuclei of some radioactive substances when protons turn into neutrons; electrons that have been accelerated to high speeds by man.

betatron A device resembling a large doughnut for speeding up electrons in a vacuum chamber by means of very strong magnetic fields; these electron projectiles, with speeds up to 30 million volts, are used for producing penetrating X-rays for atomic research.

biannual (adj.) Occurring twice a year.

biennial (adj.) Happening every two years.

big bang theory The assumption that the universe started as an explosion from a highly condensed state.

bight An indentation in the seacoast similar to a bay, but it may be larger or have a gentler curvature.

billow clouds A line of clouds formed by wave motion on the surface of the discontinuity between air layers of different temperatures and densities. They occur in a series of closely spaced bands separated by clear skies.

binary system A number system that has two as its base; e.g., in computers, 1 and 0 are used to express other numbers, letters, or symbols.

binomial nomenclature In biology, a system of classifying by both a generic and a specific name.

biochemical oxygen demand (B.O.D.) The quantity of oxygen utilized primarily in the biochemical oxidation of organic matter in a specified time and at a specified temperature.

biochemistry The science that deals with the chemistry of living matter.

biochore That part of the earth's surface having a life-sustaining climate. It is bounded on the one hand by the cryochore, or region of perpetual snow, and on the other by the xerochore, or waterless desert. Transition zones on either side are the bryochore, or tundra region, and the poechore, or steppe region. The bulk of the biochore consists of the dendrochore, or tree region.

bioclimatic (adj.) Pertaining to the relation between climate and life.

bioclimatology The science that treats of the effects of climate on human life and health.

biocoenose An assemblage of organisms that live together as an interrelated community.

biodegradable (adj.) Of an organic substance that is quickly broken down by normal environmental processes.

biogenesis The production of living organisms from other living organisms.

biogenetic theory The theory that all living matter must originate from other living matter. See: spontaneous generation

biogeochemical cycles All of the cycles involving man, animals, plants, and minerals. See: carbon cycle; phosporus cycle; nitrogen cycle

biogeography The study of the geographical distribution of plants and animals over the globe.

biological control The control of crop pests by the use of insect predators.

biomass That part of a given habitat consisting of living matter.

biome A community characterized by distinctive plant and animal life and climatic conditions.

biometrics The application of mathematical-statistical theory to biology.

biosphere That portion of the earth occupied by life forms.

biosynthesis The building of chemical substances, often complex, by living organisms, generally from simpler materials.

biota The plant and animal life of a region or period of time.

biotic community See: community

biotic potential The capacity of plants and animals to increase in numbers under optimum environmental conditions.

biotic resources Plant and animal resources.

birth control Regulation of the number of children born through the deliberate control or prevention of conception.

birth rate The proportion of births to total population, usually expressed as a quantity per 1000 persons.

bise (or bize) A cold, dry winter wind that blows from a northerly direction over the mountainous districts of southern Europe.

Bishop's ring A faint reddish-brown corona around the sun, often seen following volcanic eruptions.

bismuth A grayish-white metal, often mixed with other metals so that they will melt easily.

bison A large, shaggy, oxlike animal with short horns and a large hump. The American bison, the buffalo, was slaughtered in large numbers in the 19th century.

bituminous coal A mineral that burns with a yellow smoky flame; soft coal.

Bjerknes cyclone model An idealized meteorological model, showing the normal distribution of air masses, fronts, cloud forms, precipitation, pressure, pressure tendencies, etc., about a wave cyclone during the various phases of its existence.

black blizzard A dust storm in the Dust Bowl region of the United States. See: Dust Bowl

black body In physics, an ideal body, the surface of which absorbs all the radiation that falls upon it; i.e., it neither reflects nor transmits any of the incident radiation.

black-body radiation The maximum amount of radiation that can theoretically be emitted by unit surface of a body at a given temperature. In theory, it is the amount emitted by a blackbody.

black box A term applied loosely to experimental devices to avoid detailed descriptions not vital to the discussion.

black bulb thermometer A mercurial maximum thermometer, the bulb of which has been coated with lampblack or platinum black.

Black Death A form of the bubonic plague that spread across Europe in the 14th century and killed about a quarter of the population.

black earth A fine, fertile soil, black or dark brown in color, covering extensive areas of the U.S.S.R. as well as other grassland areas.

black (or hard) frost A condition prevailing in late autumn when both air and terrestrial objects have temperatures below freezing. Vegetation is blackened, but hoarfrost does not form.

black oak Any of several varieties of oak characterized by blackish bark.

blast furnace A furnace for melting the metals in ores.

blaze A mark, made on the trunk of a standing tree by painting or chipping off a spot of bark with an ax, to indicate a trail, boundary, location for a road, trees to be cut, etc.

blizzard A violent, intensely cold wind, laden with snow, most of which is picked up from the ground.

block harvesting In forestry, the practice of clearing extensive areas of all trees as opposed to selective cutting.

blocking The retardation or deflection of eastward-moving air pressure centers due to the stagnation of a high-pressure center in their paths.

block mountain A mountain mass formed by the uplift of land between faults or by subsidence outside the faults.

blood group (or type) One of several classes into which human blood is separated.

blood rain Rain that is tinted a reddish color by dust from nearby desert areas, leaving a red stain on the ground.

blossom showers See: bai-u

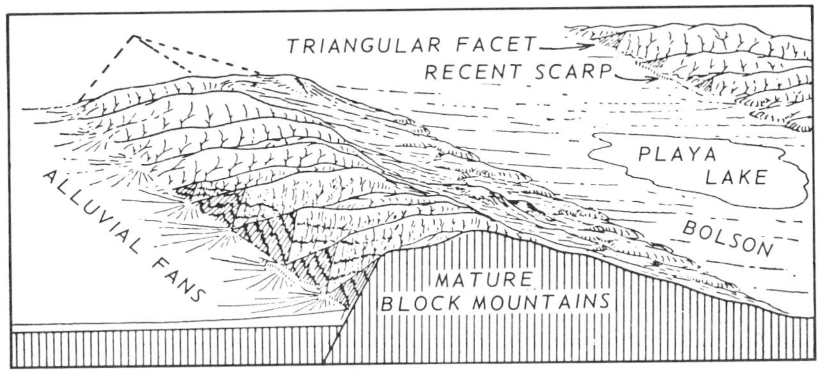

block mountain

blowhole A hole in a seacliff through which columns of spray are forced upward.

blowing snow Snow raised from the ground and carried by the wind so that the horizontal visibility becomes less than five-eighths of a mile (3,300 ft.), although no precipitation is falling.

blowout An area from which soil material has been removed by wind. Such an area appears as a nearly barren, shallow depression with a flat or irregular floor consisting of a resistant layer, an accumulation of pebbles, or wet soil lying just above a water table.

bluff A headland or cliff with a bold and almost perpendicular front.

bocage Farming country in the northwest of France that is divided by hedges and trees into small fields.

BOD The biological oxygen demand of aquatic organisms for survival.

Boer A descendant of the Dutch pioneers who settled in South Africa.

bog An area of soft, wet, spongy ground, mostly decayed or decaying moss and other vegetable matter.

bogaz A long narrow opening in a karst landscape into which a surface stream flows.

bog soil An intrazonal group of soils with mucky or peaty surface soils underlain by peat. Bog soils usually have swamp or marsh vegetation and are commonest in humid regions.

bohorok A foehn-type wind that blows in parts of Sumatra during the months of May to September.

boiling point The temperature at which the saturation vapor pressure of a liquid is in equilibrium with the external pressure on the liquid. The boiling point therefore varies with the external pressure; this explains why the higher one goes, the lower is the temperature at which water boils. The boiling point is lowered about 1.8°F. for each 1,000 feet of altitude.

bole The stem or trunk of a tree.

boll weevil A long-snouted beetle whose larvae are hatched in and damage the bolls of cotton.

bolometer An electrical device used to measure radiant energy by changes in resistance of a blackened platinum strip exposed to the radiations.

bolson A basin that drains into a central interior lake in an arid or semiarid region.

bonds Chemical forces holding atoms together to form molecules.

Bonne's projection A map projection that is a modification of the simple conical projection and resembles it except that the meridians are curved.

boom 1. Logs or timbers fastened together end to end and used to hold floating logs. The term includes also the logs enclosed. 2. Projecting arm, either swinging or rigid, of a log-loading machine, which supports the log during loading.

bora A cold northerly Adriatic wind, caused by the passage of a cyclone over Italy, which draws cold heavy air down from the Hungarian basin. It is similar to the mistral and belongs to the fallwind category.

borasco A thunderstorm, or violent squall, especially on the Mediterranean.

borax A white, water-soluble powder or crystal used as a flux, or cleansing agent, in the manufacture of glass, porcelain, etc.

border irrigation Irrigation in which the water flows over narrow strips that are nearly level and are separated by parallel, low-bordering banks or ridges.

bore A high tidal wave advancing up a narrow river estuary.

boreal forest The northern forest, also known as the taiga.

Boreas Greek name of the north, north-northeast, or northeast wind, or its personification.

botany The science of plants.

botulinum A species of anaerobic bacteria, capable of producing a highly poisonous substance in preserved foods.

boulder A large rounded block of stone lying on the ground or embedded in loose soil, which has been transported to its present position.

Bourdon tube A device that measures atmospheric pressure, and, in another form, temperature.

box canyon A canyon having steep rock sides and a zig-zag course so that an individual is virtually enclosed on four sides.

Boyle's law A thermodynamic law that the volume of a gas varies inversely as its pressure, the temperature remaining constant.

B.P. Before present; used when expressing dates determined by radiocarbon dating with the year 1950 representing the base or "present."

brachiopod A phylum of marine mollusks with two unequal shells.

brachiopod

brackish water Water containing dissolved mineral in excess of acceptable normal municipal, domestic, and irrigation standards, but less than that of sea water.

Brahman Beef cattle, native to India, raised in hot climates of the world.

Brahman

braided stream A waterway characterized by successive division and rejoining of streamflow with accompanying islands.

branch 1. A natural channel of water, part of a larger channel. 2. A limb growing from the trunk or a larger stem of a tree.

brave west winds A nautical term for the strong, often stormy, winds from the west-northwest and northwest, which blow at all seasons of the year from about 40° S. latitude to about 60° S.

breaker A wave breaking into foam as it advances toward the shore.

breakwater A barrier that breaks the force of waves, generally built in front of a harbor.

breccia A rock of coarse, sharp-cornered fragments, formed by volcanic eruption, by crushing or grinding in a fault zone, or by sedimentary deposition of materials that have not been carried far enough to become rounded.

breeder reactor A reactor that produces fissionable fuel as well as consuming it, especially one that creates more than it consumes. The new fissionable material is created by capture in fertile materials of neutrons from fission. The process by which this occurs is known as breeding.

breeze 1. A light wind. 2. In the Beaufort wind scale, a wind speed ranging from 4 to 31 miles per hour, and divided into the following Beaufort numbers: 2. Light breeze, 4-7 m.p.h.; 3. Gentle breeze, 8-12 m.p.h.; 4. Moderate breeze, 13-18 m.p.h.; 5. Fresh breeze, 19-24 m.p.h.; 6. Strong breeze, 25-31 m.p.h.

brickfielder A hot, dry, dusty north wind from the interior deserts, which sometimes visits the south coast of Australia.

brisa (or briza) 1. A breeze (Spanish). 2. Brisas: A northeast wind that blows on the coast of South America during the trades. 3. The northeast monsoon in the Philippines.

British thermal unit (BTU) A unit of heat, 1/180 of the quantity of heat required to raise the temperature of one pound of water from the melting point to the boiling point.

broadleaf A tree with two cotyledons, or seed leaves, usually deciduous, with relatively broad, flat leaves, like a maple or oak.

broken sky (or clouds) The condition of the sky when it is more than five-tenths, but not more than nine-tenths, covered by clouds.

brontide Low, thunder-like noise, of short duration, most frequent in actively seismic regions; it is the rumbling of a very feeble earthquake.

brontosaur An amphibious, herbivorous dinosaur of the Jurassic.

Bronze Age The period when men first used tools and weapons made of bronze.

brook A small stream or rivulet.

brown coal See: lignite

brown forest soils An intrazonal group of soils that have dark-brown surface horizons, relatively rich in humus, grading through lighter colored soil into the parent material. They are characterized by a slightly acid or neutral reaction and a moderately high amount of exchangeable calcium. They are commonly developed under deciduous forests from parent materials relatively rich in bases, especially calcium.

brown podzolic soils A zonal group of soils with thin mats of partly decayed leaves over thin, grayish-brown mixed humus and mineral soil. They lie over yellow or yellowish-brown, acid B horizons, slightly richer in clay than the surface soils. These soils develop under deciduous or mixed deciduous and coniferous forests in cool-temperature humid regions, such as parts of New England, New York, and western Washington.

brown snow Snow intermixed with dust particles.

brown soils A zonal group of soils having a brown surface horizon that grades below into lighter colored soil. These soils have an accumulation of calcium carbonate at 1 to 3 feet. They develop under short grasses.

brown Swiss Dairy cattle bred originally in Switzerland; they produce modest amounts of milk with middle values of butterfat content.

Brückner cycle A supposed world-wide periodic variation in temperature, rainfall, and atmospheric pressure, small in amount and quite irregular in occurrence, with an average period of 34.8 ± 0.7 years, though single periods vary from 20 to 50 years.

bruma A haze that appears in the afternoons on the coast of Chile when sea air is transported inland.

Brunton compass An accurate pocket compass with attached sights and scales for measuring angles and directions, used in rough surveying and in geologic mapping.

bryochore Tundra region.

B.T.U. See: British thermal unit

buck (v.) To saw felled trees into logs or bolts; to bring or carry, as to "buck" water.

bucker In lumbering, a workman who cuts felled tree trunks into logs.

bud An undeveloped or rudimentary stem or branch of a plant. (v) To propagate new plants by grafting a twig or a branch from one tree onto the trunk of another.

buffer strips Established strips of perennial grass or other erosion-resisting vegetation, usually on the contour in cultivated fields, to reduce runoff and erosion.

bund An artificial embankment, dike, or dam in India.

buran A violent northeast storm of south Russia and central Siberia, similar to the American blizzard.

burga (or boorga) A storm of wind and sleet in Alaska, similar to the Russian purga.

burl A hard, woody growth on a tree trunk or on roots, more or less rounded in form. It is usually the result of entwined growth of a cluster of buds, and produces a distorted and unusual (but often attractive) grain.

burlap Coarsely woven cloth made from hemp or jute.

burn Area in which fires have injured the forest.

burn off (v.) With relation to fog, to dissipate.

burster See: southerly burster

bush 1. A low woody plant, or the area where it grows. 2. A wild, uncultivated, scrub-covered region in Australia and Africa.

Bushmen Nomadic people of South Africa.

butane A colorless, flammable gas, C_4H_{10}, used chiefly as a fuel and in the manufacture of rubber.

butanol (or butyl alcohol) An alcohol derived from butane and similar to ethyl alcohol. It, together with acetone, is made by a bacterial fermentation of carbohydrates. It is widely used as a solvent for resins, waxes, fats, etc.

butte A flat-topped hill or mountain similar to a mesa but smaller in size.

Buys Ballot's law The principle governing the relation of wind direction to pressure distribution, that if one stands with his back to the wind, the pressure on his left hand is lower than on his right. Thus stated, the law applies in the northern hemisphere, but in the southern hemisphere, its reverse is true.

C horizon The relatively unmodified rock material in the lower part of the soil profile like that from which the upper horizons (or at least a part of the B horizon) have developed.

caatinga Thorn forest of northeastern Brazil.

cacao A tropical evergreen cultivated for its seeds, which are the source of chocolate.

cacimbo Heavy mists, or smokes, of the Congo Basin in Africa.

cairn A heap of stones set up as a landmark, monument, tombstone, etc.

cajú rains Light showers occuring in October in northeast Brazil.

calcareous (adj.) Made of, or containing, calcium carbonate.

calcify (v.) To become hard or strong as a result of the deposition of calcium salts.

calcimorphic soils Soils developed from limestone parent material.

calcite A common mineral, calcium carbonate, the principal constituent of limestone.

caldera A large volcanic depression, more or less circular, containing the volcanic vent or vents.

calf A piece of floating ice that has broken away from an iceberg, glacier, or ice sheet.

caliche The more or less cemented deposits of calcium carbonate very near the surface or exposed by erosion, in many soils of warm-temperate areas, as in the southwestern states. Also name for deposits of sodium nitrate in Chile and Peru.

calina A Spanish summer haze.

calm An entire or almost entire absence of wind.

calorie The amount of heat required to raise the temperature of one gram of water at 15°C. by one centigrade degree; also known as a gram calorie or small calorie.

calorimeter An instrument to measure quantities of heat; sometimes used in meteorology to measure solar radiation.

calving The breaking away of a mass of ice, called the calf, from a parent berg, glacier, floe, or barrier.

camanchaca See: garúa

cambium A soft layer, strip, or cylinder of living cells, one row thick, between the living bark and living wood of a tree. During the growing season its cells divide continuously, giving origin to the wood tissues and the bark tissues.

Cambrian See: Appendix I

campos The tropical grasslands of south-central Brazil.

Canadian Shield An area of pre-Cambrian rocks in the region surrounding Hudson Bay containing the oldest rocks of the North American continent.

canal An artificial channel of water.

Candlemas Eve winds Heavy winds that often visit England during February and March.

cannel coal A soft coal that burns with a bright but sooty flame.

canopy The cover or crown of all vegetation, formed by leaves, needles, and branches.

canyon A relatively narrow, steep-sided gorge.

canyon wind See: katabatic wind

cap cloud An apparently stationary cloud, resting on an isolated mountain peak, formed by the cooling and condensation of air forced up over the peak.

cape A headland jutting out into the sea.

cape doctor The strong southeast wind that blows on the South African coast.

capillarity 1. The degree to which a material or object containing minute openings or passages, when immersed in a liquid, will draw the surface of the liquid above the hydrostatic level. 2. The phenomenon by which water is held in interstices above the normal hydrostatic level, due to attraction of the molecules in the walls of an interstice for the molecules of the water and the attraction of the molecules of water for one another.

capillary action The rising of a liquid in a very slender (capillary) tube due to the attraction between the liquid and the material of the tube.

capillary fringe The belt of subsurface water held above the zone of saturation by capillary action.

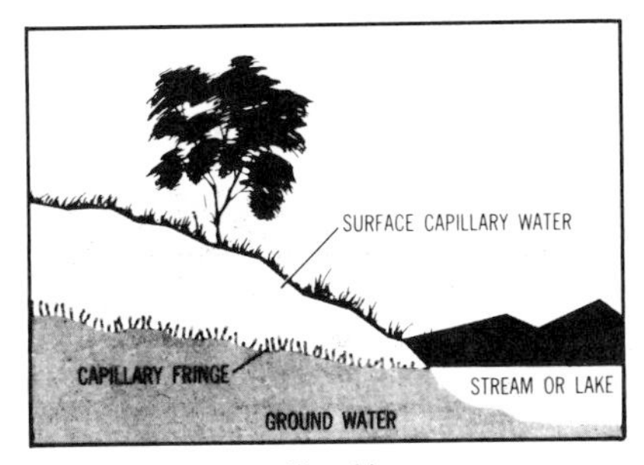

capillary fringe

capillary porosity The volume of small pores within the soil that hold water against the force of gravity.

capillary water Soil water held in the capillary spaces and as a film around soil particles.

carabao A large and powerful water buffalo.

caravan A group of traders or merchants traveling together, especially in the deserts.

carbohydrates Substances produced in all green plants by photosynthesis and having the general form CH_2O; e.g., sugar, starch, cellulose.

carbon One of the commonest chemical elements, occurring in lampblack, coal, and coke in varying degrees of purity. Compounds of carbon are the chief constituents of living tissue.

carbonaceous (adj.) Composed largely of carbon.

carbon black Carbon in the form of black powder; used in many scientific experiments.

carbon cycle The processes by which carbon is incorporated into living forms, released to the atmosphere, and returned to life forms again.

carbon cycle

carbon dioxide A gas, CO_2, comprising about 0.03 percent of the atmosphere by volume; it is necessary for plant life. It is one of the end products of the burning (oxidation) of organic matter, or carbon-containing compounds.

carbon 14 A radioactive isotope of carbon, with a half-life of 5,730 years. It can be used to determine the age of carbonaceous material younger than 30,000 years.

carbonic acid A weak, colorless acid produced when carbon dioxide dissolves in water.

Carboniferous A period of the Paleozoic era during which the major coal deposits of the world were formed. In the U.S., it is divided into the Mississippian and Pennsylvanian periods. See: Appendix I

carbon monoxide A poisonous gas with no color and little smell, formed when combustion occurs without much air.

carcinogenic (adj.) A term applied to any of the many compounds which can cause cancer.

cardinal winds Winds from the four cardinal, or principal, points of the compass, that is, north, east, south, and west.

caribou Domesticated reindeers, used as pack animals; their meat is eaten, and their hides are used for clothing.

caribou

carnauba palm A Brazilian palm tree whose leaves are coated with a wax that is used to make floor polish, candles, and other products.

carnivora Flesh-eating mammals.

carnotite An ore of uranium and vanadium.

Carolina bay An egg-shaped depression, often marshy, found in large numbers on the coastal plains of the southeastern United States.

carp A bottom-dwelling fresh-water fish, prolific and destructive, introduced into America in the 1870's.

carrier A stable isotope, or a normal element, to which radioactive atoms of the same element can be added to obtain a quantity of radioactive mixture sufficient for handling, or to produce a radioactive mixture that will undergo the same chemical or biological reaction as the stable isotope. A substance in weighable amount which, when associated with a trace of another substance, will carry the trace through a chemical, physical, or biological process. See: radioactive tracer; tracer, isotopic

carrying capacity The capability of a particular environment to support animal life; used commonly with reference to range land.

carry-over effect In cloud-seeding, the apparent persistence of the effects of silver-iodide nuclei long after seeding occurs.

Cartesian coordinates A system of coordinates for locating a point on a plane by reference to its distance from each of two intersecting lines, or in space by its distance from each of three planes intersecting at a point. In a plane, the point of intersection of the two lines is known as the origin; the vertical line is known as the y-axis; and the horizontal line is called the x-axis. Points to the right of the y-axis or above the x-axis are given positive values; those to the left of the y-axis or below the x-axis are given negative ones.

cartographer A mapmaker or chartmaker.

cartography The art of drawing maps and charts.

cascade A small waterfall, especially one of a series of closely spaced small waterfalls. A very steep rapids.

casein A protein compound in milk, the principal ingredient in cottage cheese.

cash crop A crop that is produced for sale.

cassava See: manioc

cassiterite A mineral; an important ore of tin.

caste An exclusive hereditary social class particularly observed by Hindus.

catabolism In living things, the chemical processes by which complex substances are broken down, with wastes being discarded and energy liberated.

cataclysm A violent event causing sudden change, such as a devastating flood or a huge earthquake.

catalyst A material that increases the rate of chemical reaction.

catalytic converter A device designed to remove pollutants from the exhaust stream of an automobile by promoting a chemical reaction between a catalyst and the pollutants.

cataract A large waterfall or series of waterfalls.

catastrophism The doctrine that major geological changes in the earth's history were caused by catastrophes rather than by gradual evolutionary processes.

catchment area 1. The intake area of an aquifer, and all areas that contribute surface water to the intake area. 2. The area tributary to a lake, stream, sewer, or drain. See: drainage basin; watershed

catena A group of soils within a specific soil zone, formed from similar parent materials but with unlike soil characteristics because of differences in relief or drainage.

caterpillar tractor A type of tractor that travels upon two endless metal belts.

cation An ion carrying a positive charge of electricity.

cation exchange The exchange of cations held by the soil-absorbing complex with other cations. Thus, if a soil-absorbing complex is rich in sodium, treatment with calcium sulfate (gypsum) causes some calcium cations to exchange with some sodium cations.

cation-exchange capacity A measure of the total amount of exchangeable cations that can be held by the soil. It is expressed in terms of milliequivalents per 100 grams of soil at neutrality (pH 7) or at some other stated pH value. (Formerly called base-exchange capacity.)

cat operator The driver of a caterpillar tractor.

cattle station In Australia, a cattle ranch.

cat train A string of sleds pulled by a caterpillar tractor.

Caucasian A member of the ethnic division of man, characteristically of light skin and fine hair.

caustic soda The hydroxide of sodium. A white, brittle deliquescent solid prepared chiefly by the electrolysis of a solution of common salt and by treating sodium carbonate with slaked lime. Solutions of caustic soda and water are strongly alkaline and constitute the most common type of lye found in the household and industry. Large amounts of caustic soda are used in the refining of edible oils and in the manufacture of soaps.

cave A hollow space in the earth's surface created by erosion.

ceiling The height of a cloud layer covering at least half the sky, generally in feet above the ground.

ceiling balloon A balloon used to calculate the ceiling.

ceiling height indicator See: clinometer; ceiling light

ceiling light Also called a ceiling projector or cloud searchlight; a small searchlight that is used at night to project a narrow beam of light onto the base of a cloud in order to measure its height.

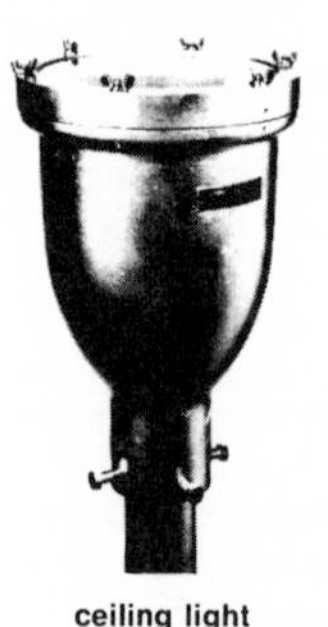

ceiling light

ceilometer An instrument for determining the height of the cloud ceiling and the rate at which the ceiling is lifting and lowering.

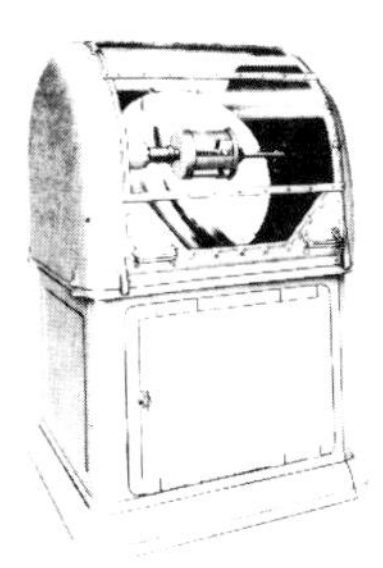

ceilometer

celestial equator A great circle of the celestial sphere lying in the same plane as the equator.

celestial sphere A sphere of infinite radius with its center at the point of the observer.

cell 1. In biology, the basic structure of which all living things are built. Consists of a membrane containing a central portion (nucleus) and surrounding cytoplasm. Living organisms may consist of one cell (the amoeba) or of billions (man). 2. In meteorology, the unit in any circulation system; thus, we speak of low- and high-pressure cells. See: Hadley cell; cyclone; anticyclone

cellular hypothesis A theory of the general circulation of the atmosphere, which seeks to explain the fact that the belt of high pressure around the earth of 30° latitude, the horse latitudes, and the belt of low pressure at 60° latitude are not continuous, even on average annual charts of pressure distribution, and are especially irregular and uncertain over the continents and the western parts of the oceans. The circulation near the horse latitudes is considered to be taking place in four cells about 90° in longitudinal width. These cells each contain an anticyclonic circulation that tilts upward from east to west, so that at high levels the anticyclones are farther west than at the surface. At about 60° latitude, there are similar cells of the same number of degrees in width but containing cyclonic circulations that tilt in the opposite direction, i.e., from west upward to east.

cellulose Glucose molecules arranged in long chains that are the principal constituent of the cell walls of higher plants. The long molecules give it a fibrous nature. Cotton fibers are almost pure cellulose. Paper is mainly cellulose separated by chemical processes from wood or other plant remains.

Celsius scale The thermometric scale in which the fundamental interval—between the temperature of melting ice and the temperature of the vapor of boiling water at 760 mm. normal atmospheric pressure — is divided into 100 equal parts, each part being called a degree. The boiling point is labeled 100°, the freezing point 0°.

cement A chemically precipitated material binding together the grains or minerals in a rock.

Cenozoic era See: Appendix I

centigrade (adj.) Designating a temperature scale divided into 100 degrees. See: Celsius scale

centimeter Metric unit equivalent to 0.3937 inch.

centimeter dyne See: erg

centimeter-gram-second A system of units based on the centimeter as the unit of length, the gram as the unit of mass, and the second as the unit of time. Abbreviated as c.g.s. See: dyne

centrifugal force The force that pulls a rotating object away from the center of rotation.

centripetal force A force pulling a rotating object toward the center of rotation.

centrosome The active center of cell division in mitosis.

centrosphere 1. Protoplasm around a centrosome. 2. The center portion of the earth.

century plant See: agave

cephalic (adj.) Of or pertaining to the head.

cephalic index The ratio of the greatest breadth of a head to its greatest length, multiplied by 100.

cephalopod A member of a group of mollusks having a distinct head surrounded by muscular tentacles (arms).

cereal grass Any grass that produces kernels used as food.

c.g.s. Abbreviation of centimeter-gram-second.

chaff The fine husks of grains such as wheat and rye.

chain A mountain system consisting of a series of more or less parallel ranges.

chain lightning Lightning in a long zigzag or broken line.

chain reaction A reaction that stimulates its own repetition. In a fission chain reaction a fissionable nucleus absorbs a neutron and fissions, releasing additional neutrons. These in turn can be absorbed by other fissionable nuclei, releasing still more neutrons.

chalcocite An important ore of copper.

chalk A soft white or greyish limestone composed of small dead marine organisms.

challiho Strong southerly winds that blow in parts of India for some forty days around the month of April.

chamsin See: khamsin

change of state The process by which a substance passes from one to another of the solid, the liquid, and the gaseous states, and in which marked changes in its physical properties and molecular structure occur. The change from the solid to the liquid state is called fusion or melting, the reverse change, freezing; the change from the liquid to the gaseous state is called vaporization, the reverse change, condensation; and the change from the solid directly to the vapor state is called sublimation, the reverse change, condensation.

channel An open conduit either naturally or artificially created that periodically or continuously contains moving water, or that forms a connecting link between two bodies of water. Rivers, creeks, runs, branches, anabranches, and tributaries are natural channels, which may be single or braided. Canals and floodways are artificial channels.

channel storage The volume of water at a given time in the channel or over the flood plain of the streams in a drainage basin or river reach. Channel storage is greatest during a flood event.

chaparral Vegetation characterized by low, dense shrubs or dwarf trees having mostly evergreen, waxy leaves, such as scrub oaks and buck brush.

Charles' law The physical law that for each rise of 1°C. in temperature, all the common gases expand by the same fraction (about 1/273) of their volume at 0°C., the pressure being kept constant.

chart A map or graph of any type.

chelate A chemical compound in which a metallic atom is firmly combined with a molecule by means of multiple chemical bonds.

chemical element See: element

chemical weathering Decomposition, the breakdown of rock material by chemical means.

chemical wood See: acid wood

chemopause The transition layer between the chemosphere and the ionosphere.

chemosphere Atmospheric zone with a high ozone concentration, about 20 to 60 miles above the earth's surface.

chemosynthesis The synthesis of organic compounds with energy derived from chemical reactions.

chernozem A zonal group of soils having deep, dark to nearly black surface horizons and rich in organic matter, which grades into lighter colored soil below. At 1.5 to 4 feet, these soils have layers of accumulated calcium carbonate. They develop under tall and mixed grasses in a temperate to cool subhumid climate.

chert A structureless form of silica, closely related to flint, which breaks into angular fragments.

cherty soil Soils developed from impure limestones containing fragments of chert and having abundant quantities of these fragments in the soil mass.

chestnut soils A zonal group of soils with dark-brown surface horizons, which grade into lighter colored horizons beneath. They have layers of accumulated calcium carbonate at 1 to 4 feet. They are developed under mixed tall and short grasses in a temperate to cool and subhumid to semiarid climate. Chestnut soils occur in regions a little more moist than those having brown soils and a little drier than those having chernozem soils.

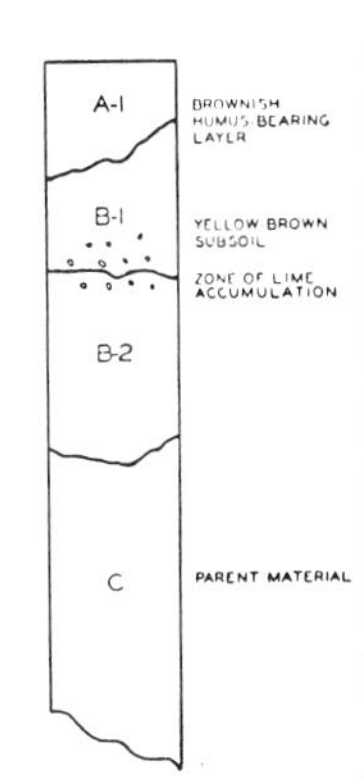

chestnut soils

chili Hot, dry, southerly sirocco wind of north Africa.

chill factor An index combining temperature, humidity, and wind, used to estimate heat loss from the human body.

chimney cloud A cumulus cloud in the tropics that has a much greater vertical than horizontal extent.

China clay See: kaolin

chinook A warm, dry, southwest wind along the eastern slopes of the Rocky Mountains in the western United States and Canada, identical with the European foehn. It may occur at any season of the year, but its effects are most marked in winter, when it may cause a rapid rise in temperature, as much as 20° to 30°F. in fifteen minutes, and cause ice and snow to disappear in a few hours.

chinook arch A bank of clouds over a range of the Rocky Mountains, heralding the approach of the chinock.

chlordane A colorless, viscous, water-insoluble toxic liquid used as an insecticide.

chlorinated hydrocarbons Organic pesticides, which include DDT, DDD, dieldrin, chlordane, etc.

chlorophyll A pigment, the constituent responsible for the green color of plants and important in photosynthesis.

chloroplasts Small bodies in cells of plants in which chlorophyll is concentrated.

chlorosis A condition in plants resulting from the failure of chlorophyll to develop, usually because of deficiency of an essential nutrient. Leaves of chlorotic plants range from light green through yellow to almost white.

chocolatero Name applied in Mexico to a milder norther.

cholesterol A solid alcohol occurring in all animal fats, thought to be one of the factors in heart disease.

chromite The principal ore of chromium.

chromium A metallic element used as an alloy, a catalyst, etc.

chromosomes Threadlike bodies that carry genes.

chronometer A precise clock used to determine longitude.

chubasco A violent wind and rain squall, attended with heavy thunder and vivid lightning, occurring during the rainy season along the west coast of Central America.

cierzo Spanish name for a mistral.

cinchona The plant from which quinine and other valuable drugs are made.

cinder cone A cone created from ejected material during a volcanic eruption.

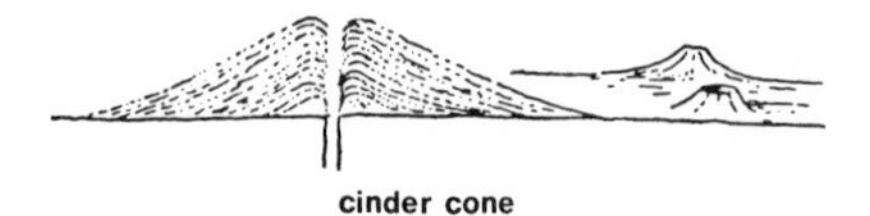

cinder cone

cinnabar The principal ore of mercury.

cion A detached shoot of a plant capable of propagation.

cirque A deep, steep-walled, bowl-shaped hollow in a mountain, open in front, a natural amphitheater created by glacial action.

cirque

cirriform (adj.) Relating to, or shaped like, cirrus.

cirro A combining form of cirrus.

cirrocumulus A cloud layer or patch of small white flakes or of very small globular masses, usually without noticeable shadows, which are arranged in groups or lines, or more often in ripples resembling those of the sand on the seashore.

cirrostratus A thin whitish veil of cloud which does not blur the outlines of the sun or moon, but often gives rise to a halo.

cirrus Detached high cloud of delicate and fibrous appearance, without shading, generally white in color, often of a silky appearance.

citrus Evergreen fruit-bearing trees native to southeast Asia; includes lemons, oranges, grapefruit, etc.

civil time Legally accepted time. Based on mean solar time, the civil day period is from one midnight to the next midnight, or 24 hours long.

clastic (adj.) Consisting of fragments of rocks or of organic materials that have been moved from their original position.

clay 1. Mineral soil material that is plastic when wet, consisting essentially of hydrated silicates of aluminum. 2. Soil material that contains 40 percent or more of clay, less than 45 percent of sand, and less than 40 percent of silt.

clay loam Soil material that contains 27 to 40 percent of clay and 20 to 45 percent of sand.

clay mineral Naturally occurring inorganic crystalline material in soils or other earthy deposits of clay size—particles less than 0.002 mm. in diameter.

claypan A compact, slowly permeable soil horizon rich in clay and separated more or less abruptly from the overlying soil, commonly hard when dry and plastic or stiff when wet.

clear (adj.) 1. Unclouded or of negligible cloudiness and good visibility. 2. The state of the sky when it is cloudless or less than one-tenth covered by clouds.

clear air turbulence (CAT) Air movements outside cloud structures, generally associated with either wind shear layers or the jet stream.

clear cutting A method of cutting that removes all merchantable trees on the area in one cut.

clear ice Ice with a glassy surface, which varies from clearness to translucency.

clearing The act of the weather in changing from cloudy or stormy conditions to a condition characterized by a nearly cloudless sky and the cessation of precipitation.

cleavage 1. In a crystal, the tendency to split along planes determined by the crystal structure. 2. In a rock, the tendency to split along definite, parallel planes. These cleavage planes may be at sharp angles to the bedding planes or strata of the rock.

cliff A high and extremely steep rock face.

climagram 1. Any graphical representation of one or more climatic features of a place or region, regardless of the quantities used as abscissae and ordinates. 2. A diagram drawn in rectangular coordinates, of which the abscissa is temperature and the ordinate is relative humidity.

climate The sum total of the meteorological elements that characterize the average and extreme condition of the atmosphere over a long period of time at any one place or region of the earth's surface.

climate change The results of climate modification, due to human or nonhuman factors.

climate control The purposeful and predictable application of weather-modification techniques.

climate divide A boundary, produced by land relief, between regions having separate types of climate. Mountain ranges produce sharp transitions (a) by the rapid increase in elevation above the sea, hence a rapid decrease in surface temperature, upward along their flanks, and (b) by preventing the free interchange of air between the regions they separate.

climate modification The purposeful or inadvertent alteration of climate.

climatic classification Classification of the climates of the different regions of the earth's surface, based on one or more of the climatic elements.

climatic controls The terrestrial and atmospheric influences which, acting upon the climatic elements, produce the changes in temperature and precipitation that give rise to varieties of weather and climate.

climatic cycles Recurrences of such weather phenomena as wet and dry years, hot and cold years, at more or less regular intervals, in response to long-range terrestrial and solar influences, such as volcanic dust and sunspots. See: Bruckner cycle

climatic elements The meteorological phenomena that comprise climate: temperature (including radiation); moisture (including humidity, precipitation, and cloudiness); wind (including storms); pressure; evaporation; and also, but of less importance, the composition and the chemical, optical, and electrical phenomena of the atmosphere.

climatic factors Those physical conditions (other than the climatic elements) which currently control climate, or may by their changes over long periods cause climatic changes: latitude, altitude, distribution of land and sea, and topography. Some climatologists include ocean currents, the semipermanent high- and low-pressure areas, prevailing wind, etc.

climatic instability The notion that the forces which control the world's climate are subject to changes that will alter the pattern of present climatic zones.

climatic optimum The period (c. 5000-2500 B.C.) when the earth's temperatures were much warmer than at present.

climatic province An area of the earth that has a definite climatological characteristic, according to a certain climatic classification.

climatic year The result obtained by substituting the values of the meteorological elements in a given year for the values of a type of climate given by one or another of the climatic classifications.

climatic zones 1. In general, any divisions of the earth's surface, large or small, based on the climatic elements. 2. In particular, the five zonal divisions of the earth's climate known to the ancient Greeks and widely used down to the present time: the torrid zone, the two temperate zones, and the two frigid zones.

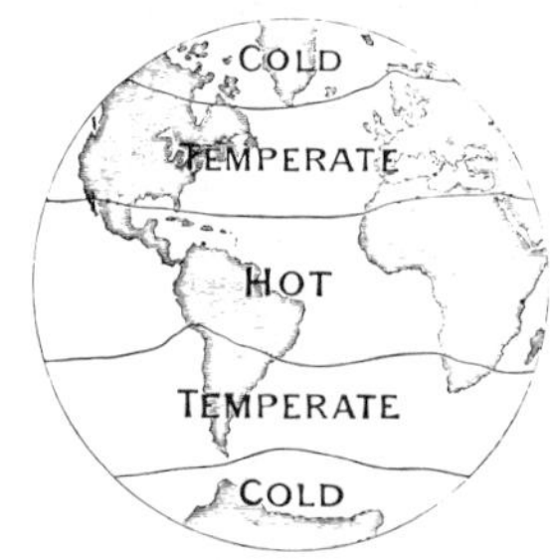

climatic zones

climatize (v.) To acclimate, or become acclimated.

climatogram See: climagram

climatography 1. The study of descriptive, statistical, and cartographical climatology. 2. An account of the climate of a particular place or region.

climatology The science, closely related to meteorology and geography, that seeks to determine and describe the various types of climate, from analysis of climatic data, and to explain the causes of these climates, their variation, geographical location, effects on plant and animal life, etc.

climax community A plant community that does not change unless there is a change in the climate. It is the culminating stage in natural plant succession. The plants in a climax community are favored by the environment that they themselves create, and so are in balance with it.

climograph A climatic diagram for showing the mean monthly values of wet-bulb temperature and relative humidity, or other pairs of elements, at any place, and for comparing such data as recorded at different places throughout the world, especially with reference to the effects of climate on mankind.

cline The gradient or pattern of variation in characteristic traits of an organism from one area to another.

clinometer A portable instrument used with a ceiling light, to measure cloud heights at night.

clone The aggregate of plants derived from a single seeding by means of vegetative propagation such as the rooting of cuttings or slips, budding, or grafting. Every member of a clone has the same heredity, so that under uniform environment a group of plants from a single clone is quite uniform.

close (adj.) Descriptive of the weather, when the air is still, moist, and hot.

closed basin A basin draining to some depression or pond within its area, from which water is lost only by evaporation or percolation.

closed loop system A pollution control system consisting of monitoring stations that measure the effluent of industrial operations, especially smelters. When emissions exceed allowable standards the loop is closed by the transmission of this information to the smelter, which then ceases its operations.

cloud A mass of small droplets or ice crystals formed by condensation of water in the atmosphere; fog is a cloud at ground level.

cloud bank A mass of clouds stretching across the sky and usually of considerable vertical extent.

cloudburst A sudden and heavy downpour of rain.

cloud camera A photographic camera used in, or specially designed for, obtaining cloud pictures.

cloud classification A scheme of distinguishing and grouping clouds according to their appearance, elevation, or method of formation.

BASE ALTITUDE	CLOUD TYPE	ABBREVIATION	SYMBOL
BASES OF HIGH CLOUDS USUALLY ABOVE 20,000 FEET	CIRRUS	Ci	
	CIRROCUMULUS	Cc	
	CIRROSTRATUS	Cs	
— 20,000 FEET —			
BASES OF MIDDLE CLOUDS RANGE FROM 6,500 FEET TO 20,000 FEET	ALTOCUMULUS	Ac	
	ALTOSTRATUS	As	
— 6,500 FEET —			
BASES OF LOW CLOUDS RANGE FROM SURFACE TO 6,500 FEET.	*CUMULUS	Cu	
	*CUMULONIMBUS	Cb	
	NIMBOSTRATUS	Ns	
	STRATOCUMULUS	Sc	
	STRATUS	St	
— SURFACE —			

*CUMULUS AND CUMULONIMBUS ARE CLOUDS WITH VERTICAL DEVELOPMENT. THEIR BASE IS USUALLY BELOW 6,500 FEET BUT MAY BE SLIGHTLY HIGHER. THE TOPS OF THE CUMULONIMBUS SOMETIMES EXCEED 60,000 FEET.

cloud classification

cloudiness The amount of sky covered by clouds, irrespective of the thickness of the clouds; usually measured in tenths of the sky canopy.

cloudless (adj.) Clear; unclouded.

cloud searchlight See: ceiling light

cloud seeding The injecting of a substance into a cloud to influence its subsequent development; silver-iodide, carbon black, and dry ice are commonly used substances.

cloud street A line of individual cumuliform clouds, often formed downwind of small islands in the tropics.

cloudy (adj.) 1. In general, the state of the weather when clouds are present in the sky. 2. The state of the sky when from six- to nine-tenths of it is covered with clouds.

coagulation The act or state of becoming jellylike or of uniting into a coherent mass. The change from a liquid to a thickened curdlike state, not by evaporation, but by chemical reaction or heat; as the coagulation of blood, milk, or egg white.

coal Generally, a rocklike derivative of fossilized vegetation, composed of carbon and various compounds. See: bituminous coal; anthracite

coalescence Uniting two or more objects into one, as in the growth of cloud and rain drops by collisions between drops.

coast That part of a land mass that borders an ocean.

coastal plain A plain that borders the sea and extends inland to the nearest elevated land.

coastline (or shoreline) The edge of a land mass bordering the sea, which follows major bays and indentations but crosses inlets and estuaries.

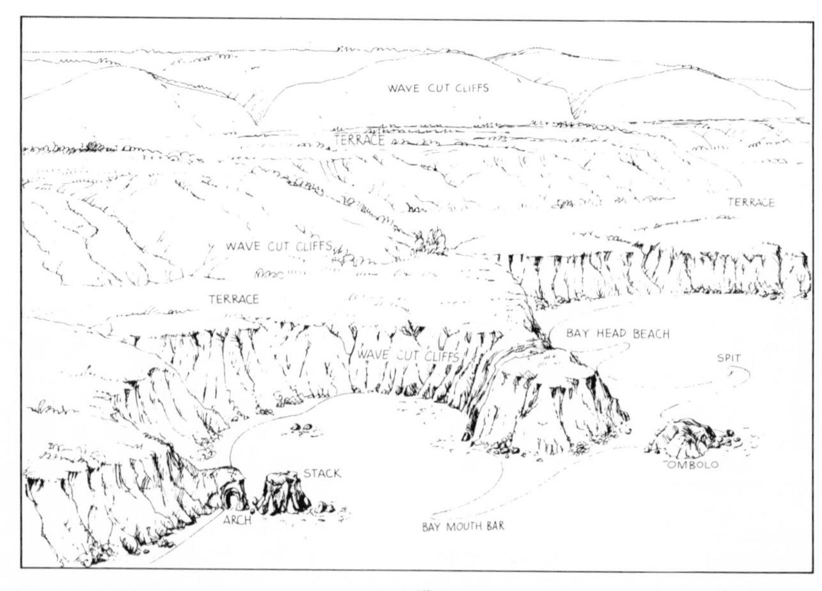

coastline

cod Carnivorous commercial fish, important in the New England fishing industry but found in both the Pacific and Atlantic oceans.

coefficient of correlation A number, between the limiting values of +1 and –1, which expresses the degree of linear relationship between two variables, and which can be determined from any one of several formulas. A high positive value of the coefficient, +0.8 or greater, indicates a close relationship between the variables such that large values of the one are associated with large values of the other; conversely, a low negative value, –0.8 to –1, means that the two quantities vary closely in the opposite sense. A value of the coefficient of correlation near 0 shows that there is little relationship between the variables.

coffee Tropical evergreen trees grown for their berries, which are roasted and ground and used to make a beverage.

coffee

coke A modified form of bituminous coal from which the volatile materials have been driven off by heat.

col 1. A neck of relative low pressure between two anticyclones; also called a saddle or neutral point. 2. A pass between two peaks.

cold air mass Broadly, an air mass that is cold relative to neighboring air masses, implying that it originated in higher latitudes.

cold box An insulated temperature-controlled box used to make estimates of the number of freezing nuclei in a sample of air.

cold fog Term generally applied to radiation fog.

cold front The line along which a wedge of cold air underruns and displaces a warmer air mass.

cold-front occlusion An occlusion formed when the air in the rear of a cold front is colder than the air in advance and hence will underrun the latter.

cold-front thunderstorm A type of thunderstorm that occurs in series along a line usually 100 miles in advance of a cold front.

cold poles The places in the northern and southern hemispheres that have the lowest annual mean temperatures.

cold wave A rapid and marked fall of temperature during the winter.

coliform Resembling or designating a certain group of bacteria, some species of which are normally present in the large intestine and feces of all warm-blooded animals. Other species are present on grains, grasses, and other plant materials.

colla In the Philippines, a period of stormy weather lasting for several days, with rain and winds mostly from the southwest.

collada A strong wind (35 to 50+ mph) from the north or northwest in the upper Gulf of California and from the northeast in the lower Gulf of California.

collagen A gelatinlike protein occurring in vertebrates. It is the chief constituent of the fibrils of connective tissue and an important part of hides to be tanned.

collective farm A large farm managed and worked by people under government supervision.

colloid A gelatinous substance that when dissolved in a liquid will not diffuse through membranes.

colloidal system An intimate mixture of two substances, one of which is uniformly distributed in a finely divided state throughout the other.

colluvium Mixed deposits of soil material and rock fragments accumulated near the base of steep slopes through soil creep, slides, and local wash.

colony A group of individuals (human or animal) living together and having some form of organization for the provision of their needs.

Colorado low A low pressure area that makes its first appearance as a definite center near the eastern slopes of the Rocky Mountains in eastern Colorado and deepens rapidly over the Great Plains Region.

columnar jointing Jointing that breaks a large mass of rock into parallel columns, commonly found in basaltic rocks.

columnar jointing

combine A machine that cuts, threshes, and cleans grain while moving across a field.

combusion nucleus A condensation nucleus formed as a result of combustion processes.

comet A heavenly body revolving around the sun in an eccentric orbit; it consists of a solid nucleus and a luminous, gaseous tail.

comet

comfort zone The combination of humidity, temperature, and air movement under which most people feel comfortable.

commons Land belonging to a local community and open for public use.

community 1. The assemblage of plant and animal populations occupying a given area. 2. A social group of any size whose members reside in a specific locality and share common interests and form of government.

compage A structure in which the many parts function as a whole. A system similar to an ecosystem.

companion crop A crop grown with another crop, usually a small grain with which alfalfa, clover, or other forage crops are sown.

compass An instrument consisting of a free-swinging magnetized needle used to find the north magnetic pole.

compass

compass card The piece of paper on which the directions of the compass are printed.

compass card

compass rose A device imprinted on a map to indicate compass directions.

compass rose

competition The struggle among organisms for food, space, and other requirements for existence.

composite cone A volcanic cone, generally large, built by alternate eruptions of lava and cinders.

compost Rotted organic matter made from waste plant residues, with inorganic fertilizers, especially nitrogen, and soil added.

compound A substance with its own distinct properties, formed by the chemical combination of two or more elements in fixed proportion.

compression The process or state of being condensed in size or volume.

compressional wave See: p wave

concentration time 1. The period of time required for storm runoff to flow from the most remote point of a catchment or drainage area to the outlet or point under consideration. It is not a constant, but varies with depth of flow and condition of channel. 2. The time when the rate of runoff equals the rate of rainfall of a storm of uniform intensity.

concretion A hard mass, usually rounded, in a sedimentary rock. Formed by deposition of successive layers around a nucleus, such as a grain of sand or a fossil.

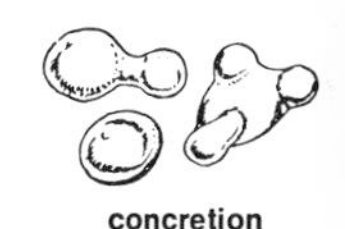

concretion

condensation The process by which a vapor becomes a liquid or a solid

condensation adiabat See: wet adiabat

condensation level The level at which air becomes saturated when it is lifted adiabatically.

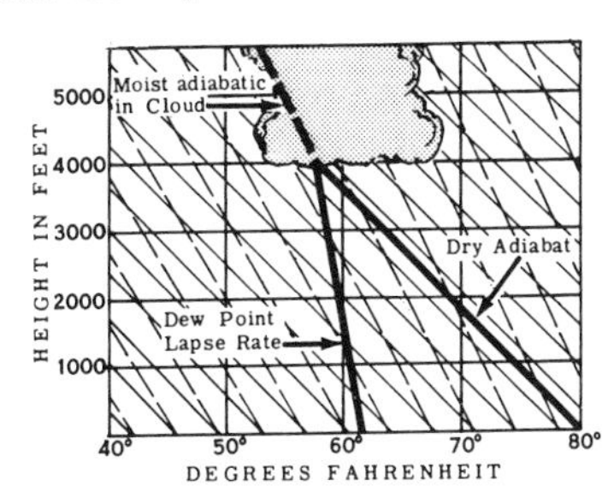

condensation level

condensation nucleus A particle upon which condensation of water vapor begins in the free atmosphere, commonly sea salt, products of combustion, or dust blown up from the earth's surface. See: hygroscopic particles

condensation trails Small artificial clouds observed frequently behind aircraft flying in air at a low temperature.

conditional equilibrium See: stability

conditional instability The state of moist unsaturated air in which the lapse rate is intermediate between the dry- and the moist-adiabatic; stability is conditional on the water-vapor content and the occurrence of lift.

conduction The transference of heat within and through a substance by means of internal molecular activity and without any obvious external motion.

conductivity (electrical) A measure of the readiness with which a medium transmits electricity. Commonly used for expressing the salinity of irrigation waters and soil extracts because it can be directly related to salt concentration. It is expressed in mhos per centimeter (or millimhos per centimeter or micromhos per centimeter) at 25°C.

cone 1. Anything shaped like a cone. 2. The fruit of conifers bearing seeds.

cone of influence (or depression) The depression, roughly conical in shape, produced in a water table by the extraction of water from a well at a given rate.

configuration An arrangement or pattern of the parts of something.

confined aquifer An aquifer that is bounded above and below by formations of impermeable or relatively impermeable material.

confluence The point where two streams converge and unite.

conformal map A map in which all angles between lines on the surface of the earth are preserved on the map, and consequently the shape of any small area on the map is the same as the shape of the corresponding area upon the earth; also known as an orthomorphic map.

conglomerate Rock composed of rounded pebbles cemented together in a matrix of finer material.

conical projection A map projection in which the earth's surface is produced on a paper cone, so that latitudes appear as concentric circles and meridians as radiating lines.

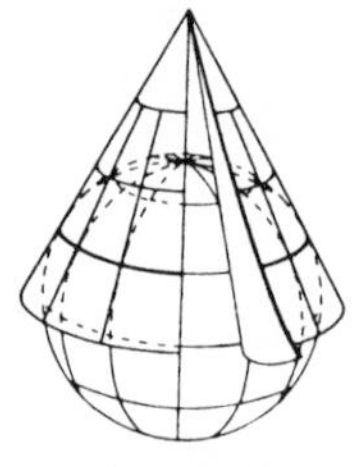

conical projection

conifer A plant, generally evergreen needleleafed, bearing naked seeds singly or in cones.

coniferous forest A forest of conifers, which provides most of the softwoods used for lumber and pulp.

conjunction The apparent proximity of two heavenly bodies, generally as seen in the same direction from earth.

connate water Water present in rock at the time of its formation, frequently saline.

consecutive mean The mean of a small group of consecutive values, especially annual means, assigned to the middle time-term of the group, in the process of smoothing.

consequent stream A stream that flows over the initial surface of the land.

conservation Rational use of the environment.

conservation of angular momentum The principle of the constancy of moment of momentum in a system on which no external force acts. The angular momentum of a body revolving about an axis is the product of its mass, its linear velocity, and the distance of its center of mass from the axis of rotation. As long as the torque acting on the system is balanced, this product remains constant.

conservation of energy A principle that energy is never created and never destroyed, so that the total amount of energy in the universe remains constant, though transformations from one form to an equivalent amount of some other form are continually occurring.

conservative property Any property the nature or value of which is affected to a comparatively small degree by the various modifying influences; e.g., specific humidity.

constant level chart A chart that represents the synoptic distribution of one or more meteorological elements at any fixed geometric elevation (including zero) above sea level. The sea level or ten-thousand-foot pressure maps are examples of constant level charts.

constant pressure chart A chart that contains the synoptic contour lines of the height above sea level of any selected isobaric surface in the free atmosphere.

constellation Any of a number of groups of fixed stars, about ninety of which have been identified.

constructional landforms Landforms created by forces of earth movement; i.e., folding, faulting, and volcanism.

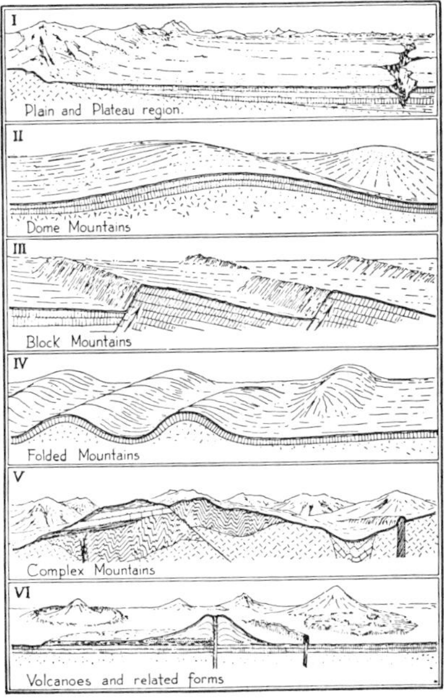

constructional landforms

consumption See: consumptive use; domestic use; industrial consumption

consumptive use The quantity of water discharged to the atmosphere or incorporated in the products of vegetative growth, food processing, or an industrial process.

contamination (water) Impairment of the quality of water sources that creates hazard to public health.

continent One of the earth's principal land masses: Europe, Africa, Asia, Australia, Antarctica, North America, South America.

continental air mass An air mass, generally with a low moisture content, that has its source over a continent.

continental climate The climatic type characteristic of the interiors of great land masses, the distinctive features of which are large annual and daily range of temperature and cold winters and warm summers.

Continental Divide On the North American continent, the line formed by crests of the Rocky Mountains that separates streams flowing toward the Atlantic from those flowing toward the Pacific.

continental drift The concept that the continental land masses, composed of lighter rock materials, float or drift on the denser material beneath them. See: Gondwana

UPPER CARBONIFEROUS

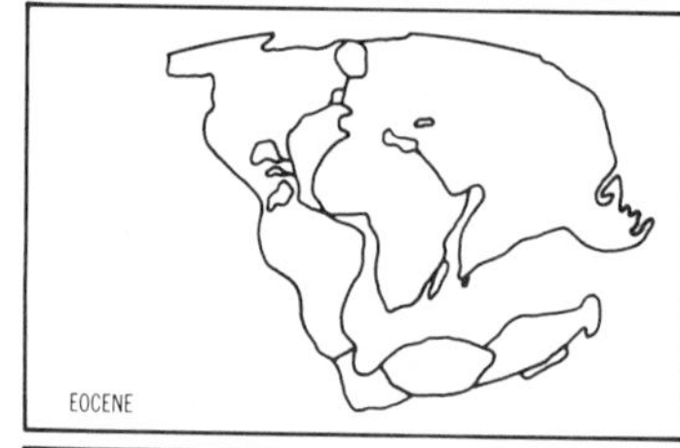

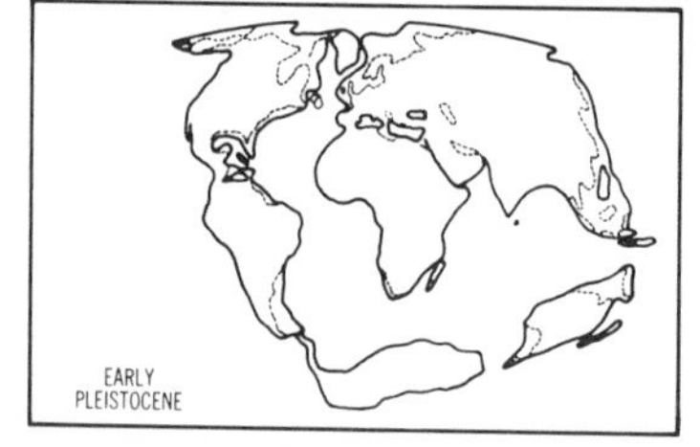

continental drift

continental glacier An ice sheet covering a large part of a continent.

continentality A measure of the degree to which the climate of a particular region approaches the typical continental climate.

continental platform That portion of the earth's crust consisting of the continents and the continental shelves.

continental shelf The sea bed bordering the continents, covered by shallow water 100 fathoms or less in depth.

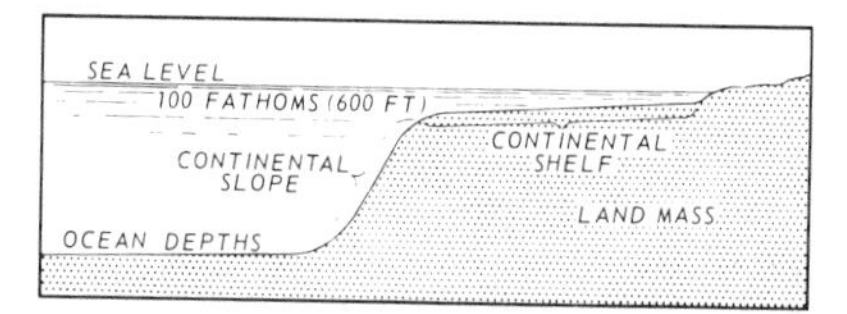

continental shelf

continental slope The slope that descends from the edge of the continental shelf to the deep ocean bed.

contour (or contour line) A line on a map drawn through points having the same elevation above or below sea level.

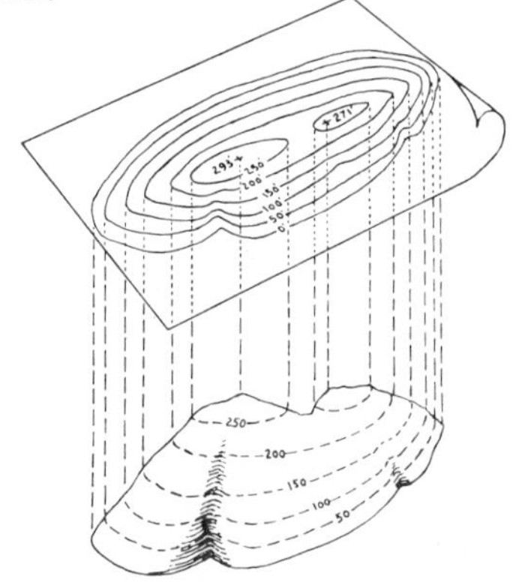

contour

contour map A map showing by means of contour lines the shape of the land surface and the elevation of each point.

contour plowing Plowing at right angles to the slope of the land to prevent erosion and conserve water.

control days Days whose weather is supposed to indicate the weather for a certain number of following days. Groundhog Day (February 2) and Saint Swithin's Day (July 15) are well-known examples.

conurbation A metropolitan area formed by the growth and coalescence of several neighboring but formerly separate cities and towns.

convection 1. In physics, the circulation in a fluid of nonuniform temperature, owing to differences in density and the action of gravity. With radiation and conduction it is one of the three methods for the transfer of energy. 2. In meteorology, the process whereby a circulation is created and maintained within a layer of the atmosphere, due either to surface heating of the bottom of the layer or to cooling at its top, and consisting in the sinking of relatively heavy air and the consequent forcing up of air which, volume for volume and under the same pressure, is relatively light.

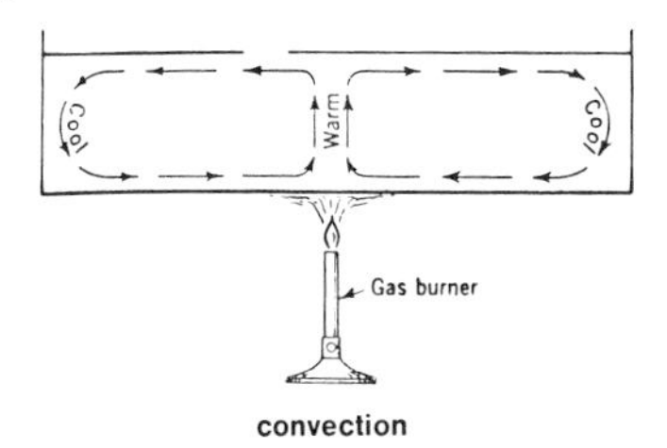

convection

convection cell See: Benard cell

convection theory The theory that slow convection currents inside the earth carry heat generated in the interior by radioactive elements to the surface. The movements of plastic rock material may produce continental drift and give rise to the earth's magnetic field.

convectional rain Rain resulting from convectional processes in the atmosphere.

convective cell An individual air system involving an updraft and surrounding weaker downdrafts.

convective condensation level The level to which air, if heated sufficiently from below, will rise before it will become saturated; accordingly, the height at which the base of cumuliform clouds may form by thermal convection from the surface. See: condensation level

convective instability The condition of an unsaturated layer of air having a stratification of humidity such that, upon being lifted, the lower part of the layer becomes saturated first, and hence cools thereafter at a slower rate than does the upper, drier portion, until the lapse rate of the whole layer becomes equal to the saturation adiabatic and any further lifting results in instability. Also called potential instability.

convective rain Rain that is caused by the adiabatic cooling of moist air that rises by reason of the vertical thermal or convective instability of the atmosphere.

convective showers Intermittent convective rain.

convective thunderstorm Convective showers that develop to thunderstorm intensity, i.e., that build up electrical space charges of sufficient potential to produce lightning discharges.

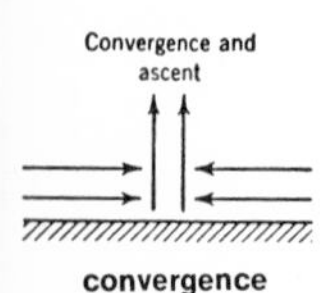

convergence

convergence The increase of mass within a given layer of the atmosphere when there is a net horizontal inflow of air into the layer. The accumulation of mass leads to vertical motion that limits the amount of the excess. Hence, if there is convergent flow at the surface, there must be upward vertical motion; if there is horizontal convergence at any level aloft, there must be upward or downward motion.

conversion factor A number by which a quantity expressed in one system of units may be converted to another system.

converter reactor A reactor that produces some fissionable material, but less than it consumes. In some usages, a reactor that produces a fissionable material different from the fuel burned, regardless of the ratio. In both usages the process is known as conversion. (Compare breeder reactor.)

conveyance loss The loss of water from a conduit due to leakage, seepage, evaporation, or evapo-transpiration.

cooling processes Processes that bring about a decrease in temperature. See: nocturnal cooling; advective cooling; evaporative cooling; dynamic cooling

coordinate One of a set of two or more lines used in reckoning the position of a point, line, or plane. See: geographic coordinates; Cartesian coordinates

copper A metallic element, an excellent conductor of heat and electricity; used in alloys such as brass and bronze.

copra The dried meat of the coconut, valued for its oil.

coral polyp A small marine creature with a hard skeleton that lives in shallow water; skeletons accumulate to form coral reefs and islands.

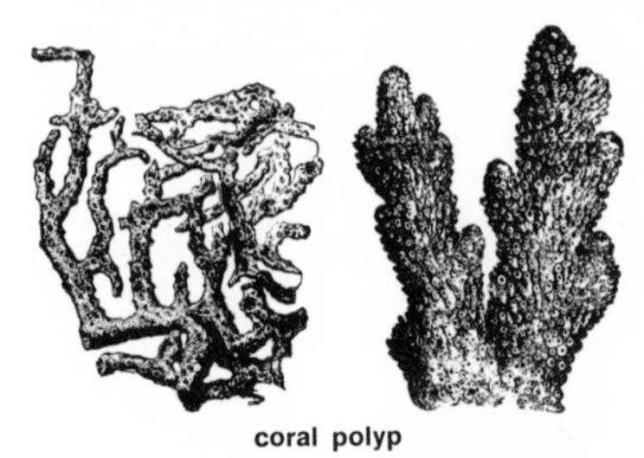
coral polyp

coral reef The skeletal remains of coral polyps built into a series of rock-like edifices that lie at or near the surface of the sea.

cordillera A group or system of mountain ranges including the valleys and plains between them.

cordonazo A violent southerly gale on the west coast of Mexico, generally occurring between June and November, and probably due to the passage of a tropical cyclone offshore.

corduroy road A road built of logs or poles laid side by side across the roadway, usually in low or swampy places.

Coriolis force The effect of the earth's rotation on wind direction, expressed as an acceleration by French scientist G. G. Coriolis in 1844. The Coriolis acceleration becomes a force when applied to a moving mass of air.

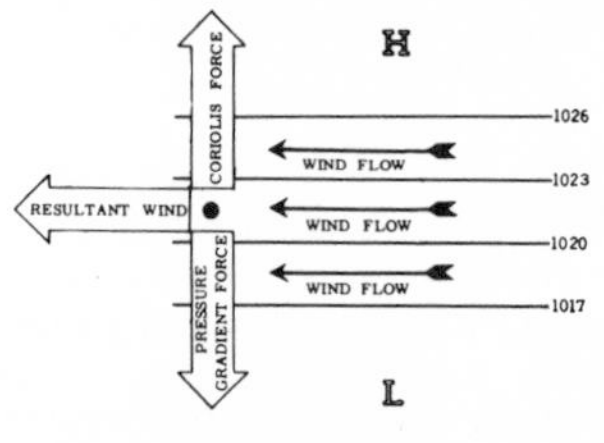

Coriolis force

cork oak Oak trees native to Asia from the outer bark of which cork is obtained; grown chiefly in Mediterranean countries.

corn belt The region in the central United States (Iowa, Illinois, Nebraska) where corn is the principal crop.

corneous endosperm See: endosperm

corona A set of one or more rainbow-colored rings of small radii concentrically surrounding the sun, moon, or other source of light when covered by a thin cloud veil.

coronary (adj.) Pertaining to either of two arteries (right and left) that are connected to the aorta and supply the heart tissues with blood.

corralines 1. Any red algae that are impregnated with lime. 2. Any coral-like animals.

corrasion The breakup of rock by abrasion; physical weathering.

correlation 1. The process of establishing a relation between a variable and one or more related variables. Correlation is simple if there is only one independent variable; multiple, if there is more than one independent variable. 2. The amount of similarity in direction and degree of variations in corresponding pairs of observations in two series of variables.

correlation coefficient See: coefficient of correlation

corrosion The wearing away of rocks by chemical action.

cosmic dust Fine particles thought to be spread thinly in space except where they are gathered into enormous clouds.

cosmic radiation Corpuscular radiation of great energy, penetrating power, and speed, which apparently travels through space equally in all directions.

cosmology The branch of philosophy that deals with the origin and general structure of the universe.

cost-benefit analysis A method used to determine the feasibility of pursuing a particular project by balancing estimated costs against expected benefits; used especially in evaluating water projects.

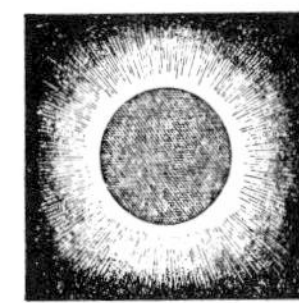
corona

cotton belt The region in the American South in which most of the nation's cotton has traditionally been grown.

cotton gin Machine for separating cotton fibers from cotton seed.

cotyledon The rudimentary leaf of the embryo of seed plants.

coulee A deep gulch in the Western U.S.

countertrade See: antitrade

country rock Rock that surrounds igneous intrusions such as batholiths, dikes, or veins.

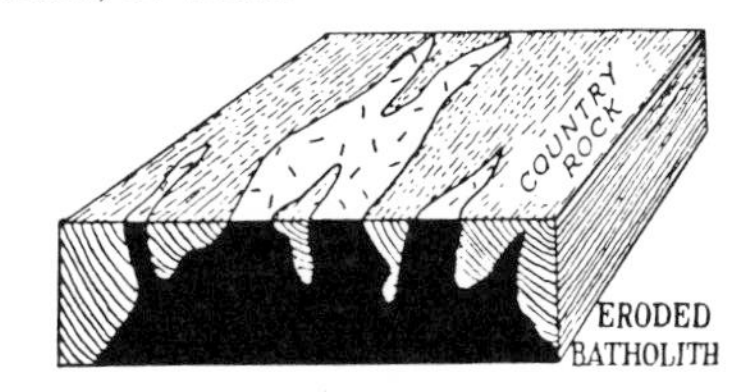

country rock

cove A small inlet or bay.

cover crop Crop, generally legume, planted to protect the soil from leaching and erosion.

crag A rough, steep rock or a projecting rock point.

crankcase smog controls (PCV System) The Positive Crankcase Ventilation System allows the engine crankcase of an automobile to breathe without discharging the products of combustion into the atmosphere. Blow-by gases leaking past the piston rings into the engine crankcase are drawn into the intake manifold by means of a tubing and valving system. These gases are then reburned in the engine cylinders.

crater The funnel-shaped hole at the top of a volcanic cone.

creek A small stream of water that serves as the natural drainage course for a drainage basin of nominal or small size. Its size is relative: some creeks in the humid section would be called rivers if they occurred in the arid portion.

creep, soil See: soil creep

Creole 1. In Latin America, one born in the region but of European ancestry. 2. In Louisiana, those of French ancestry. 3. In some cases, those of French and Negro ancestry.

creosote bush A resinous desert shrub dominant in many parts of the American southwest.

crepuscular rays Clearly defined rays of the sun that shine through breaks in heavy cloud layers, seen at sunrise and sunset; sometimes referred to as "the sun drawing water."

crest 1. The top of a dam, dike, spillway, or weir, to which water must rise before passing over the structure. 2. The summit or highest point of a wave.

crest cloud See: cap cloud

crest stage The highest elevation, or stage, reached during a rise by flood waters flowing in a channel.

Cretaceous See: Appendix I

crevasse A deep, nearly vertical fissure in the ice of a glacier.

criador In northern Spain, the rain-bringing west wind.

critical mass The smallest mass of fissionable material that will support a self-sustaining chain reaction under stated conditions.

critical point The highest possible temperature at which the liquid and gaseous phases of a substance coexist in equilibrium. For water the point is 374°C.

crivetz In Rumania and parts of southern Russia, a cold, bora-like wind from the north-northeast.

Cro-Magnon The prototype of European man of Upper Paleolithic times.

crop rotation The practice of growing two or more crops in succession on the same field in order to preserve fertility.

cross-bedding Arrangements of small layers at angles to the main planes of strata in a sedimentary rock.

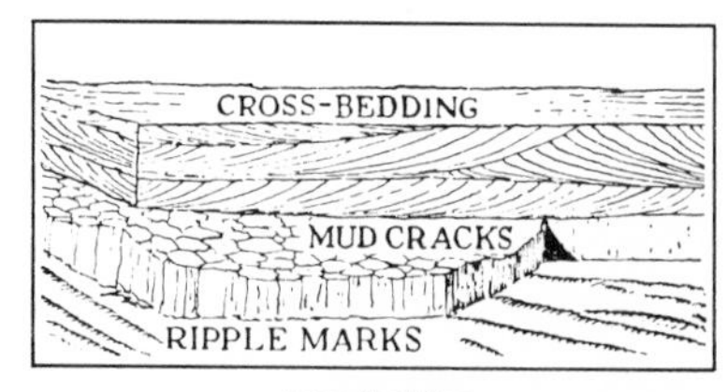

cross-bedding

cross pollination The transfer of pollen from the flower of one plant to another with a different genetic constitution.

cross section A section made by a plane cutting anything at right angles to the longest axis.

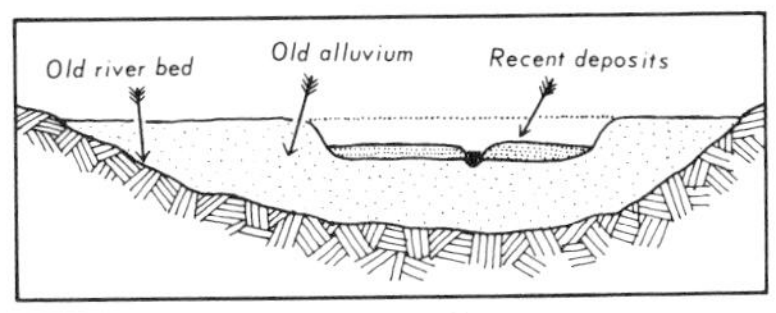

cross section

crown The upper part of a tree, including the branches with their foliage.

crude birth rate The number of live births per 1000 persons.

crude death rate The number of deaths per 1000 persons.

cruise A survey of forest lands to locate and estimate volume and grades of standing timber; also, the estimate obtained in such a survey.

crumb A unit or particle of soil composed of many small grains sticking together.

crumb structure Porous granular structure in soils.

crust 1. A hard layer at the surface of softer material. 2. The outer layer of the earth, thought to be of varying thicknesses, probably not over 20 to 30 miles.

crustal movements The movements of the outer portions of the earth, which produce folds, faults, etc., in the crust.

cryochore Region of perpetual snow.

cryology The science of the physical aspects of snow, ice, hail, and sleet and other forms of water produced by temperatures below 0°C.

crystal A particle formed by an element or a compound so that it is bounded by plane surfaces arranged symmetrically; e.g. quartz crystals, snow crystals.

crystalline (adj.) Having a regular molecular structure evidenced by crystals.

crystallography The science dealing with crystals and, more generally, with the interatomic arrangement of solid matter.

crystallization The forming of crystals.

cube A solid with six equal square sides.

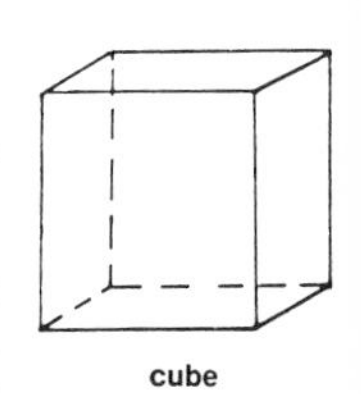
cube

cubic feet per second (CFS) A unit expressing the rate of discharge of a liquid through a rectangular cross section one foot wide and one foot deep, flowing at an average velocity of one foot per second.

cuesta A ridge or belt of hilly land with a long gentle slope in one direction and a steep escarpment in the other.

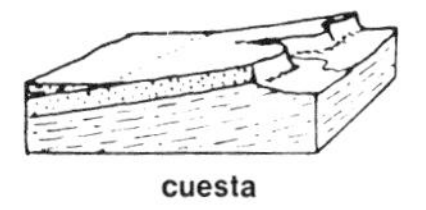
cuesta

cull 1. A tree or log unmerchantable because of defects. 2. The deduction from gross volume made to adjust for defect. (v.) 1. To cut a small portion of a stand by selecting one or a few of the best trees. 2. To reject a tree, log, or board in scaling or grading.

cultigen A cultivated plant whose taxonomic origin is unknown or obscure.

cumulative temperature See: accumulated temperature

cumuliform (or cumuliformis) (adj.) Pertaining to clouds having dome-shaped upper surfaces that exhibit protuberances, the bases of such clouds being generally all on the same horizontal plane. Such clouds are characteristically distinct and separated from one another by clear spaces.

cumulonimbus A cumulus cloud of great vertical development (often four miles or more deep from base to summit) and comparable horizontal extent, the top of which is composed of ice crystal clouds, its distinguishing characteristic.

cumulus A cloud with vertical development, the mean lower level of which is 1,600 feet. The summit of the cloud, in general dome-shaped, shows rounded bulges, while the base is usually horizontal. It appears variously white, shaded, or dark, according to its position with reference to the sun. Its base is generally of a gray color.

curie The unit used in measuring radioactivity, amounting to a decay rate of 3.7×10^{10} disintegrations per second.

current 1. The vertical component of air motion, distinguished from wind, which is the horizontal component. 2. That portion of a stream of water that is moving with a velocity much greater than the average or in which the progress of the water is principally concentrated. See: ocean current

current meter A device for determining the velocity of water, flowing in a stream, open channel, or conduit, by ascertaining the speed at which elements of the flowing water rotate a vane or series of cups.

curvature effect In small water droplets in a cloud, a factor of surface tension that affects their rate of growth. External vapor pressures must be greater than those in the water droplet if growth is to occur. As the drop grows in size so that its radius is 2 to 3 microns, the curvature effect decreases and becomes negligible.

cut The yield, during a specified period, of products that are cut, as of grain, timber, or, in sawmilling, lumber.

cutoff A new and relatively short channel formed when a stream cuts a new course through a horseshoe bend or meander.

cycle A regularly recurring succession of events, such as the cycle of the seasons.

cycle of erosion A concept to explain the more or less systematic series of changes in a landscape as it passes from youth to maturity to old age.

cyclic compound Organic compound containing rings of carbon atoms instead of chains.

cyclogenesis The process that creates a new cyclone or that intensifies the circulation around a pre-existing one.

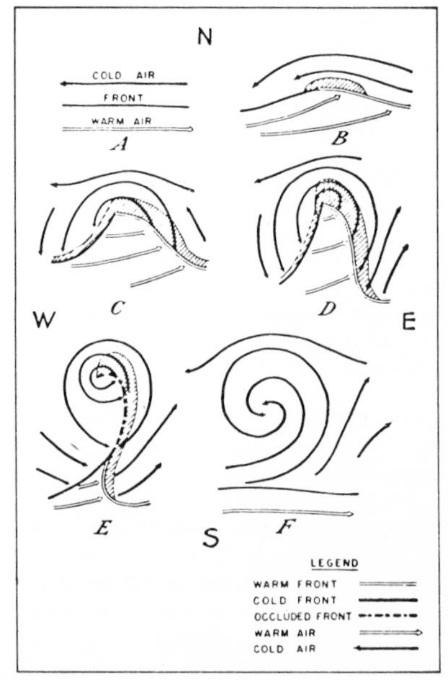

cyclogenesis

cyclolysis The decrease and eventual extinction of the circulation around a low pressure center.

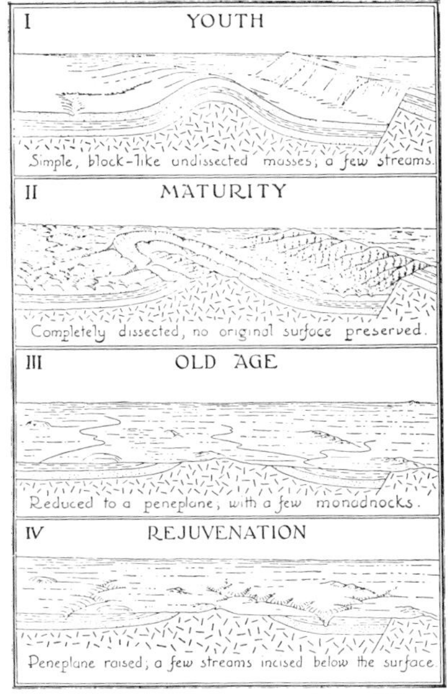

STAGES IN THE GEOMORPHIC CYCLE OF A HUMID REGION

STAGES IN THE GEOMORPHIC CYCLE OF AN ARID REGION

cycle of erosion

cyclone A circular or nearly circular area of low atmospheric pressure around which the winds blow counterclockwise in the northern hemisphere, clockwise in the southern.

cyclone family A series of extratropical cyclones, usually four, which occurs in the interval between two successive outbreaks of polar air, and which travels along the polar front toward the east or northeast. The time interval may vary from three to seven days, and the number of cyclone waves from three to six.

cyclone wave The first stage in the development of an extratropical cyclone.

cyclonic rain Any rainfall connected with and caused by the passage of a barometric depression; the immediate

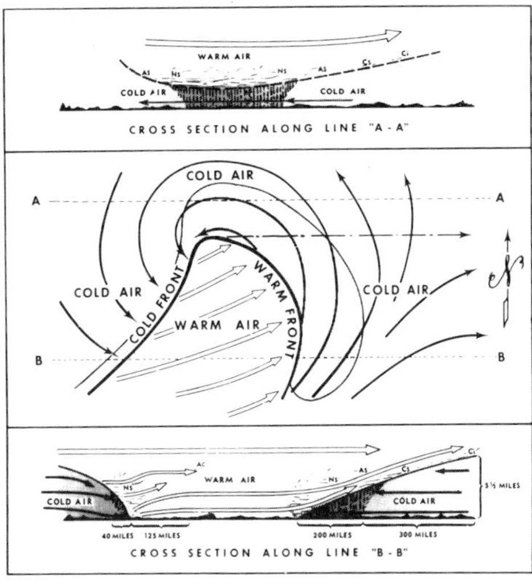

cyclone

causes of the rain are rising air currents, vertical currents or warm air masses overrunning colder air masses.

cyclotron A device for producing high-speed projectiles for disintegrating the nucleus of atoms; protons and deuterons move in spiral paths as they are accelerated twice per revolution, until they attain speeds up to 350 million electron-volts.

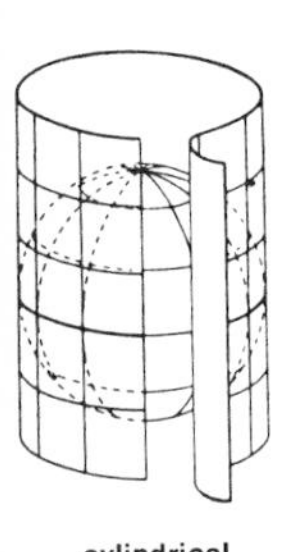

cylindrical projection

cylindrical projection 1. A map projection produced by placing a cylinder around the globe and projecting the earth grid upon it. 2. Also applied to most map projections having straight lines representing meridians and parallels.

cypress A genus of evergreen trees with dense, brushy foliage and ball-shaped cones, used in many places as windbreaks.

cytogenesis The formation and development of cells.

cytology The branch of biology concerned with the study of cells.

cytoplasm In biology, all of the protoplasm lying outside the nuclear membrane of a cell.

daily extreme See: extreme

daily mean See: mean

daily range See: range

Dalton's law A physical law that states that in a mixture of gases, the total pressure is equal to the sum of the pressures that would be exerted by each gas if it alone were present.

dam Any barrier that controls the movement of water.

dam

damp haze Small water droplets or hygroscopic particles in the air which reduce the horizontal visibility, but not to less than 1¼ miles. Damp haze is similar to a very thin fog, but the droplets or particles are more scattered than in light fog and presumably also smaller.

damping-off A fatal disease of seedling plants caused by microorganisms.

Darwinism The theory of evolution that organisms develop, by a process of natural selection, through small changes from one generation to the next that increase their ability to survive.

daughter A nuclide formed by the radioactive decay of another nuclide, which in this context is called the parent.

DDD An insecticide closely related to DDT in chemical composition and in other properties.

DDT A white, crystalline, water-insoluble solid used as an insecticide.

death rate The number of deaths per unit of population, generally per 1000, in a given place and time.

decade Usually a group of 10 years; often in meteorology a group of 10 days.

decibel The unit of measuring the intensity of sound. Zero on the decibel scale is the slightest sound that can be heard by humans—rustling leaves, breathing. The scale: eardrum ruptures (140 decibels—jet taking off); deafening (100 decibels—thunder, car horn at three feet, loud motorcycle, loud power lawn mower); very loud (eighty decibels—portable sander, food blender.

deciduous (adj.) Applied to plants that lose their leaves every year.

declination 1. In astronomy, the angular distance of a celestial body north or south of the celestial equator, measured along an hour circle. 2. In terrestrial magnetism, the angle between true north as measured along a meridian and north as shown by the needle of a magnetic compass pointing toward the magnetic pole. Declination is either east or west of true north depending on the location of the observer. See: isogonic line

decomposers Living plants and animals, chiefly fungi and bacteria, that live by extracting energy from the tissues of dead plants and animals.

decomposition The chemical breakdown of organic or inorganic matter.

decreasers Tall, nutritious climax grasses that generally decrease in number under moderate grazing because they are preferred by livestock.

deduction A process of reasoning in which a conclusion follows necessarily from the stated premises.

deep A steep-walled depression on

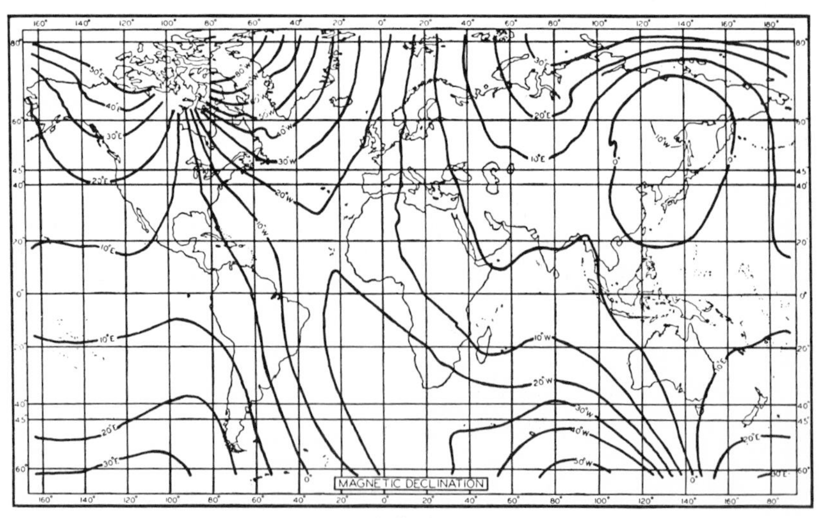

declination

an ocean floor representing one of the deepest areas in the ocean.

deepening The process by which the central pressure system of a low pressure system decreases.

deep-sea plain An extensive, nearly level portion of the ocean floor 2,000 to 3,000 fathoms below the surface.

deep sea zone The ocean bottom lying beyond the area of the continental shelf and the continental slope generally at depths of 6,000-18,000 feet (2000-6000 m).

deflation The removal of loose material by the wind, leaving rocks bare to the continuous attack of the weather. Deflation is responsible for the uniformly sandy or stony surfaces of deserts.

deflecting (or deflective) force See: Coriolis force; Ferrel's law

deflection of the wind The deviation of wind to the right of its original direction in the northern hemisphere, to the left in the southern, due to Coriolis force.

deflocculate (v.) To separate or to break up soil aggregates into the individual particles; to disperse the particles of a granulated clay to form a clay that runs together or puddles.

defoliate (v.) To shed leaves; to lose leaves; to cause a tree to lose its leaves.

deforestation The removal of forests from the land.

deformation Any change in the original form or volume of rock masses by folding or faulting.

degradation 1. The geologic process by which parts of the surface of the earth are worn down and carried away. 2. The change of one kind of soil to a more highly leached kind, such as the change to a chernozem to a podzol.

degree 1. On the Celsius thermometer scale, 1/100 of the interval from the freezing point to the boiling point of water under standard conditions; on the fahrenheit scale, 1/180 of this interval. 2. A unit of latitude or longitude, 1/360 of a circle.

degree day The departure, per degree per day, in the mean daily temperature, from an adopted standard reference temperature, usually 65°F. The number of degree days for an individual day is the departure of the mean temperature from the standard; and the number of degree days in a month, or other interval, is the sum of all the daily values.

dehydration An abnormal removal of water from an organism.

delta A deposit of alluvium at the mouth of a river or smaller stream, often triangular, resembling the Greek letter delta.

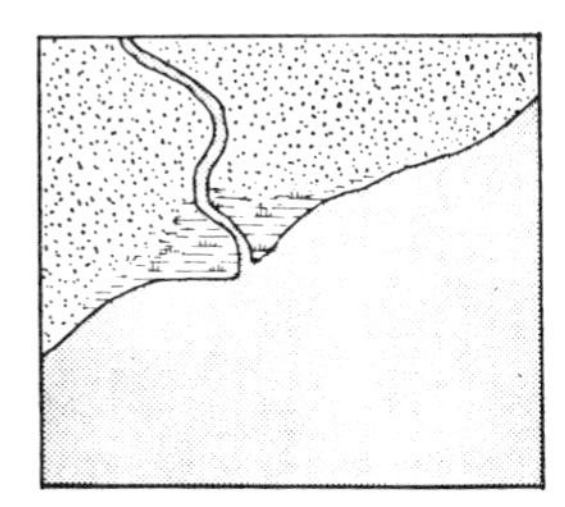

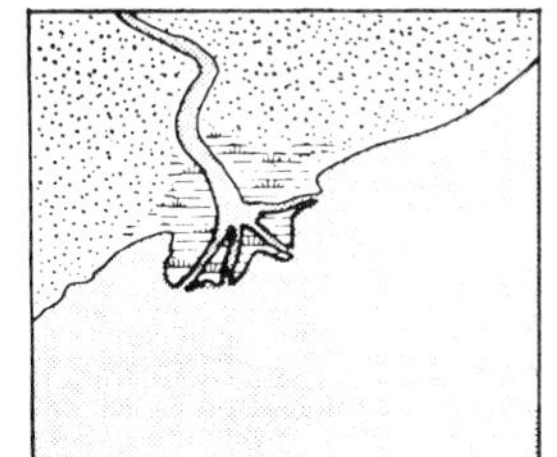

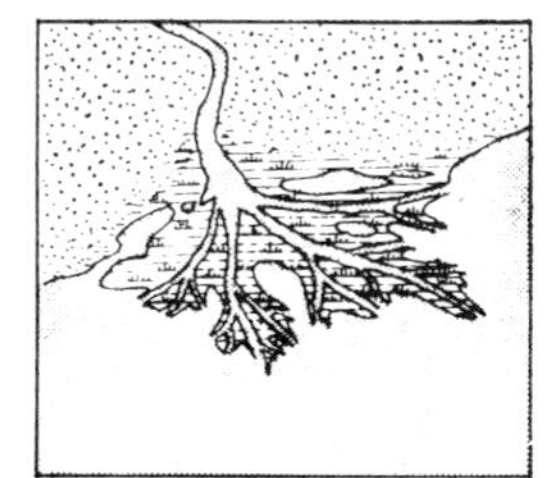

delta

demographic revolution The abrupt change in the rate of natural increase of a population, associated with a marked decrease in death rate, mainly as a result of the decreased effects of famine, plague, and infant mortality.

demographic transition (or cycle) The succession of changes in birth, death, and natural increase rates that occur as a country passes through various stages of economic development.

demography The statistical study of population.

dendritic drainage The form of the drainage pattern of a stream and its tributaries when it follows a treelike shape, with the main trunk, branches, and twigs corresponding to the main stream, tributaries, and subtributaries, respectively, of the stream.

dendrochore A tree region.

denitrification The process by which nitrates or nitrites in the soil or organic deposits are reduced to ammonia or free nitrogen by bacterial action. The process results in the escape of nitrogen into the air and is therefore wasteful.

density The amount of mass per unit volume of any substance (pound per cubic foot, gram per cubic centimeter, kilogram per cubic meter, etc.). Heating causes a substance to expand, decreasing density. Cooling increases the density of a substance.

density current A flow of water maintained by gravity through a large body of water, such as a reservoir or lake, and retaining its unmixed identity because of a difference in density.

density of snow The ratio, expressed as a percentage, of the volume a given quantity of snow would occupy if it were reduced to water, to the volume of the snow.

denudation The wearing away of the land by various natural agencies.

denude (v.) To make bare; to expose by erosion.

departure The amount by which the value of a meteorological element (either the instantaneous value or the mean over a brief period) differs from the value taken as normal for the given time.

dependency ratio The proportion of the population in the relatively unproductive ages (under 20 and over 65) with those of working age (20-64).

depletion (ground water) The withdrawal of water from a ground water source at a rate greater than its rate of replenishment, usually over an extended period of several years.

deposition The laying down or precipitation of mineral matter that may eventually form rocks or that creates secondary landforms such as deltas, sand dunes, etc.

depression 1. An extensive area of relatively low barometric pressure. 2. A low place in a landscape; a basin.

depression contours Contour lines that indicate an enclosed basin or depression on a map.

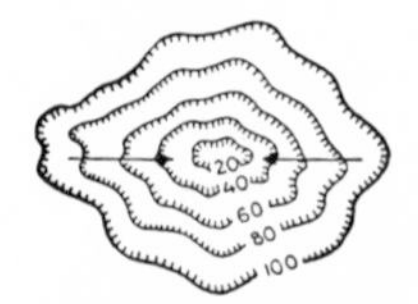

depression contours

depression of the dew point The difference in degrees between the prevailing temperature and the current temperature of the dew point.

depression of the wet-bulb The difference in degrees between the current temperature of the dry- and the wet-bulb thermometers of a psychrometer.

depth of runoff The total runoff from a drainage basin, divided by its area. For convenience in comparing runoff with precipitation, the term is usually expressed in inches of depth during a given period of time over the drainage area or in acre-feet per square mile.

desalination (or desalinization) The process of removing salts from brackish or ocean waters or from soil.

descriptive climatology That branch of climatology that treats of the component elements of weather and climate as revealed by climatological statistics and discusses the interaction of these elements at any place upon the life and health of man and upon his various activities. See: physical climatology

desert A region so devoid of vegetation as to be incapable of supporting any considerable population. Three kinds of deserts may be distinguished: (a) the polar ice and snow deserts, marked by perpetual snow cover and intense cold; (b) the middle latitude deserts, in the basin-like interiors of the continents, such as the Gobi, characterized by scant rainfall and high summer temperatures; and (c) subtropical deserts, such as the Sahara, the distinguishing features of which are negligible precipitation and large daily temperature range.

desert pavement The accumulation of pebbles at the surface of the desert as a consequence of the removal of finer particles by the wind.

desert soil A zonal group of soils that have light-colored surface soils and usually are underlain by calcareous material and frequently by hard layers. They are developed under extremely scanty scrub vegetation in warm to cool, arid climates.

desert varnish The shiny brown or black coating that covers many exposed rock surfaces in the desert, caused by a surface stain or crust of iron oxide or manganese oxide.

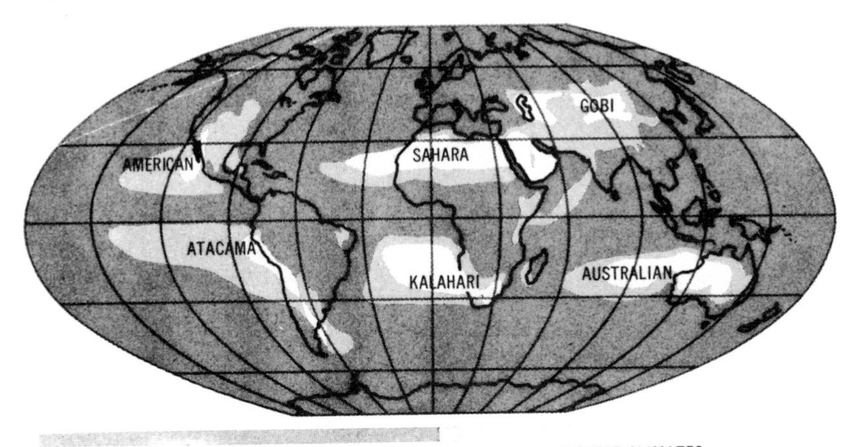

desert

desiccation The process by which a region suffers a complete loss of its water, due to decrease of rainfall, increase of evaporation, or to changes in other climatic controls. Desiccation is manifested by the drying up of streams and lakes, the destruction of vegetation, the loss of surface soil, etc. See: exsiccation

detention dam A dam constructed for the temporary storage of flood flows where the opening for release is of fixed capacity and not manually operated.

detergent A synthetic preparation that acts like soap to emulsify oils and hold dirt in suspension; some of the chemicals used have caused pollution problems.

detritus 1. The heavier mineral debris moved by natural watercourses, usually in bed load form. 2. The sand, grit, and other coarse material removed by differential sedimentation in a relatively short period of detention.

deuterium An isotope of hydrogen whose nucleus contains one neutron and one proton and is therefore about twice as heavy as the nucleus of normal hydrogen, which is only a single proton. Deuterium is often referred to as heavy hydrogen.

deuteron The nucleus of deuterium. It contains one proton and one neutron.

deviation 1. The algebraic difference between the mean of a series of data and an individual member of the series. See: standard deviation. 2. The difference in degrees between true north and the magnetic north indicated by a compass, which varies with geographical location. See: declination

devil grass See: Bermuda grass

Devonian See: Appendix I

dew Water condensed onto objects near the ground whose temperatures have fallen below the dew point of the adjacent air due to radiational cooling during the night, but are still above freezing.

dew point The temperature to which air must be cooled, at constant pressure and constant water vapor content, in order for saturation to occur.

diagenesis The processes involving physical and chemical change in sediments that convert them to consolidated rocks.

dialysis The separation of small molecules (usually crystalloids) from large molecules (colloids) in solution by taking advantage of their unequal rates of diffusion through natural or artificial membranes; the small molecules diffuse readily, whereas the colloids diffuse very slowly or not at all.

diameter limit A specified diameter at breast height (4½ feet above the ground) above which all trees are cut, under a diameter-limit cutting agreement.

diastase An enzyme that hastens the hydrolysis of the starches in grains to form fermentable sugars.

diastrophism All the processes that change the shape of the earth's surface, producing mountains and valleys, continents and ocean basins, etc.

diatom A member of a group of microscopic, single-celled plants found

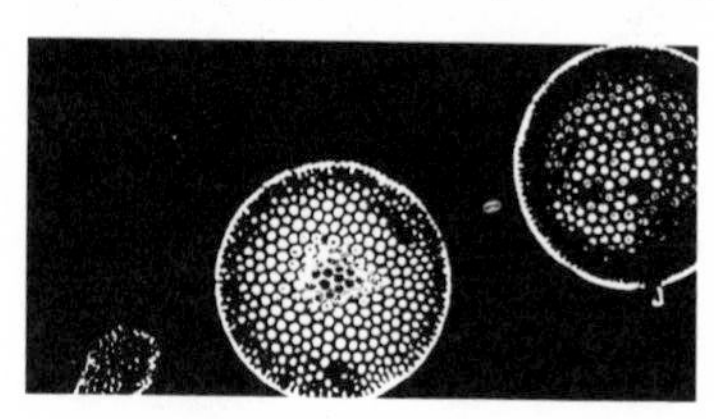

diatom

in both fresh and salt water. The limey or siliceous cell walls (shells) of diatoms may accumulate in enormous numbers in sediments.

diatomaceous earth A friable rock material of nearly pure silica derived from diatoms.

diatomic (adj.) Having two atoms per molecule.

dicotyledon An angiosperm with two cotyledons present in the seed; beans belong to this subclass.

dieldrin A light tan crystalline, poisonous solid used as an insecticide.

differentiate (v.) To become more specialized during development, used especially of cells or tissues.

diffraction 1. The bending of a wave or ray as it passes the edge of an obstacle. 2. In seismology, a pulse produced by reflection of a seismic wave from a fault or sharp fold in a reflecting surface.

diffuse reflection The reflection of light by large particles in the atmosphere such as dust, water droplets, and ice crystals.

diffuse sky radiation The scattered solar radiation received by the earth from the atmosphere as distinguished from the radiation incident in direct sunlight.

diffusion The process by which a fluid permeates its environment. See: eddy diffusion

dihybrid The offspring of parents differing in two specific pairs of genes.

dike 1. A relatively thin, flat-surfaced mass of igneous rock that cuts across the structure of the enclosing rocks or extends through massive rocks. Usually formed by the intrusion of magma into a fissure. 2. An embankment, usually of natural materials.

dike

dinkey A small logging locomotive.

dinosaur An extinct reptile of the Mesozoic era, certain species of which were among the largest known land animals.

dip The downward angle that a stratum or any planar feature makes with the horizontal plane.

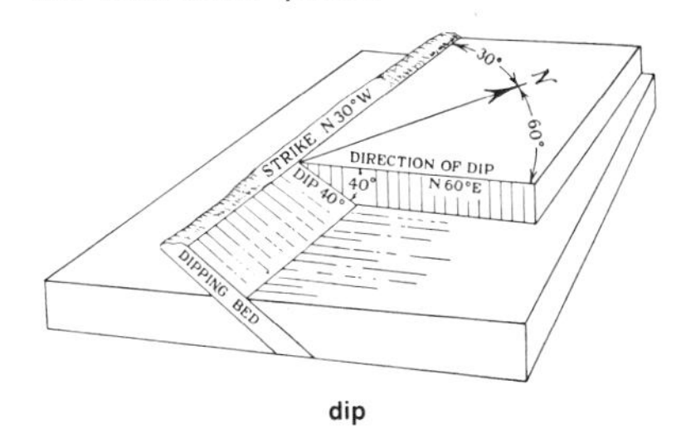

dip

diploid An organism with the usual paired chromosomes.

dipolar (adj.) Having two poles as a result of separation of electric charge.

dip slope A slope of the land surface

that is roughly parallel to the dip of the underlying rocks.

direction of the wind The point of the compass from which the wind blows, not that toward which it is moving. Wind direction is also expressed in degrees measured clockwise from north: thus an east wind has a direction of 90°, a northwest wind, of 315°.

direct runoff The runoff entering stream channels promptly after rainfall or snow melt. Superposed on base flow, it forms the bulk of the hydrograph of a flood.

discharge The rate of volume flow of water in a given stream at a given time and place, usually expressed in cubic feet per second (abbreviated to "second feet"), liters per second, gallons per minute, or million gallons per day.

disconformity An unconformity between parallel layers.

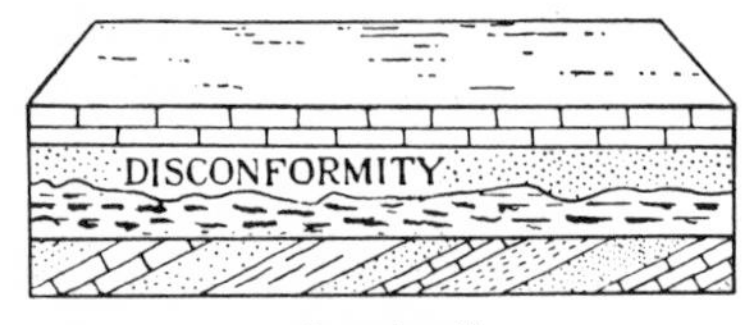

disconformity

discontinuity A zone or layer in the atmosphere within which there is a comparatively rapid transition of any of the meteorological elements from one value to another. See: frontal zone

disintegration The breakdown of rock material by mechanical or physical means. See: decomposition; chemical weathering

dispersal The spread of organisms from their centers of origin to other parts of the world.

dispersion 1. In chemistry, a suspension of particles in a medium. 2. In physics, the separating of a wave into its components.

dissected plateau A plateau into which a number of valleys have been carved by erosion.

dissection The cutting of ravines and valleys into a relatively flat surface.

dissolved oxygen The amount of free (not chemically combined) oxygen in water, usually expressed in milligrams per liter.

distillation A process in which a substance or substances are heated in a closed container, with the liberation of a vapor or gas, which is conducted from the container and is later converted entirely or in part to a liquid by cooling.

distillation wood See: acid wood

distributary A channel carrying water away from the main stream and across a delta or alluvial fan.

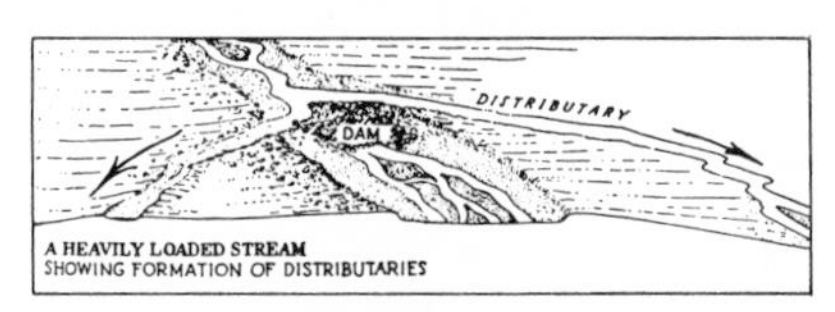

distributary

disturbance 1. Any local departure from the normal wind pattern. 2. A cyclone, both tropical and extratropical, as in the phrase "a tropical disturbance."

diurnal (adj.) Daily.

diurnal range The amount of variation between the maximum and minimum of any element within a 24-hour period.

diurnal variation The daily change in both wind direction and speed, due to the heating of the earth's surface by the sun.

divalent (adj.) Having the ability to enter chemical combination simultaneously with two other atoms or combining units. A divalent element is capable of losing two electrons to, or acquiring two electrons from, the elements with which it combines.

divergence 1. In fluid motion, a net outflow of mass across a closed surface bounding a limited volume of the fluid. 2. In meteorology, the expansion of winds within a given layer of air so that there is a net horizontal outflow of air from the region and a compensating upward or downward movement of air. 3. The changes in a group of animals from ancestral type to divergent descendent types. See: adaptive radiation

divide The line of separation between drainage systems.

divining rod A forked branch, supposed to indicate the presence of water or minerals under the soil.

D layer 1. Any stratum underlying the soil profile that is unlike the material from which the soil has been formed. 2. An ionized region or layer of the atmosphere approximately 70 kilometers above the earth's surface, which partly absorbs radio waves and also reflects long radio waves.

DNA (deoxyribonucleic acid) A constituent of the nucleus of cells, that functions in the transfer of genetic characteristics and in the synthesis of protein.

doab In India, the tongue of land lying between the confluence of two or more rivers.

doctor 1. A sea breeze in tropical climates. 2. The harmattan trade wind.

dodo An extinct flightless bird related to the pigeon.

dodo

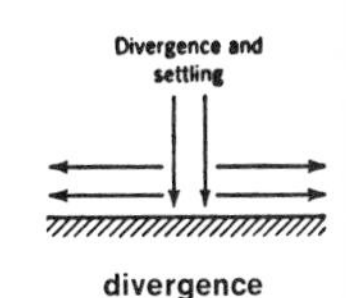

divergence

dog days A period of hot weather, supposedly extending from late July to early September. It derives its name from Sirius, the Dog Star, the rising of which was associated with the coming of dry, hot, and sultry weather.

doldrums The equatorial belt of calms or light fitful winds, lying between the northeast trade winds of the northern hemisphere and the southeast trades of the southern.

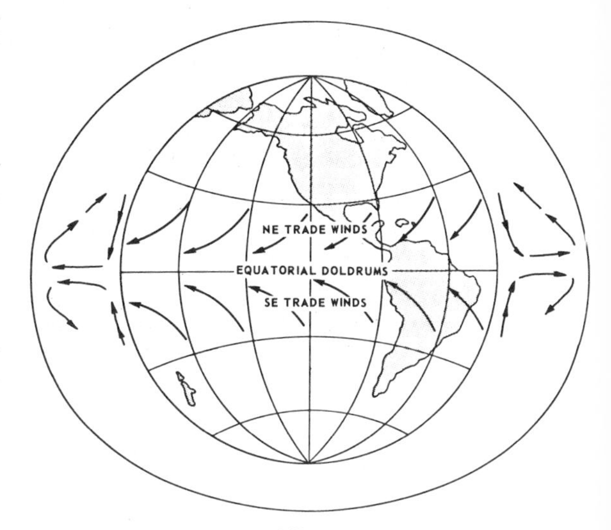

doldrums

doline (dolina) A depression in a karst region created by the subsidence of the roofs of caverns created by the solution of limestone.

dolomite A relatively resistant rock that forms ridges, especially in arid regions.

dome A symmetric land upfold with the beds dipping in all directions from the highest point.

domestic (adj.) 1. Pertaining to household or home use. 2. Tame.

domestication The process of taking a wild plant or animal and converting it to a product easily harvested or used by man.

domestic use (of water) The use of water primarily for household purposes: the watering of livestock, the irrigation of gardens, lawns, and shrubbery surrounding a house.

dominant 1. The gene that masks the effect of its paired gene so that a particular characteristic is represented. 2. Any plant or animal that because of its abundance, size, or habits exerts an influence on the condition of an area so important as to determine what other organisms may live there.

Doppler effect The phenomenon that the relative motion between an observer and the source of waves results in an apparent change in frequency of the waves. If the two are approaching, the frequency is increased; if they are receding, it is decreased.

dormant (adj.) In a state of rest or inactivity; this state may be produced by light, temperature, or moisture conditions.

double cropping The practice of raising two crops on a field in one year.

double-mass curve A plot on arithmetic cross-section paper of the cumulated values of one variable against the cumulated values of another or against the computed values of the same variable for a concurrent period of time.

doubling rate The period of time it takes a given population to double; it is about 70 years for the United States at present.

Douglas fir An evergreen tree, not a true fir. It is one of the principal sources of wood for the construction industry, used in dimension lumber and plywood.

down 1. An area of open, treeless, hilly land; in particular, the chalk hills of southern England. 2. The soft underplumage of birds.

downwarp An area that has been gently folded downward.

dowse (v.) To search for deposits of ores or water with the aid of a divining rod.

drainage The removal of excess surface water or excess water from within the soil by means of surface or subsurface drains.

drainage basin A part of the surface of the earth that is occupied by a drainage system, which consists of a surface stream or a body of impounded surface water together with all tributary surface streams and bodies of impounded surface water.

drainage density The relative density of natural drainage channels in a given area. It is usually expressed in terms of linear miles of natural drainage or area, and obtained by dividing the stream channel per square mile of total length of stream channels in the area in miles by the area in square miles.

drainage divide The boundary line, along a topographic ridge or along a subsurface formation, separating two adjacent drainage basins.

drainage patterns The form taken by the tributaries of a stream, generally determined by the underlying rock structure. See: annular, radial, dendritic, trellis, rectangular, parallel drainage

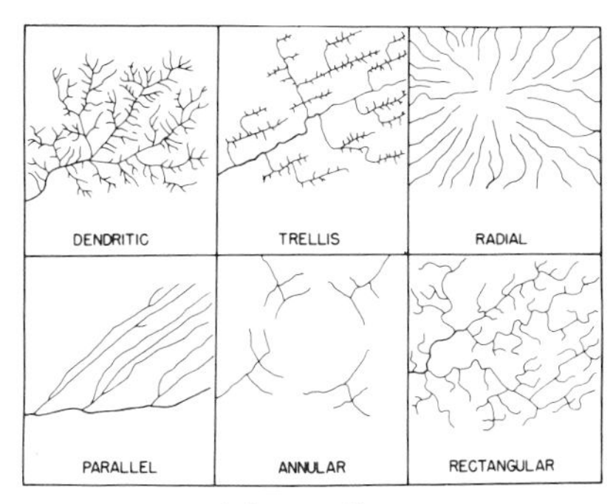

drainage patterns

Draper catalog A system for classifying stars according to their observable characteristics, such as color.

draw A tributary valley or coulee that usually discharges water only after a rainstorm.

drift 1. Snow driven by the wind along the surface, sometimes rising to heights of 100 feet or more. 2. Snow lodged in the lee of surface irregularities under the influence of the wind. 3. The unresisting motion of sea ice or of vessels, resulting from ocean currents. 4. Boulders, gravel, sand, clay, or rock material that was transported by a glacier.

drift ice Detached portions of ice carried by currents into the open sea beyond the limit of the pack ice.

drifting snow Snow raised from the ground and carried by the wind, so that the horizontal visibility becomes less than five-eighths of a mile (3,300 feet), although no precipitation need be falling.

driftless area A region lying mostly in southwestern Wisconsin, not reached by the Pleistocene continental glaciers, though surrounded by areas covered with glacial drift.

drip Moisture from fog or cloud condensed on leaves, twigs, etc., and falling therefrom. See: fog drip

drizzle Precipitation from stratiform clouds, in the form of small and numerous drops of water, which apparently float in the air and bring poor visibility.

drought An extended period of dry weather; or, a period of deficient rainfall that may extend over an indefinite number of days sufficient in length and severity to cause at least partial crop failure.

drowned valley A valley that has been submerged by the advance of the sea or a lake, owing to the sinking of the land or the rising of sea level.

drumlin A ridge or teardrop-shaped hill or drift originally deposited beneath and shaped by an advancing glacier or ice sheet. The long axis of the drumlin normally parallels the direction in which the glacier flowed.

drumlin

drupe A simple, fleshy, or pulpy fruit; a stone fruit, as peach, plum, and cherry.

dry adiabat A temperature-height or temperature-pressure curve along which a rising or sinking air parcel will move, provided no condensation occurs and no heat is given to or taken from the air during its motion.

dry adiabatic lapse rate The rate at which dry air warms or cools during adiabatic descent and ascent, respectively; for absolutely dry air, it is 1°C per 102.39 meters of vertical distance when the lapse rate is neutral.

dry bulb thermometer An ordinary thermometer. See: psychrometer

dry farming A method of farming in arid and semiarid regions involving strip farming, fallowing, and the use of drought-resistant plants.

dry haze Dust or salt particles in the air that are dry and so small that they cannot be felt or discerned individually by the unaided eye.

Dry Ice A trademark for solidified carbon dioxide. Valuable as a refrigerant because the solid carbon dioxide turns into a gas at room temperature without becoming liquid.

dry-ki Trees killed by flooding. Often found in areas flooded by beaver dams.

dry season The season of the year characterized by scanty rainfall.

dry spell A period, usually of not less than two weeks duration, during which no measurable rainfall occurs at a certain place or region. When extended into a month or more, a dry spell is considered a drought.

duff The partly decomposed organic material on the forest floor.

dune A ridge of sand formed by, and constantly changed by, the wind.

dusk That part of morning twilight from complete darkness to the beginning of civil twilight, and that part of evening twilight from the ending of civil twilight to the beginning of complete darkness. See: twilight

dust Solid material suspended in the atmosphere in the form of small particles, many of them microscopic.

Dust Bowl A name given early in 1935 to the region in the south-central United States then afflicted with droughts and dust storms. It included parts of five states: Colorado, Kansas, New Mexico, Texas, and Oklahoma.

dust-cloud hypothesis The theory that our solar system grew out of a great cloud of dust and gases, parts of which coalesced to form the sun and each of the planets.

dust devil See: dust whirl

dust storm A strong wind carrying large clouds of dust or sand, common in desert and plain regions.

dust whirl A rotating column of air, about 100 to 300 feet in height, carrying dust, straw, leaves, and other light material, best developed on a calm, hot afternoon with clear skies, and in desert regions.

duty of water In irrigation, the quantity of water required to satisfy the irrigation water requirements of land.

dynamic climatology See: air-mass climatology

dynamic cooling The adiabatic loss of heat by ascending air which, in rising to a region of lower pressure, undergoes expansion, requiring the expenditure of energy and consequently leading to a depletion of internal heat.

dynamic heating The adiabatic addition of heat by the work done upon air undergoing compression in descending to a level of a higher pressure.

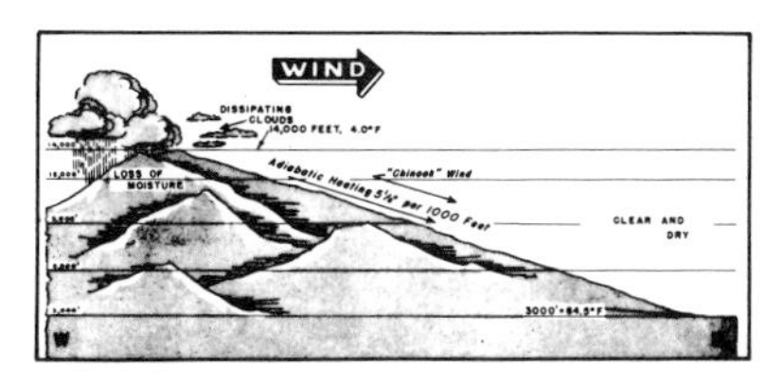

dynamic heating

dynamic meteorology A branch of meteorology, also known as theoretical meteorology, that seeks to explain, by the application of mathematical physics and the laboratory data of classical physics, the forces that create and maintain atmospheric motions and the heat transformations associated therewith.

dynamic trough A barometric trough formed on the lee side of a north-south mountain range when a west wind is blowing; often seen on U.S. weather maps east of the Rocky Mountains.

dyne The absolute, c.g.s. unit of force, defined as that force which, acting upon a free mass of 1 gram, would impart to it an acceleration of 1 cm/sec^2.

dysgenics The retrogressive evolution of a population through the reproduction, in disproportionately large numbers, of its genetically inferior elements.

Earth One of eight major planets in the solar system, fifth in size.

earth-air current The steady electrical current that flows between the earth and the atmosphere because of their difference in electrical charge. Its average value for fair weather over the earth is about 2×10^{-16} amperes per square centimeter.

earth currents Electrical currents circulating within the earth's crust, differing markedly from place to place and subject to variations in intensity and direction.

earth grid The system of parallels and meridians that is used to locate the position of any place on the earth's surface. See: latitude, longitude

earthlight (earth-shine) The faint illumination of the dark part of the moon's disk caused by sunlight reflected from the earth's surface and atmosphere.

earthquake A movement of the earth's crust, as a result of slippage along faults or volcanic action.

earthquake intensity A rating of the effects of an earthquake on man and his environment; it varies with the distance from the epicenter, with the nature of the ground, and with the observer's perception of events. See: Mercalli scale

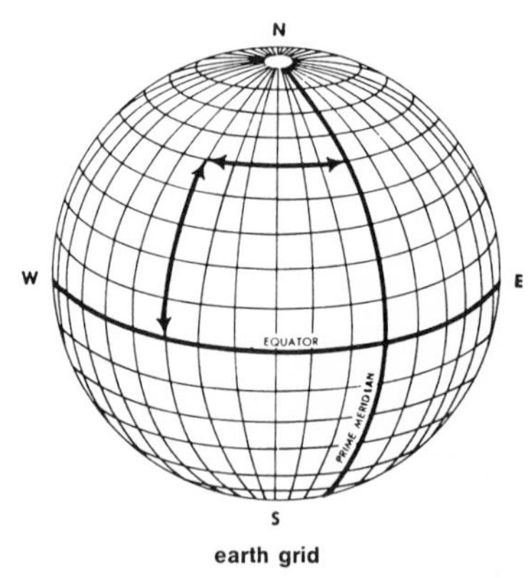

earth grid

earthquake magnitude A measure of the energy released by an earthquake, derived from seismograph records; it is usually expressed in units on the Richter scale.

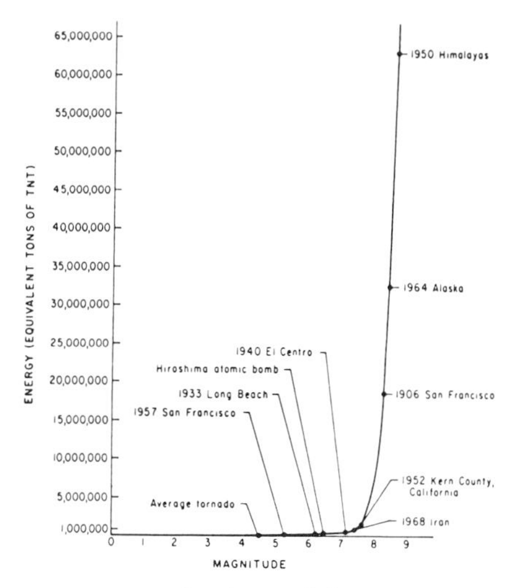

earthquake magnitude

earth radiation See: nocturnal radiation

earth temperature The temperature of the surface of the earth. Heat is received by direct radiation from the sun, by radiation and conduction from the atmosphere, and by conduction from below; heat is lost by radiation and conduction; the net balance between gain and loss determines the temperature.

easterly wave A wave in the pressure patterns of the atmosphere moving from east to west; it is preceded by clear weather and followed by clouds and rain.

eastern hemisphere The half of the earth's surface lying east of the prime meridian.

eastern hemisphere

ebb tide The receding tide following high tide.

echinoderm Marine animals with body walls strengthened by calcareous plates.

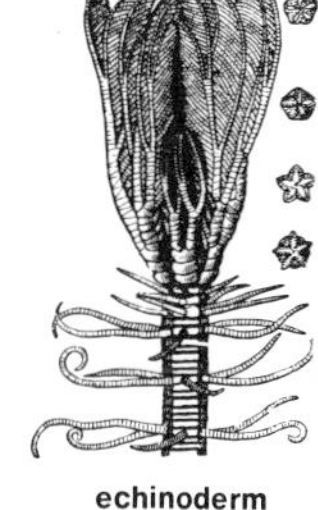
echinoderm

echo sounding A method of determining the depth of water by measuring the time required for sound to travel to the bottom and return.

eclipse Any cutting off of light, particularly the total or partial obscuring of the sun or of the moon. See: solar eclipse; lunar eclipse

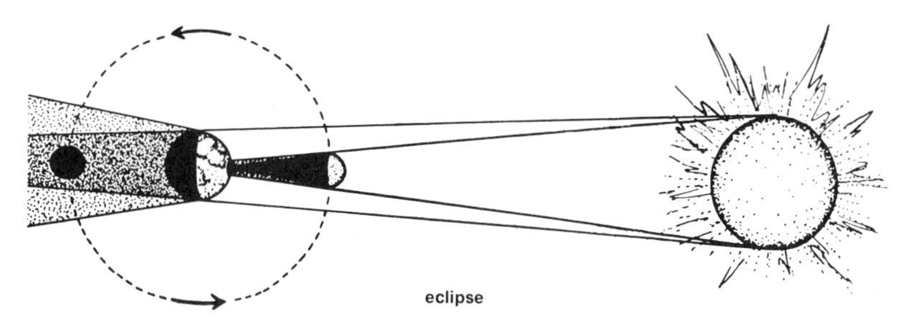
eclipse

ecliptic The great circle of the celestial sphere in which the apparent annual motion of the sun among the fixed stars takes place; it is the intersection of the plane of the earth's orbit with the celestial sphere, inclined to the celestial equator by approximately 23½°. The zodiac extends 8° on each side of the ecliptic.

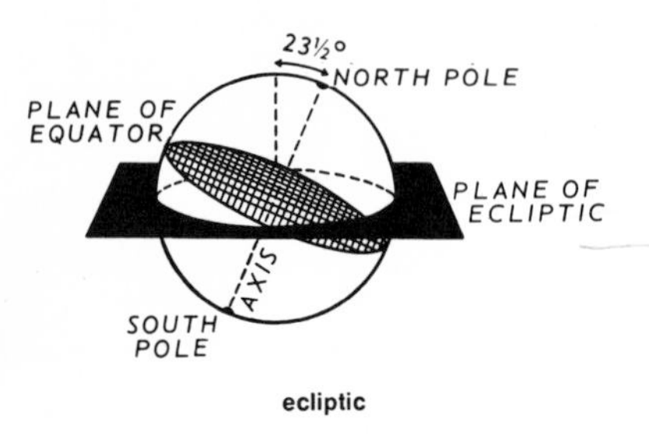

ecliptic

ecocide A substance which enters, permeates, and kills an entire system, be that system an insect pest, a lake, or a man-environment community.

ecological equilibrium See: equilibrium

ecological isolation The separation of one group of organisms from other related organisms so that there can be no interbreeding.

ecological niche The position that an organism occupies in relation to its environment.

ecology The study of organisms in relation to their environment.

economies of scale The savings that occur with large-scale production or consumption; e.g., assembly-line production techniques.

ecosphere That part of the atmosphere in which it is possible to breathe normally.

ecosystem A system formed by the interaction of a group of organisms and their environment.

ecotone The transition zone between two adjacent ecosystems.

ectoplasm The outer portion of the cytoplasm of a cell.

ecumene 1. The permanently occupied area of earth. 2. That part of a country having the greatest concentration of people and activity.

edaphic (adj.) Relating to soil.

eddy The more or less circular motion produced by an obstruction in the path of a moving fluid, such as the wind.

eddy conductivity The transfer of a physical property, such as heat, by eddies, from one layer to another of a fluid in turbulent motion.

eddy diffusion Atmospheric mixing brought about by eddy or turbulent motion, in which heat, water vapor, and the momentum of the wind are transported from one layer to another.

effective radiation The difference between the radiant energy received from the sky by any surface and that emitted by the surface due to its own thermal energy.

effective rainfall The portion of the total rainfall that reaches streams and rivers.

effective temperature An arbitrary, empirically determined index that unites into a single value the effects of temperature, humidity, and air movement on the degree of warmth or cold felt by the human body, subjected to various combinations of these elements.

effluent The outflow of water from a subterranean storage space. Also used generally for gases and other liquids.

ejido A tract of land owned jointly and farmed by all the inhabitants of a Mexican village.

Ekman spiral A diagrammatic representation of the theoretical change of vector wind velocity with height, from the surface of the earth to the gradient level, about 1,500 feet, where the wind blows parallel to the isobars in the absence of the deviating effect of surface friction.

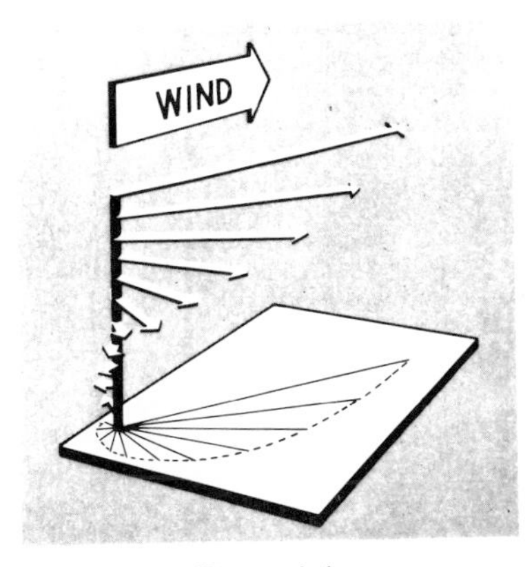

Ekman spiral

elastic rebound theory The theory that earthquakes occur when rocks that have been gradually strained or bent over a long period of time suddenly break, and rocks on each side of the break spring back to a position of no strain.

E layer A rarefied and highly ionized stratum of the atmosphere some 20 miles thick, its base at about 50 miles above the earth's surface. Commonly called the Kennelly-Heaviside.

electric storm 1. Name sometimes given to a thunderstorm because of its accompanying electrical phenomena. 2. A condition of highly electrified air, in which electrical discharges may be seen in the air or may take place through the body, in fine weather without clouds or precipitation, and often with dry, dusty winds.

electrolyte Any conductor of electric current in which chemical change accompanies the passage of the current and the amount of the change is proportional to the amount of current passed. Usually electrolytes are solutions of substances in a liquid, such as salt in water.

electromagnetic spectrum See: spectrum

electron A negatively charged particle in every atom. In a neutral atom, the number of electrons is equal to the number of protons.

element 1. Any one of the properties or conditions of the atmosphere that specify its physical state at a given place for any particular period of time. See: climatic elements, meteorological elements. 2. "The elements": the weather, especially when stormy. 3. A chemical element, that cannot be separated into constituents other than itself by chemical means, such as iron, mercury, oxygen, etc.

elevation The height of any point above mean sea level.

El Niño A warm-water current that in some years flows south along the Peruvian coast around Christmastime; it upsets the equilibrium of the area because it brings copious rains to an otherwise arid land.

eluvial soil Soil that has lost material through eluviation.

eluviation The movement of material from one place to another within the soil in either true solution or colloidal suspension. See: illuviation

embryo An organism in the early stages of development.

emigration The movement of people or organisms away from their native habitats.

emission standards Allowable automobile pollutant emission levels, set by federal and state legislation.

emphysema A respiratory disease associated with excessive smoking and smog, causing difficulty in breathing.

empirical Provable through observation or experiment.

endemic Indigenous or native in a restricted locality; confined naturally to a certain limited area or region, in contrast to epidemic.

end moraine The pile of glacial drift that marks the point of farthest advance of a glacier.

endosperm The main portion of a grain kernel, the nutritive material on which the embryo feeds when germinating.

endrin An insecticide, resembling dieldrin in toxicity and other properties.

energy Capacity to do work.

energy balance See: heat balance

energy-balance climatology The study of climatology involving energy flows, surpluses, and deficiencies, which stresses energy budget calculations.

energy budget The flow of energy of an object or locality expressed in terms of incoming and outgoing radiation.

energy diagram See: thermodynamic diagram

entisol An azonal soil lacking a distinctive soil profile.

entrenched (or incised) meander An old meander that has been deepened by rejuvenation.

entropy A measure of the capacity of a system to undergo spontaneous change in its quantity of heat at a given temperature.

environment The sum total of all the conditions and influences that affect the development of an organism.

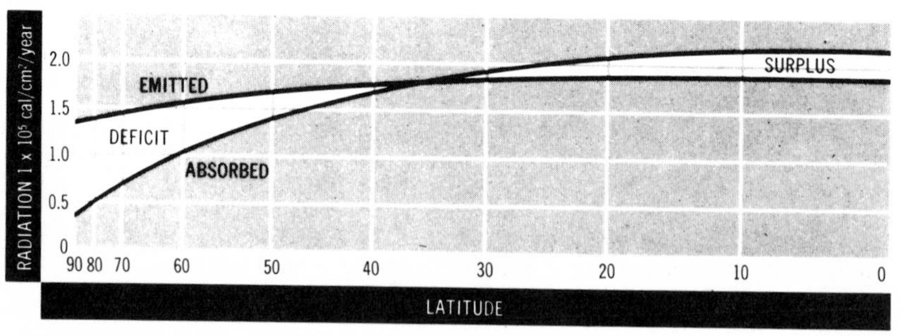

energy budget

environmental lapse rate The actual variation with height of temperature and moisture conditions in the atmosphere.

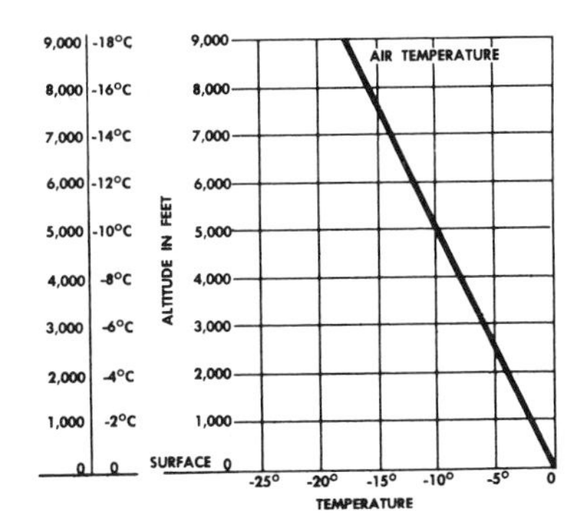

environmental lapse rate

environmental limits Those conditions that limit the range of an organism; they include temperature, moisture, soil, etc.

environmentalism (environmental determinism) The idea that environmental factors are the most important determinants in the kind of development that occurs in an individual or group.

enzyme A complex organic substance capable of producing chemical changes in other organic substances; it is thought that each enzyme in a cell acts as a catalytic agent for only one kind of substance.

Eocene See: Appendix I

eolation The wearing away of land surfaces by wind-driven dust or sand.

eolian See: aeolian

eon The largest division of the geologic time scale. See: Appendix I

epeirogenesis Large-scale elevation or depression of portions of the earth's crust.

ephemeral stream A stream, with its channel above the level of the water table, that flows only in direct response to precipitation, and discontinues its flow during dry seasons.

ephemeris A table showing the positions of a heavenly body on a series of dates.

epicenter The point on the earth's surface directly above the focus of an earthquake.

epicontinental sea A shallow sea resting on a continental shelf.

epidemic (adj.) Spreading or increasing rapidly; used most often of disease. See: endemic

epidemiology The science that deals with the incidence, distribution, and control of disease.

epidermis The outermost living layer of cells of any organism.

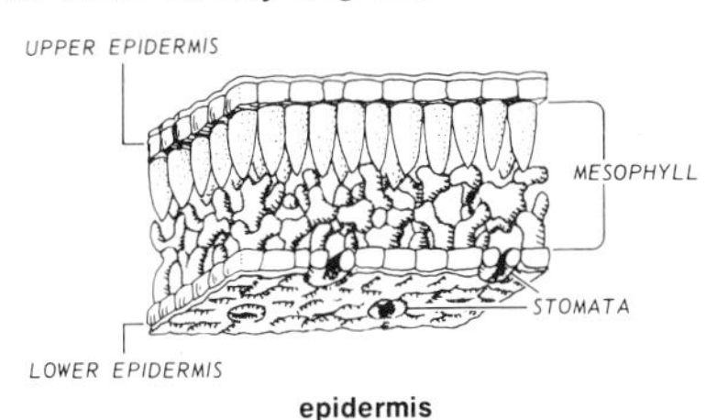

epidermis

epiphyte A plant growing on another plant but obtaining food from the atmosphere.

epithelium The tissue, consisting wholly of living cells of varying form and arrangement forming a practically unbroken sheet, or membrane, which covers the free surfaces of the skin, mucous membranes, and various organs, and lines, tubes, or cavities in the body. It encloses and protects the various parts of the body and facilitates the absorption of nutrients and the excretion of waste products.

epoch 1. A particular period of time marked by distinctive features, activities, etc. 2. Geological time period See: Appendix I

equal area map A map drawn so that a square mile in one portion of the map is equal to a square mile in any other portion.

equal area projection See: homolographic projection

equation of time The difference between mean solar time and apparent solar time; it is zero, four times a year, about April 15, June 15, August 31, and December 24. Between April 15 and June 15 and between August 31 and December 24, mean time is ahead of apparent time; and between June 15 and August 31 and between December 24 and April 15, mean time is behind apparent time.

equator See: terrestrial equator

equatorial air Warm and moist air, originating in equatorial regions.

equatorial calms See: doldrums

equatorial forest The hot, wet evergreen forest of the tropics.

equidistant projection An azimuthal projection in which distances are true outward from the center of the map.

equilibrium A state of balance due to the offsetting effects of opposing forces. In the environmental system, the balance of nature.

equinoctial (adj.) Pertaining to the equinox. See: equinoctial storm

equinoctial rain Rains regularly occurring in the equatorial regions about the time of the equinoxes.

equinoctial storm A rainstorm that, according to long-standing popular belief, is likely to occur in the latter half of September and also of March, the times of the equinoxes.

equinox The moment, occurring twice each year, when the sun, in its apparent annual motion among the fixed stars, crosses the celestial equator, so called because then the night is equal to the day, each being twelve hours long over the whole earth. The autumnal equinox occurs on or about September 22, when the sun is traveling southward; the vernal equinox on or about March 21, when it is moving northward.

equivalent-potential temperature The temperature to which air would come if it were brought adiabatically to the top of the atmosphere (i.e., to zero pressure), so that all its moisture content were condensed and precipitated and the latent heat of condensation given to the air, and then lowered and compressed to a level having the standard pressure of 1,000 millibars.

equivalent temperature The temperature to which air would come if sub-

equidistant projection

jected to a pseudoadiabatic process until all its water vapor content had been condensed, and then returned dry-adiabatically to its initial pressure.

era A period of time marked by a distinctive character. See: Appendix I

Eratosthenes Greek scientist who first calculated the circumference of the globe by measuring the distance from Syene to Alexandria (500 miles) and measuring the angle made by the shadow of a wall at Alexandria. He found the angle equal to 7.2° or one-fiftieth of a circle. Therefore 50 x 500 = 25,000 miles—the circumference of the globe.

erg 1. The absolute c.g.s. unit of energy and work, whose length and force factors are the centimeter and the dyne; i.e., the centimeter dyne. 2. Desert area marked by sand dunes.

erosion The wearing away of the land surface by detachment and transport of soil and rock materials through the action of moving water, wind, or other geological agents.

erratic A rock fragment, usually a boulder, lying on bedrock different from itself, and usually transported by a glacier or by floating ice.

error In the theory of measurements, the difference between the true value of a quantity and the observed value. An error may be due to several causes, such as an error of scale, a defect in the instrument, or personal equation.

escarpment The steep-faced edge of a cuesta or a plateau.

esker A long narrow ridge of sand and gravel, thought to have been formed by streams flowing under glaciers.

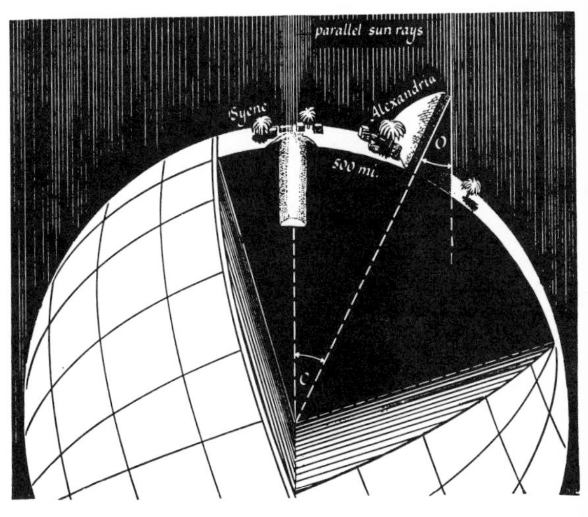

Erastosthenes

estancia Spanish for cattle ranches in parts of South America.

estival (adj.) Pertaining to summer.

estivation A state of dormancy of certain animals, fishes, reptiles, and insects during the summer. See: hibernation

estrogen A hormone produced by the ovary and used in birth control pills.

estuary That part of a drainage channel emptying into the sea in which tidal currents flow.

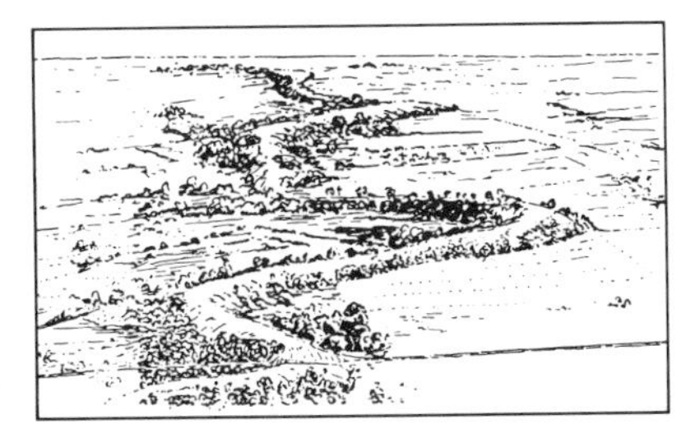

esker

etesian The northerly summer winds of the Mediterranean, especially over the Aegean Sea. They blow intermittently from mid-May to mid-October, having their greatest frequency during July and August.

ethane A colorless, odorless, flammable gas, used chiefly as a fuel.

ethanol See: ethyl alcohol

Ethiopian Realm One of the regions identified by Wallace in his classification of animal distributions. It includes all of Africa except the area lying north of the Sahara.

ethnic (adj.) Pertaining to the cultural characteristics of a people.

ethnography The scientific description of individual cultures.

ethnology The study of cultures with reference to their origin and development and a comparison of their similarities and differences.

ethology The scientific study of animal behavior.

ethyl alcohol An alcohol made by the yeast fermentation of carbohydrates and found in all alcoholic beverages such as beer, wine, and whisky. It has many uses, in medicine as an antiseptic, in automobile radiators as an antifreeze, and in chemical products as a solvent for a wide variety of materials. Also known as grain alcohol or ethanol.

eugenics The science of improving the human race, especially through genetic control.

Eulerian wind A class of winds, such as tornadoes and tropical cyclones, in which the chief controlling influence to balance the pressure gradient is acceleration (terrestrial rotation and friction being relatively negligible).

European corn borer The larvae of a moth, a serious pest introduced into the United States from Europe in 1910.

euryhaline An organism that can tolerate a wide range of salinity.

eustatic movement A large-scale rise or fall of sea level.

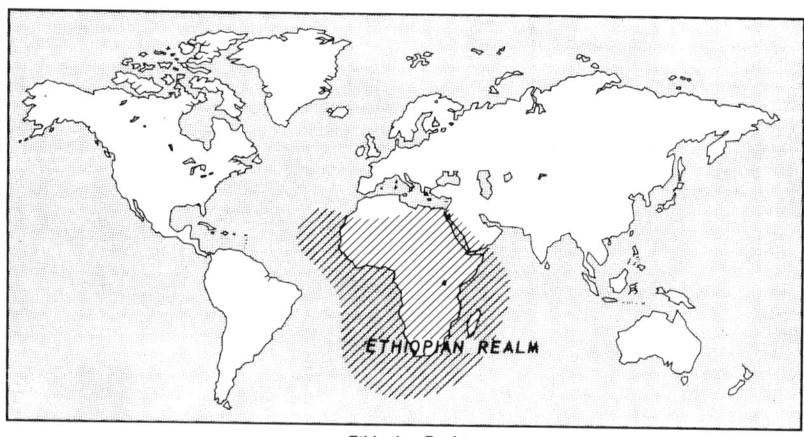

Ethiopian Realm

eutrophication The process of over-fertilization of a body of water by nutrients that produce more organic matter than the self-purification processes can overcome.

evaporation The process by which a liquid changes to the gaseous state; in meteorology, ordinarily understood to refer to the change of liquid water into water vapor, which process continues, under the proper conditions, until saturation is reached.

evaporation pan A pan used to hold water when determining the quantity of evaporation at a given location. Such pans are of varying sizes and shapes, the most commonly used being circular or square.

evaporation pan

evaporation power An index to the degree to which a region is favorable or unfavorable to evaporation. For instance, the evaporation power is greater in the deserts of California than elsewhere in the United States.

evaporation rate The quantity of water, expressed in terms of depth of liquid water, that is evaporated from a given surface per unit of time.

evaporative cooling Cooling of air by the evaporation of water drops falling in a rain. Similar cooling occurs whenever liquid water is changing to vapor, thereby taking latent heat energy from the environment.

evaporativity The potential rate of evaporation as distinguished from the actual rate; defined as the rate of evaporation under the existing atmospheric conditions from a surface of water that is chemically pure and has the temperature of the atmosphere.

evaporimeter See: atmometer

evaporite A rock deposit resulting from the evaporation of an aqueous solution.

evapotranspiration The volume of water evaporated and transpired from water, soil, and plant surfaces per unit land area.

evening star A bright planet generally Venus, seen in the west after sunset.

Everglades A large tract of swampy land in south-central Florida.

evolution 1. The process of formation or development. 2. The continuous genetic adaptation of organisms or species to the environment.

ewe A female sheep.

Ewing-Donn theory The idea that open Arctic seas would greatly affect the earth's climate, acting as a catalyst to produce another ice age.

exchangeable (adj.) Describes the ions in the absorbing complex of the soil that can be exchanged with other ions. For example, when acid soils are limed, calcium ions exchange for hydrogen ions in the complex; when alkali soils are treated with gypsum, calcium ions exchange for sodium ions that can be leached away.

exchangeable sodium Sodium that is attached to the surface of soil particles which can be exchanged with other positively charged ions in the soil solution, such as calcium and magnesium.

exfoliation The peeling or spalling off of thin layers from bare rock surfaces.

exhaust manifold reactors Redesigned exhaust manifolds into which additional air is introduced. The added air re-ignites the unburned fuel in the exhaust, resulting in a lower level of pollution. The reactor is designed to promote complete burning.

exosphere The highest region of the atmosphere, where molecules both join and escape the atmosphere.

exotic (adj.) Of foreign origin or character—Anything exotic, as a plant.

expanding universe theory The theory that all other galaxies are constantly moving away from us, based on observations that light from them is moved in the direction of the red end of the spectrum.

exposure The location of meteorological instruments with respect to the sun, altitude, and natural and artificial surroundings.

exsiccation In climatology, the drying up of an area due to some change that decreases the amount of moisture without reducing the rainfall. Draining a marsh is an example of exsiccation. See: desiccation

extended family A family group consisting of a fundamental social group (spouse and children) and their near relatives; e.g., cousins, uncles, aunts, and grandparents.

extensive cultivation A system of agriculture involving low expenditures of capital and labor per acre, generally resulting in low returns per acre.

extinct species Any species no longer living on earth.

extratropical cyclone A barometric formation of middle and higher altitudes, characterized by an extensive low pressure area around which the winds blow in a counterclockwise direction in the northern hemisphere, clockwise in the southern; it is in general circular or oval in shape, approximately 500-1,000 miles in diameter, causing precipitation, cloudiness, and moderate to strong winds over hundreds of thousands of square miles. It is so named because these formations are a characteristic feature of the weather over most of the earth poleward from 30° north or south latitude.

extreme (adj.) Term applied to the highest and the lowest temperatures (or other meteorological element) that had occurred over a long record for each month and for the year.

extrusive rocks Rocks formed by solidification of magma above the earth's surface.

eye of the storm An area in the center of a tropical cyclone with no precipitation, very light winds, and sometimes complete calm and clear sky.

eye of the wind Windward, the direction from which the wind is blowing.

F layer An ionized region of the atmosphere, some 60 miles thick, the base of which is at about 125 miles above the earth's surface. It is divided into the lower layer, the F_1, or Appleton layer, and the F_2 layer, at about 190 miles elevation.

Fahrenheit A widely used temperature scale that registers the freezing point of water at 32° and the boiling point at 212° under standard atmospheric pressure.

fair (adj.) Of weather, cloudless.

fall 1. A decrease in the reading of an instrument. 2. A season of the year, also called autumn; in the popular mind it usually includes September, October, and November in the North Temperate Zone; 3. A waterfall. 4. (v.) To cut a tree.

fall line The line along which the waterfalls of a number of approximately parallel rivers are found; generally this is associated with a change in the nature of the underlying rock material.

fallout 1. Both the process and the material involved in the deposition of solid material on the earth's surface. 2. Radioactive particles associated with a nuclear explosion.

fallow Cropland that is plowed but left idle in order to restore productivity, mainly through accumulation of water, nutrients, or both.

fallwind A cold wind blowing down a mountain slope; warmed adiabatically during its descent, but still cold relative to the surrounding air.

false cirrus Cirrus proceeding from a cumulonimbus cloud, and composed of the debris of the upper frozen parts of the cloud.

famine Scarcity of food, producing hunger and starvation.

fat Solid or liquid esters of glycerol appearing as whitish to yellowish greasy substances in an organism's body.

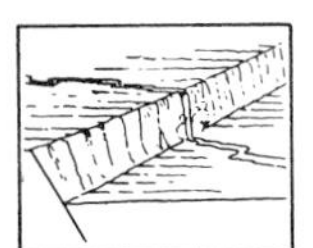
fault scarp

fathom A unit, equal to six feet, used for determining ocean depths.

fathometer An instrument used to determine the depth of the sea.

fault A fracture in soil or in a rock mass along which movement has occurred, causing one side to be displaced in relation to the other.

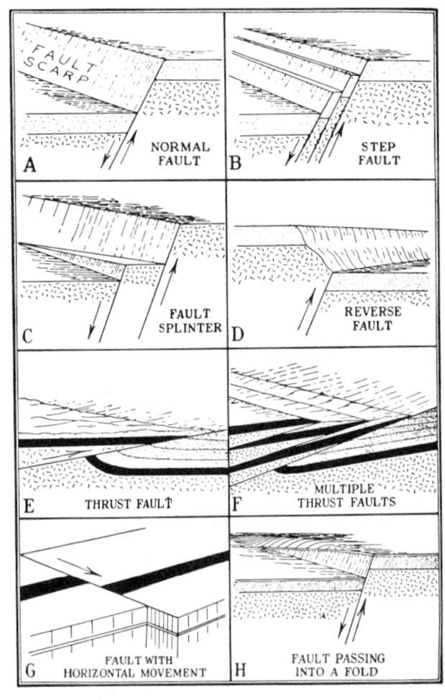

fault

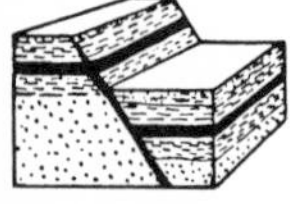
fault block

fault block A portion of the earth's crust, bounded by faults, which has moved relative to the surrounding landscape.

fault movement (or faulting) The movement that causes the displacement along a fault.

fault plane The surface along which faulting occurs.

fault scarp An escarpment along a fault line.

fault zone A portion of the earth's crust in which a number of breaks or faults create a zone of fractured rock.

fauna The animal life of a region or of a geological period.

fazenda Plantations in Brazil, usually coffee plantations.

fecundity The capacity for reproduction in great numbers.

feldspar Any of a group of minerals composed of silicates of aluminum combined with potassium, sodium, and calcium, which make up a large portion of the earth's crust.

fen A tract of low-lying marshy land in eastern England.

feral (adj.) Refers to plants or animals that are no longer cultivated or domesticated and revert to a wild state.

fermentation Any transformation induced by microorganisms, or by their enzyme products that reduce complex organic substances into simpler substances. Organic substances are fermented to a wide variety of products, such as ethyl alcohol, lactic acid, penicillin, and vitamin B_2.

ferralite Soil formed under hot and wet climates.

Ferrel's law The principle that relates to the deflection of winds by the earth's rotation, an effect of the Coriolis force.

ferric iron An oxidized or high-valence form of iron responsible for red, yellow, and brown colors in soils.

ferroalloy An alloy of iron and some other metal used in the making of steel.

ferromagnesium A mineral containing mostly iron and magnesium, typical of basaltic rocks.

ferrous iron A reduced or low-valence form of iron, imparting a blue-gray appearance to some wet subsoils on long standing.

fertile (adj.) 1. Capable of re-producing. 2. Of soil, rich in material for plant growth.

fertility The birth performance of a population or an individual as evidenced by the number of live births. 2. The quality of a soil that enables it to provide compounds, in adequate amounts and in proper balance, for the growth of specified plants, when other growth factors are favorable.

fertility ratio The ratio of the number of children under five years of age per 1000 women of child-bearing age (15-44 years).

fertilizer A natural or artificial substance added to soil to increase productivity.

fertilizer grade The percentages of plant nutrients in a fertilizer. Thus a 10-20-10 grade contains 10 percent nitrogen, 20 percent phosphoric oxide, and 10 percent potash.

fetch 1. The distance waves have traveled in open water, from their point of origin to the point where they break. 2. The distance of the water or the homogeneous-type surface over which the wind blows without appreciable change in direction.

fetus The later stages in the development of the embryo of a mammal when the body structures become identifiable.

fiber Generally, any thread-like material; specifically, those threadlike tissues having sufficient toughness for use in textile or similar industries.

fibril A small thread or fiber; a fiber-like chain of molecules arranged in one direction.

fiducial points Points (or lines) of any kinds of scale used for reference or comparison.

field A region of space at each point of which a given physical or mathematical quantity has some definite value. Thus, one may speak of a gravitational or magnetic field.

field ice Large tracts of ice floating in the polar seas.

field moisture capacity The quantity of water that can be permanently retained in the soil in opposition to the downward pull of gravity.

filament A single continuous fiber of indefinite length, such as rayon, nylon, or silk, in contrast to staple fibers, such as cotton or wool, whose lengths are limited to one or several inches. A filament often extends for several thousand yards.

filaria A group of parasitic round worms usually requiring two hosts to complete their life cycles; one phase occurs in the body of a biting insect and the other in the body of the bitten organism.

filling The process, opposite to deepening in which the central pressure of a cyclone increases.

filtration The process of separating solid particles from a liquid by passing the liquid, by gravitation, through a sheet or continuous and compact layer of porous material. Filtration may also be effected by centrifugal force, by hydraulic or air pressure applied to the liquid, or by sucking the liquid through the filter with a partial vacuum.

fiord (fjord) A long narrow inlet along the seacoast with steep sides created by valley glaciers.

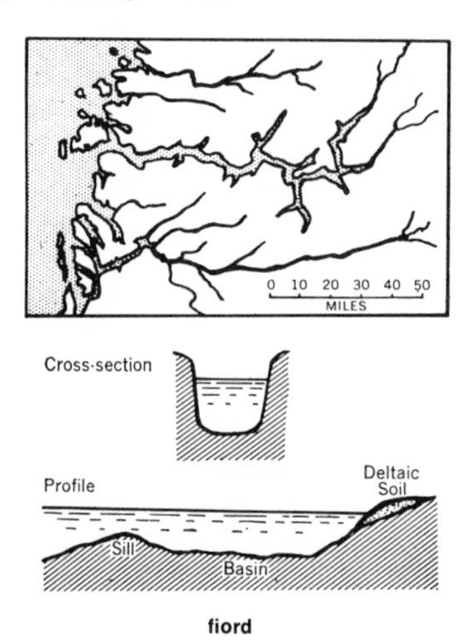

fiord

fire edge The line, usually irregular, to which a fire has burned at a given moment; the boundary of a fire at a given moment.

firebreak A barrier, natural or constructed, from which all or most of the inflammable materials have been removed, to stop or check creeping or running but not spot fires, or to serve as a line from which to work and to facilitate the movement of men and equipment in fire suppression.

fire weather Weather conditions that are favorable for the starting of fires, particularly in forests. The elements that are given in a report on fire weather include weather, temperature, relative humidity, wind direction and velocity, and visibility, to which are added advice and information on the degree of danger.

firm yield The maximum annual supply of a given water development that is expected to be available on demand, with the understanding that lower yields will occur in accordance with a predetermined schedule or probability.

firn (or firn snow) 1. Snow that has become granular and compacted. New firn snow lies with its grains fairly loose, but as it grows old the grains become more and more firmly held together by a cement of ice originating from a film of thaw water surrounding them. Also called névé. 2. Old snow on top of glaciers, but not yet converted into ice.

firnification The process by which newly fallen snow is converted into firn.

firn line The highest level to which the fresh snow on a glacier's surface retreats during the melting season. The line separating the accumulation area from the ablation area.

first law of thermodynamics The law of conservation of energy, that whenever heat energy is transformed into any other kind of energy, or vice versa, the quantity of energy that disappears in one form is exactly equivalent to the quantity produced in another form.

firth A long, narrow inlet in Scotland, usually in the lower part of an estuary.

fish ladder An underwater step-like structure commonly built near a dam to permit fish to migrate upstream.

fission 1. The division of any unit into parts. 2. In biology, the division of an

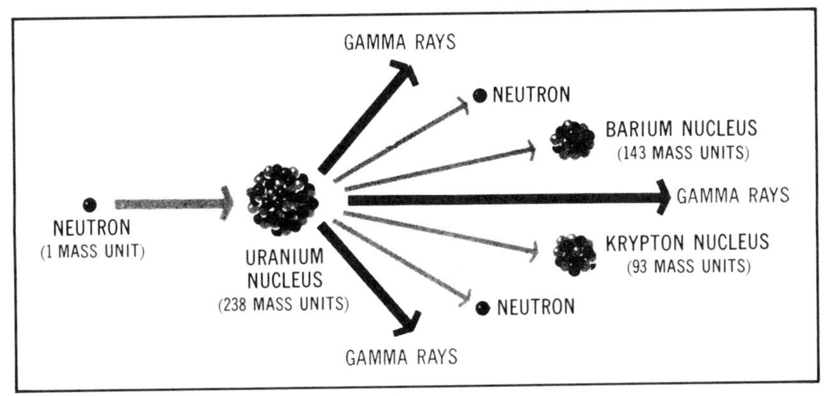

fission

organism into new organisms. 3. In physics, the splitting of an atom into nuclei of lighter atoms.

fissure A narrow opening or crack.

fixation (in soil) The conversion of a soluble material, such as a plant nutrient like phosphorous, from a soluble or exchangeable form to a relatively insoluble form.

fjard A low irregular coastline marked by a lack of cliffs but with many long, branching bays.

fjord See: fiord

flagellate A microscopic form of protozoan which propels itself by means of a long lashlike appendage.

flaring The burning of excess gas associated with oil production, now seen only rarely.

flash flood A local flood that rises and subsides rapidly.

flatirons A triangular, sloping layer of rock generally occurring in series along the flanks of mountains and created by the erosion of tilted sedimentary beds.

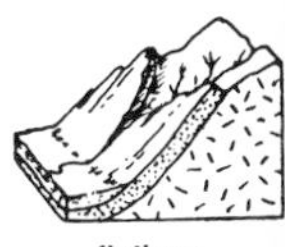

flatiron

flocculate (v.) To aggregate or clump together individual tiny soil particles, especially fine clay, into small groups or granules. The opposite of deflocculate, or disperse.

floe A sheet of floating ice; a detached portion of an ice field.

flood 1. A rise of the sea above its normal tidal height, due to storm or volcanic action. 2. A rise of a river or stream above its banks, generally on account of a heavy snowfall or excessive rainfall in the watershed through which it passes, and most frequently in spring.

flood

flood control capacity That part of the gross reservoir capacity that, at the time under consideration, is reserved for the temporary storage of flood waters.

flood control storage Storage of water in reservoirs to abate flood damage.

flood frequency The average interval of time between floods equal to or greater than a specified discharge or stage, generally expressed in years.

flood-frequency curve 1. A graph showing the number of times per year on the average that floods of a given magnitude are equaled or exceeded. 2. A similar graph but with recurrence intervals of floods plotted as one of the variables.

flood irrigation Irrigation by running water over nearly level soil in a shallow flood.

flood, maximum probable The largest flood for which there is any reasonable expectancy in a particular climatic era.

flood peak The highest value of the stage or discharge attained by a flood; thus, peak stage or peak discharge.

flood plain A strip of relatively smooth land bordering a stream built of sediment carried by the stream and dropped in the slack water beyond the influence of the swiftest current in times of flood.

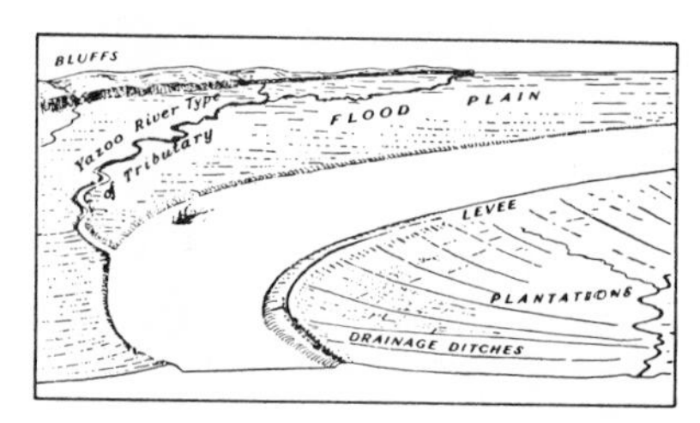

flood plain

flood profile A graph of elevation of the water surface of a river in flood, plotted as ordinate, against distance, measured in the downstream direction, plotted as abscissa. A flood profile may be drawn to show elevation at a given time, crests during a particular flood, or stages of concordant flows.

flood routing The process of determining progressively the timing and shape of a flood wave at successive points along a river.

flood stage The elevation, or stage, at which overflow of the natural banks of a stream begins to cause damage in the reach in which the elevation is measured.

floodway The channel of a river or stream and those parts of the flood plains adjoining the channel that are reasonably required to carry and discharge the floodwater or flood flow of any river or stream.

flood zone The land bordering a stream that is subject to floods with some frequency.

flora The plant life of a region or of a geological period.

floriculture The cultivation of flowers or flowering plants.

flotation A method of mineral separation involving the use of a reagent to make some minerals float while others sink.

floury endosperm See: endosperm

flow-duration curve A cumulative-frequency curve that shows the percentage of time that specified discharges are equaled or exceeded.

flow (flowing) resource See: renewable resource

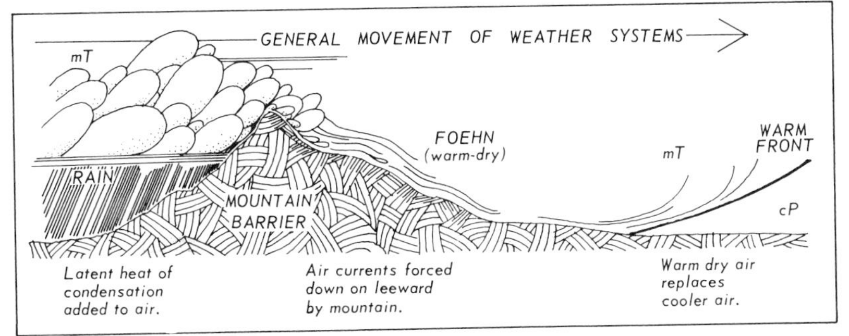

foehn

flowing well A well from an artesian aquifer in which the water is under sufficient pressure to cause it to rise above the ground surface.

fluvial (adj.) Pertaining to flowing water.

flux 1. The flow of a substance across an interface, as water from an ocean to the atmosphere. 2. State of change.

fly ash Particulates that are emitted from smoke stacks, especially coal-fueled power plants.

flyway A route along which birds customarily migrate between their breeding grounds and wintering areas.

focus The center point of an earthquake from which earthquake waves radiate in all directions.

fodder Coarse food, such as hay and cornstalks, for domesticated animals.

foehn A dry wind with a strong downward component, warm for the season, characteristic of many mountainous regions, notably the Alps, and also the Rockies, where it is known as the chinook.

foehn cloud (or wall) A cloud formation on the leeward side of a mountain range over which a foehn wind is passing.

fog A cloud formed at the surface of the earth by the condensation of atmospheric water vapor into a multitude of minute water droplets (average diameter about 40 microns) or, less frequently, tiny ice crystals, and interfering to varying degrees with the horizontal visibility at the surface.

fog dispersal The process of causing fog to dissipate, generally by mixing, heating, or seeding.

fog drip Water dripping to the ground from trees or other objects that have collected moisture from wind-blown fog.

föhn See: foehn

fold A bend in rock layers caused by compression of the earth's crust.

folded (or fold) mountains Mountains created by folding, which consist of tilted sedimentary rock structures.

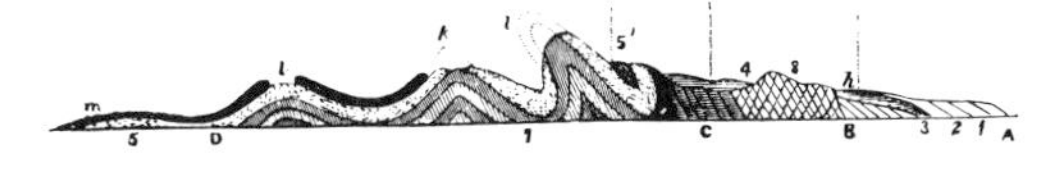
folded mountains

foliage Plant leaves.

foliation 1. A layerlike structure found in some metamorphic rocks, such as schist and gneiss, resulting from the separation of different minerals into parallel layers. 2. Bearing foliage.

food chain The dependence of one type of life on another, each in turn eating or absorbing the next organism in the chain.

food web The complex set of energy and matter transformations in an ecosystem involving a number of food chains.

footwall The rock immediately below a fault plane or mass of ore.

forage Plant material that can be used as feed by domestic animals, by being either grazed or cut for hay.

foraminifera Tiny one-celled animals, mostly marine, with hard shells of calcium carbonate or of cemented sedimentary grains.

ford The shallow part of a river or other body of water that may be crossed by wading.

forebay reservoir A reservoir used to regulate the flow of water to a hydroelectric plant; it may serve other purposes, such as recreation.

forest An extensive area covered by trees.

formation A group of strata or layers having certain common characteristics that more or less readily distinguish them from adjacent layers.

fossil Remains of animals or plants that have been preserved in the earth's crust since before the beginning of historic time. The word originally referred to anything, such as ores, coal, or rocks, dug up from the earth.

fossil fuels Those fuels derived from the fossils of organisms; i.e., gas, oil, coal.

Foucault pendulum An apparatus used to demonstrate the rotation of the earth on its axis.

fraction A compound separated into components by distillation or crystallization.

fractional distillation A physical method for separating liquid mixtures of chemical compounds by distilling them at gradually increasing temperatures and collecting any uncondensed gas and the condensed liquid and residue as a number of fractions, each one of which represents a definite range of boiling points. The process is repeated with individual fractions to narrow the boiling ranges of the final fractions as much as possible, and eventually to obtain fractions having fixed boiling points (characteristic of pure compounds).

fracto- A prefix to indicate a basic cloud form with a torn, ragged, and scattered appearance caused by strong winds.

fractocumulus A low, small, scudding, ragged variety of cumulus cloud, lacking the clear-cut outlines of ordinary cumulus, generally seen in bad weather under a sheet of altostratus or nimbostratus.

fractostratus A remnant of a broken-up layer of stratus or a shred-like wispy cloud of independent formation, that may develop into a thin, low layer below a sheet of altostratus or nimbostratus, that may be visible through the interstices of the fractostratus.

fracture A break in rock material due to intense folding or faulting.

fragipan A dense and brittle pan or layer in soil that owes its hardness mainly to extreme density or compactness rather than to high clay content or cementation. Removed fragments are friable, but the material in place is so dense that roots cannot penetrate and water moves through it very slowly because of small pore size.

fragmentation nucleus A tiny ice nucleus that has broken from a larger one and acts as a growth center within a cloud.

frazil (frazil ice) Fine needlelike ice, resembling cinders, composed of small particles that, when first formed, are colloidal and not seen in the water in which they are floating.

free fatty acids A group of organic, aliphatic-type, monobasic acids occurring in the uncombined state in fats and oils or their derivatives.

freeze The condition that exists when over a widespread area the surface temperature of a whole air mass remains below 0°C or 32°F for a sufficient time to constitute the characteristic feature of the weather. (v.) To change state from liquid to solid.

freezing level The height at which a temperature of 0°C is reached in a column of rising air.

freezing nucleus Any particle that when present within a mass of supercooled water will initiate the growth of an ice crystal.

freezing point In meteorology, the temperature at which pure water freezes or pure ice melts, i.e., 0°C or 32°F.

freezing rain Precipitation in the form of rain, a portion of which freezes and forms a smooth coating of ice (or glaze) upon striking exposed objects.

frequency 1. The time rate of vibration or the number of complete cycles per unit time. 2. In climatology, the number of times a certain phenomenon of the weather occurs in a given time.

frequency curve A curve that expresses the relation between the frequency distribution plotted, usually with the magnitude of the variables as abscissas and the number of occurrences of each magnitude in a given period as ordinates.

frequency distribution The quantitative distribution of objects on the basis of some variable characteristic.

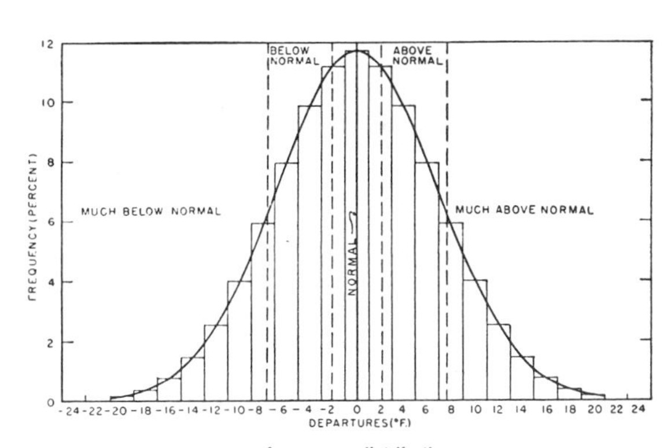

frequency distribution

freshet 1. A flood or overflowing of a stream or river caused by heavy or long-continued rains or melted snow. 2. A stream of fresh water.

friable (adj.) Easily crumpled.

friagem An occasional period of cool weather on the middle Amazon in Brazil, lasting five or six days in May or June, and brought about by the penetration of a relatively cold south wind to this usually uniform equatorial climate.

frigid zone One of the climatic divisions of the earth based on astronomical considerations, rather than on actual climatic conditions. See: Antarctic Zone; Arctic Zone

front The line of intersection of a frontal surface with a horizontal plane (e.g., the earth's surface); or the line on the earth's surface, or at higher levels, where two different air masses meet.

frontal (adj.) Pertaining to a front, as frontal surface.

frontal surface The surface of separation between two different and adjacent air masses.

frontal zone The region of transition between two air masses; a sloping layer of the atmosphere, separating air of different temperature, density, or wind velocity, in which values of such elements change gradually.

frontogenesis The creation of a new front or regeneration of an old one, generally in a region where a large temperature gradient and a converging wind system combine to bring into close proximity air of different temperature, density, and speed and direction of movement.

frontolysis The process by which a front weakens or dissolves, owing to the fact that the air masses it separates have become one homogeneous whole.

frost 1. A light, feathery deposit of ice caused by the condensation of water vapor, directly in the crystalline form, on terrestrial objects whose temperatures are below freezing. 2. The atmospheric condition when the temperature is below 32°F.

frost-free season The growing season; the average number of consecutive frost-free days occurring each year at a given place.

frost heaving The lifting of a surface by the internal action of frost. It generally occurs after a thaw, when the soil is filled with water droplets and when a sudden drop of the temperature below freezing changes the droplets into ice crystals, which involves expansion, and, consequently causes an upward movement of the soil.

frostless zone Same as thermal belt.

frost smoke A thick fog or mist, most common in high latitudes, rising from the surface of the sea when the relatively warm water is exposed to an air temperature much below freezing; also called sea smoke or frost mist.

fuel cell An experimental method of producing electrical energy on demand. It has been used in powering automobiles, farm and construction equipment. To date, this system has proven heavy, expensive and unreliable.

fugitive resource See: nonrenewable resource

fumarole A hole or crack in the ground in volcanic regions from which hot gases escape.

fungus A low form of plant life having no chlorophyll, reproducing by spores, having a mycelium, and living as a parasite or saprophyte on organic

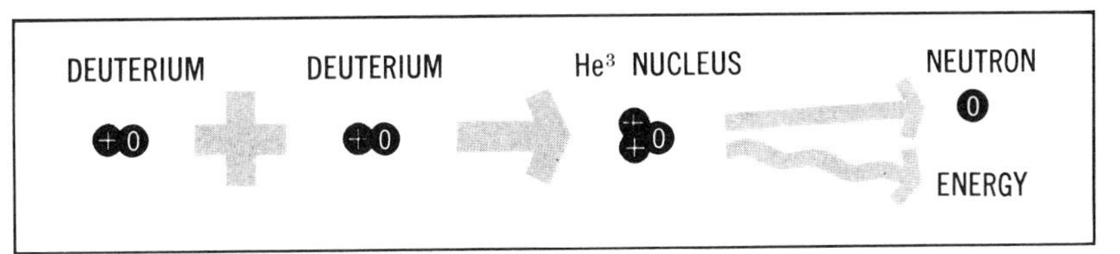

fusion

matter. Numerous on and in soil, fungi aid in breaking down organic debris to humus.

funnel cloud The characteristic tornado cloud, often shaped like a funnel, that develops in a low cumulonimbus cloud mass, and reaches down toward the earth. See: tornado

fusion 1. In general, the transition of a substance from the solid to the liquid state, or the melting together of two substances. 2. In meteorology, the melting of ice, at the temperature of 0°C or 32°F under standard pressure. 3. The formation of a heavier nucleus from two lighter ones with attendant release of energy.

gabbro Coarse-grained igneous rock consisting essentially of feldspar and one or more of the ferromagnesian minerals.

gage See: gauge

gauge (gage) A device for measuring changeable magnitude or position in specific units; e.g. the elevation of a water surface, the velocity of flowing water, the pressure of water, the amount or intensity of precipitation, the depth of snowfall. (v.) To register or measure the magnitude or position of a thing when these characteristics are undergoing change.

gauge height The water-surface elevation, referred to some arbitrary gauge datum.

gauging station A particular site on a stream, canal, lake, or reservoir where systematic observations of gauge height or discharge are obtained.

galaxy One of the systems of billions of stars, nebulae, gases, and dust that move through space together.

gale A wind with a velocity exceeding 30 miles an hour; the precise limiting velocities vary among the different meteorological services.

galena A dark gray or black ore bearing silver and lead.

gamete A mature sexual reproductive cell, a sperm or an egg, which unites with another cell to form a new organism.

gamma ray A ray given off by the nucleus of some radioactive substances. A form of radiant energy similar to the X ray and to visible light, but having a much shorter wavelength than either.

gang plow A set of plows hooked side by side.

gangue The material that surrounds an ore in a lode or vein.

garigue (garrigue) The scrub bushes and trees characteristic of the drier lands around the Mediterranean Sea, comparable to the chaparral of the American southwest.

garúa (or camanchaca) A thick, damp fog on the coasts of Ecuador, Peru, and Chile, which creates a raw, cold atmosphere that may last for weeks in winter and supplies most of the moisture in the area around Lima.

gas A fluid of low density and high compressibility having neither independent shape nor volume and tending to expand indefinitely.

gas constant The constant in the equation of state for a perfect gas which relates the variables of pressure, volume and temperature. For dry air, the characteristic gas constant may be taken in c.g.s. units as 2.87×10^6 cm^2/sec^2 per degree.

gas equation of state A relation between the variables that specify the physical state of a gas, which results from combining Boyle's law and Charles' law. See: gas laws

gas laws Laws governing, strictly, the behavior of an ideal gas, but sufficiently valid for the purposes of meteorology when applied to the gases of the atmosphere. See: Boyles' law, Charles' law

gastropod A member of a group of mollusks having a one-piece coiled shell; e.g., a snail.

gaucho South American cowboy, famous for horsemanship, usually descendant of both Spanish and Indian peoples.

geest A sandy region in the glaciated northern lowlands of Germany.

gene The unit that controls hereditary characteristics transmitted in the chromosome.

gene pool All of the genes in a given population that are available for exchange in interbreeding.

general circulation The average or prevailing large-scale movements of the atmosphere as represented by the yearly means of all available records of surface and upper-air wind velocities, which fit into the average annual pressure patterns. Thus, the general circulation is the mean condition and, to some extent, an idealized picture of the atmospheric circulation.

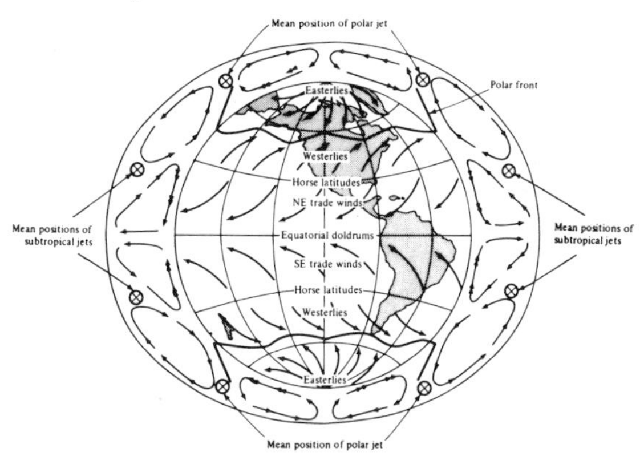

general circulation

general gas law The law that the volume of a given mass of an ideal gas varies directly with its absolute temperature and inversely with its pressure.

gastropod

General Land Office Survey The system of dividing the public domain of the United States into townships and sections, initiated by the U.S. Land Ordinance of 1785.

genetics The science of heredity.

genotype A group sharing a specific genetic constitution.

genus A major subdivision in the classification of living things, usually consisting of more than one species.

geoanticline (or geanticline) A large anticline extending over several miles.

geochemical cycle The set of processes involved in rock formation, weathering, transportation, and redeposition.

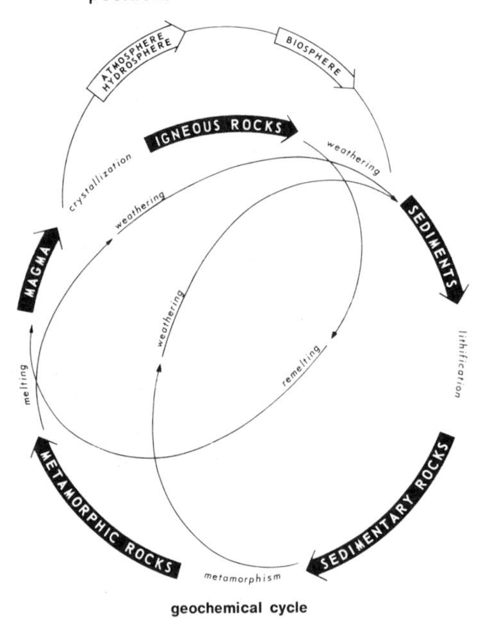

geochemical cycle

geodesy The science of measuring the earth's shape, size, weight, etc.

geography The study of the earth and its people involving an analysis of spatial processes, distributions, and relations.

geographic coordinates Lines on a map by which points on the earth's surface are located with reference to previously determined lines.

geohydrology That branch of hydrology relating to subsurface or subterranean waters.

geoid The equipotential surface of gravity that coincides with the mean surface of the oceans of the earth.

geologic erosion The normal or natural erosion caused by geological processes acting over long geologic periods and resulting in the wearing away of mountains and the building up of flood plains, coastal plains, etc. See: erosion

geologic time scale See: Appendix I

geology The study of the composition, structure, and history of the physical changes that have occurred in the earth.

geomorphology The study of the evolution of landforms.

geophysics The study of the physical characteristics and properties of the earth; including geodesy, seismology, meteorology, oceanography, atmospheric electricity, terrestrial magnetism, and tidal phenomena.

geopotential The potential energy of unit mass relative to sea level, numerically equal to the work that would be done in lifting the unit mass from sea level to the height at which the mass is located.

geoscience Any of the sciences that deal with planet earth.

geosphere The solid portion of the earth, synonymous with the lithosphere.

geostrophic wind A steady horizontal air motion along straight, parallel isobars in an unchanging pressure field, with gravity the only external force, and in a direction perpendicular to that in which the Coriolis force (due to the earth's rotation) and the pressure gradient force are acting equally and oppositely.

geosyncline A large trough-like depression that sinks slowly over a long period of time, while a thick sequence of sedimentary and possibly volcanic rocks accumulates in it.

geotropism Movement or growth directed by gravity.

germ That part of the grain kernel, also called the embryo, that becomes a seedling in the process of germination. It is embedded in the endosperm and is covered only by the pericarp, or seed-coat layer.

germinate (v.) To come into existence.

gestate (v.) To carry in the womb until born.

geyser A hot spring or fountain that at intervals shoots forth, often violently, hot water and steam.

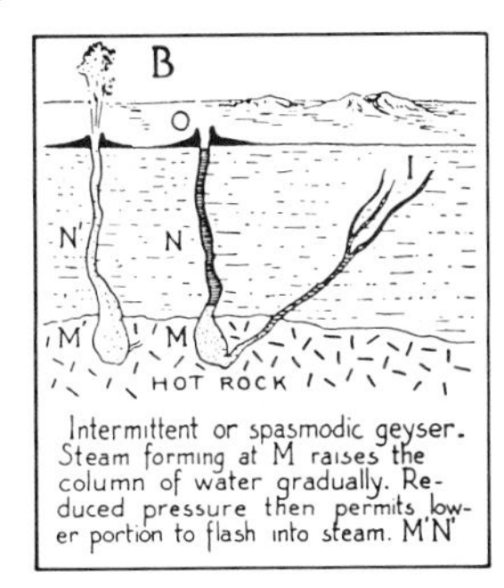

Intermittent or spasmodic geyser. Steam forming at M raises the column of water gradually. Reduced pressure then permits lower portion to flash into steam. M'N'

geyser

geyserite The siliceous materials formed around hot springs and geysers.

gibli The hot, dry, southerly sirocco wind of North Africa.

gibbous Convex on both sides; as the moon when more than half full.

girdling (v.) The act of encircling the stem of a living tree with cuts that completely sever bark and cambium and often are carried well into the outer sapwood, usually killing the tree.

glacial (adj.) Pertaining to a glacier; applied to the periods of widespread glacier activity, and to the climatic conditions prevailing in those times.

glacial drift Material that has been deposited by a glacier or in connection with glacial processes. It consists of rock flour, sand, pebbles, cobbles, and boulders. It may occur in a heterogeneous mass or be reasonably well sorted, depending upon its manner of deposition.

glacial striations Fine lines or scratches carved into rock surfaces by rock fragments carried in a glacier.

glaciation The covering of an area with glacial ice.

glaciation

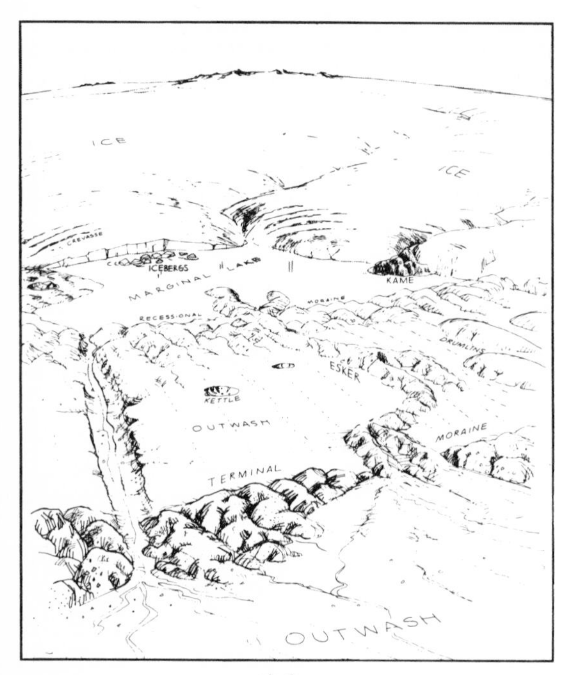

glacier

glacier An extensive, slowly flowing body of ice, formed on land by snow transformed into ice by pressure and frozen liquid water.

glaciology The study of ice and the action of ice in all its forms.

glade An open space in a wood or forest.

glaze A transparent or translucent coating of ice having a smooth glossy appearance. It is formed by the freezing of rain on terrestrial objects. See: freezing rain

glazed frost In the classification of hydrometers adopted by the International Meteorological Organization and used by the British Meteorological Office, the name given to glaze.

Gley soil A soil horizon in which waterlogging and lack of oxygen have caused the material to be a neutral gray in color. The term "gleyed" is applied, as in "moderately gleyed soil," to soil horizons with yellow and gray mottling caused by intermittent waterlogging.

glucose A sugar occurring in many fruits.

gluten The tough nitrogenous substance remaining when the flour of wheat or other grain is washed to remove the starch.

glycerol (or glycerin) A colorless, odorless, sweet liquid used for sweetening and preserving food; it also has a number of other industrial uses.

gneiss A banded metamorphic rock.

gnomonic projection An azimuthal projection with the point of projection the center of the earth; the plane upon which the earth's grid is reproduced is tangent at any point on the globe.

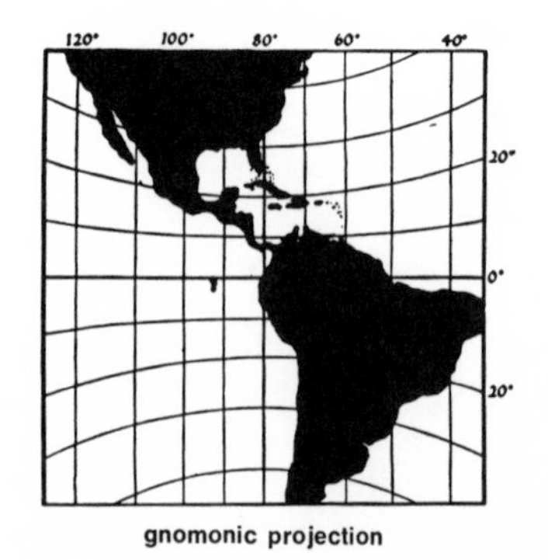

gnomonic projection

Gondwana (Gondwanaland) A continent believed to have existed before the Permian period. It is supposed to have broken up and the pieces to have drifted apart, forming modern India, Australia, Antarctica, and parts of Africa and South America.

Goode's interrupted homolosine projection An equal-area projection in which each segment is centered on the central meridian of each continent. It is based on the sinusoidal and homolographic projections.

gores The pieces of paper attached to a sphere to make a globe. The earth grid and details of the continents are imprinted on them.

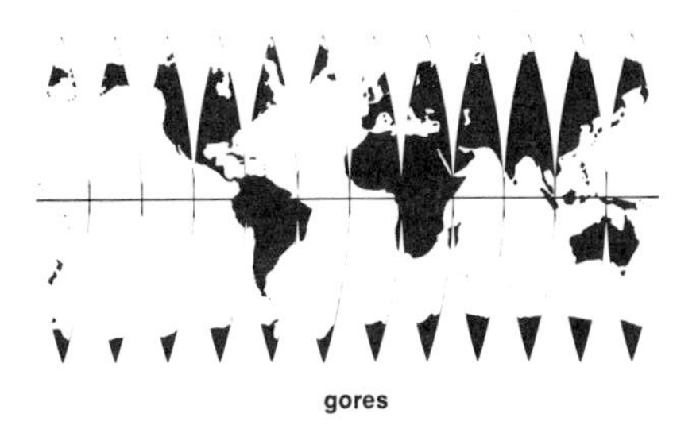

gores

gorge A deep, narrow, and steep-walled valley.

graben A portion of the earth's crust, between faults, which has dropped lower than the surrounding area.

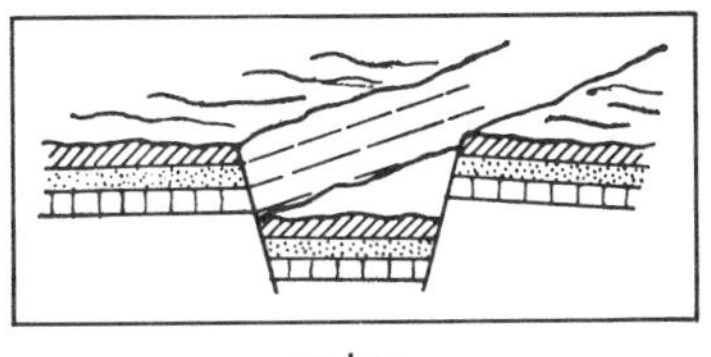

graben

gradation 1. A process taking place in a series of stages or steps. 2. The process of bringing a stream to grade.

grade 1. The continuous slope of a stream bed necessary to carry both water and sediment. 2. The measure of a slope, expressed in percent.

graded stream A stream in equilibrium, so that it neither aggrades nor degrades its bed.

grade level The level to which a stream is eroding.

gradient The rate of change in the value of any quantity with distance in any given direction; in practice, usually the rate of change in a horizontal or vertical plane, in the direction of the maximum rate of change. Horizontal pressure gradients are usually measured along a direction normal to the isobars, positive toward lower pressures. When the isobars are close together, the gradient is said to be steep, and when far apart, the field of pressure is said to be "flat," or the gradient "weak."

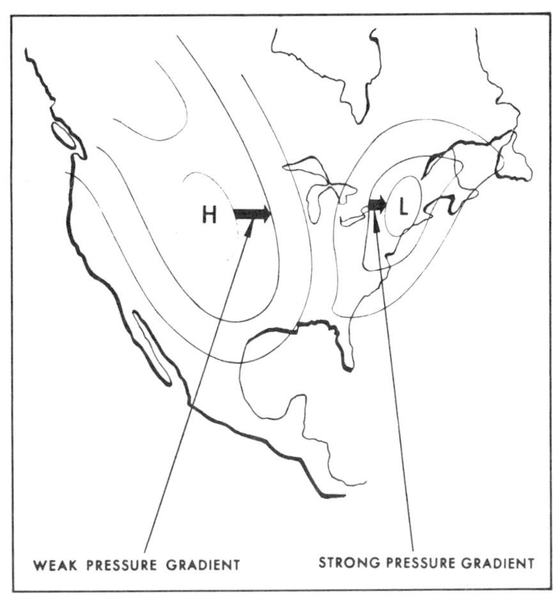

gradient

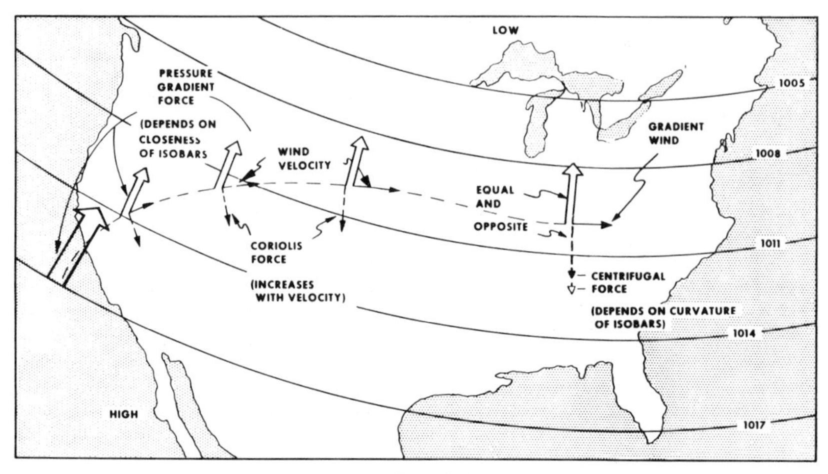

gradient wind

gradient wind Steady, horizontal, functionless, atmospheric motion in which the wind blows parallel to curved isobars in an unchanging pressure field, and the centrifugal, Coriolis, and pressure-gradient forces balance. The gradient wind velocity is the wind speed attained under these conditions. It does not occur at the earth's surface, due to friction, but is closely realized at a height of about 1,500 feet.

grafting The act or process of inserting a cion (less strictly a bud), of a specified variety into a stem, root, or branch of another plant so that a permanent union is effected, especially for purposes of propagation.

grain alcohol See: ethyl alcohol

gram A unit of mass, originally defined as the mass of 1 cubic centimeter of water at 4° C., but now taken as the one-thousandth part of the standard kilogram.

gram calorie See: calorie

granite A course-grained igneous rock containing quartz, feldspar, hornblende, and other minerals in varying quantities.

granular snow Precipitation of white, opaque, snowlike grains, similiar to soft hail but more or less flattened or oblong in shape and generally less than 1 mm. in diameter.

granular structure Soil structure in which the individual grains are grouped into spherical aggregates with indistinct sides. Highly porous granules are commonly called crumbs. A well-granulated soil has the best structure for most ordinary crop plants.

graphite One of the softest minerals, composed of pure carbon; used in pencils.

grasslands Subhumid and semiarid regions of the earth where the vegetation consists mostly of grass. See: savanna, prairie, steppe

graupel German name for the hydrometeors called "grésil" in French and "soft hail" in English and formed by the freezing of water droplets on a snow crystal falling through a cloud. Graupel looks like crumbly pellets of snow–miniature snowballs–and ranges in size from that of coarse shot to that of small peas.

gravel A mixture of rounded stones and finer material such as sand or clay.

gravel envelope In well construction, the uniform gravel poured several inches thick into the annular space between the well casing and the drilled hole.

gravitation A universal force by which particles of matter attract each other. Often refers to the strong attraction near the surface of the earth that pulls bodies toward the ground. See: gravity

gravitational water The water in the large pores of the soil that drains away under the force of gravity with free under-drainage. Well-drained soils have such water only during and immediately after rains or applications of irrigation water. In poorly drained soils, this water accumulates in the pores at the expense of air. Under such conditions, the soil lacks oxygen for the roots of most crop plants and is said to be waterlogged.

gravity The apparent force per unit mass with which the earth attracts bodies near its surface, as measured by the acceleration of a freely falling body relative to the surface of the earth.

gravity correction A quantity that must be applied to the readings of all mercurial barometers not located at sea level at 45° north or south latitude, where the value of gravity is taken as a standard of reference. The correction is made in order to obtain comparable pressure values all over the world, since the height of the mercury column varies with gravity, and gravity in turn varies with elevation and latitude.

gravity fault In geology, a fault along an inclined plane in which the upper rock has moved downward with respect to the rock mass below the fault plane.

gravity wind See: katabatic wind

gray body See: grey body

gray-brown podzolic soils A zonal group of soils having thin organic coverings and thin organic-mineral layers over grayish-brown leached layers that rest upon brown B horizons richer in clay than the soil horizon above. These soils have formed under deciduous forests in a moist-temperate climate.

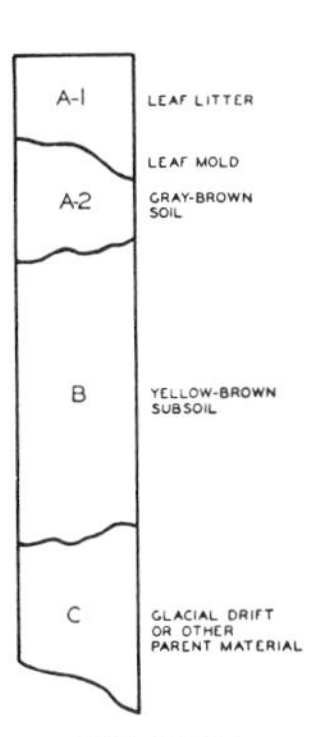

gray-brown podzolic soils

gray desert soils Soils formed under conditions of low precipitation with little or no vegetation cover. They are light in color and contain no humus.

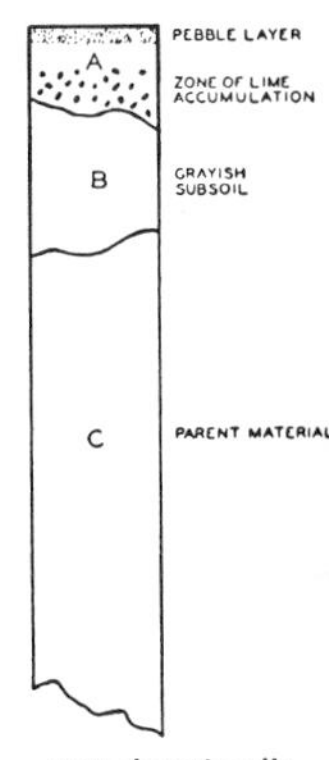

gray desert soils

Great Basin high An area of high barometric pressure, centered over the Great Basin region of Nevada, formed by the piling up and stagnation of air owing to the stability of the surface layers and the natural barriers afforded by the topography of the Basin.

great circle A circle on the earth's surface created by a plane passing through the center of the earth, dividing it into hemispheres.

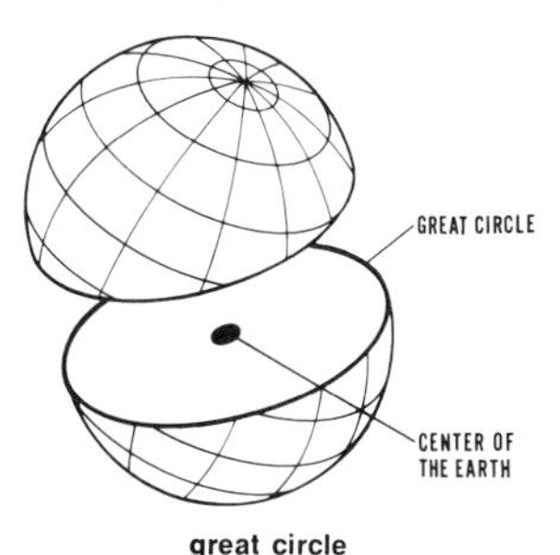

great circle

great circle route A route between any two points on the earth's surface that follows the great circle passing through them.

Great Plague The bubonic plague of 1665 in London. It killed about 15 percent of the population.

Great Salt Lake A lake lying in northwest Utah which is what remains of ancient Lake Bonneville.

Great soil group Any one of several broad groups of soil with fundamental characteristics in common. Examples are chernozem, gray-brown podzolic, and podzol.

green belt In a metropolitan region, open farmland or park areas that separate individual communities from each other.

green flash A brilliant green coloring of the upper edge of the sun as it is about to disappear at sunset below a distant clear horizon. It is due to refraction by the atmosphere, which disperses the last spot of light into a spectrum and causes the colors to disappear in the order of refrangibility.

greenhouse effect The thermal result of the fact that solar radiation, of comparatively short wave length, penetrates the atmosphere rather freely, only to be largely absorbed near and at the earth's surface, whereas terrestrial radiation, of long wave length, passes upward with great difficulty. This effect is due to the fact that the absorption bands of water vapor, ozone, and carbon dioxide are more prominent in the wave lengths occupied by terrestrial radiation than in the short wave lengths of solar radiation. Hence the lower atmosphere is almost perfectly transparent to incoming radiation, but partially opaque to outgoing long-wave radiation.

green revolution The name given to the current increase in agricultural yields due to the development of high-yielding plants, increased irrigation and use of fertilizer, and new technology.

Greenwich Mean Time (G.M.T.) Standard time established at the prime meridian at the Greenwich Observatory in England. It is used for navigational purposes throughout the world.

gregale A strong northeast wind of the central Mediterranean.

grey body A radiating surface which is such that while its radiation has the same spectral energy distribution, its emissive power is less at any temperature than that of a black body, and such that while not black, its absorptivity is nonselective.

grid A systematic arrangement of points or lines along which measurements are made or objects located, as on a map, chart, or aerial photograph.

grit Rock particles, consisting mostly of quartz.

groin A low wall built out into the sea, more or less perpendicular to the coast line, to minimize shore erosion by trapping sand being carried along the shore.

gross reservoir capacity The total amount of storage capacity available in a reservoir for all purposes.

ground fog A shallow but often dense fog of the radiation type, through which the stars may be observed at night and the sun in daytime. It appears first at the ground and remains there even after it thickens. It is characterized by a temperature inversion whose base is at the ground and whose upper edge is at the top of or above the fog.

ground frost A freezing condition considered to have occurred when a minimum thermometer exposed to the sky and just above a grass surface records a temperature of 30.4°F. or below. It is injurious to vegetation.

ground inversion In the layer of air just above the earth's surface, an increase of temperature with height.

ground moraine Material transported by a glacier and deposited on the ground as the glacier stagnates and melts.

ground swell A broad, deep rolling of the sea generated by a distant storm.

ground water All water below the land surface. It has its origin in the downward seepage of surface water to a layer of impervious material.

ground-water basin A ground-water storage area more or less separate from neighboring ground-water storage areas, so that it can be considered a hydrologic unit.

ground-water divide A line on a water table on either side of which the water table slopes downward; analogous to a drainage divide between two drainage basins on a land surface.

ground-water hydrology The branch of hydrology that treats of ground water.

ground-water mining Pumping water from a basin in which it has accumulated over a long period of time. The safe yield is very small. Withdrawals exceed replenishment, which may be negligible.

ground-water outflow That part of the discharge from a drainage basin that occurs through the ground-water process.

ground-water overdraft Pumpage of ground water in excess of safe yield.

ground-water recharge Inflow to a ground-water reservoir.

ground-water reservoir An aquifer or aquifer system in which ground water is stored. The water may be placed in the aquifer by artificial or natural means.

ground-water runoff Ground water which has been discharged as spring or seepage water into a stream channel.

ground-water storage capacity The reservoir space available in given volume of deposits. Also, the optimum volume of water that can be alternately extracted and replaced in a deposit, within specified economic limits.

growing season The season that is warm enough for the growth of plants, the extreme average limits of duration being from the average date of the last killing frost in spring to that of the first killing frost in autumn. On the whole, however, the growing season is confined to that period of the year when the daily mean temperatures are above 42°F.

growth ring See: annual ring

guano The excement of seabirds or bats, valuable as a fertilizer because it is rich in phosphates and nitrates.

Guernsey One of a breed of dairy cattle originally bred in the British Isles. Its milk has a high butterfat content.

guinea corn A cereal grass similar to millet.

gulch A narrow, deep ravine with steep sides formed by rushing water.

gulf A large bay or extensive inlet.

gully A long, narrow channel worn by the action of water. Smaller than a ravine.

gum The raw product (oleoresin) that exudes from the wood of a living tree when a wound is made through the bark into the living tissues.

gumbo Soils that form a sticky mud when wet.

Günz First glacial period in Europe, during the Pleistocene.

gust A sudden brief increase in the force of the wind, of more transient character than a squall and followed by a lull.

gustiness Irregularity in the velocity of the wind, caused either by the mechanical effect of surface irregularities, which create eddy currents that disrupt the smooth air flow, or by convection currents due to surface heating.

gustiness factor A measure of the gustiness of the wind. A gustiness factor may be defined as follows: gustiness factor=range of fluctuation in gusts and lulls—mean wind. The range is determined from the highest gusts and lowest lulls over a period of time, such as an hour, ignoring any very exceptional variation. The factor is conveniently given as a percentage; for example, if the mean wind is 30 miles per hour with gusts up to 45 miles per hour and lulls down to 15 miles per hour, the range is 30 miles per hour and the factor is 100 percent.

guyot A seamount, flat-topped and generally 100 fathoms below the surface of the water.

gymnosperm A plant having its seeds exposed and not enclosed in an ovary; e.g., the conifer.

gypsum A common mineral of evaporites, used in the manufacture of wallboard and plaster.

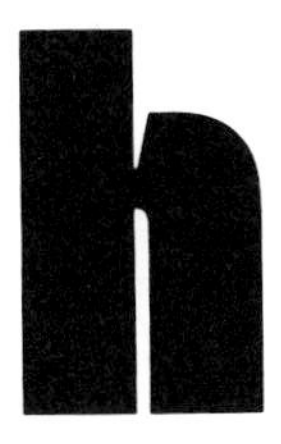

haar A local name in eastern Scotland and parts of eastern England for a kind of wet sea fog that at times invades coastal districts. Haars occur most frequently in summer months.

habitat The natural environment of a plant or animal.

haboob A dense dust or sand storm on the deserts of Egypt and Arabia or on the plains of India. The dust cloud may rise to 5,000 feet, and the wind may be violent enough to unroof verandas and overturn light streetcars. This type of storm lasts about two hours and is sometimes followed by heavy rain. Other spellings are haboub, habub, and hubbub.

hachures Short lines drawn on a map to represent differences in the slope of the land.

hacienda A Spanish word for a large farm or estate.

Hadley Cell The convectional cell in the atmosphere that consists of rising air in equatorial regions, descending air near 30°N., and the return surface flow of the trade winds.

hail Precipitation in the form of balls or lumps of ice more than 0.2 inch in diameter, formed by alternate freezing and melting as the lumps are carried up and down in highly turbulent air currents.

hail stage The part of the adiabatic process that theoretically begins when the ascending air has cooled to 0°C and its liquid water content has begun to freeze.

hailstorm A storm (generally a severe or prolonged thunderstorm) in which hail occurs.

halcyon days The name given by the ancients to the seven days preceding and the seven days following the winter solstice. Also calm, pleasant weather.

half-life The time required for half of the atoms of a radioactive substance to disintegrate. See: carbon 14

halide A general term for any salt of chlorine, bromine, iodine, or fluorine. The halides include the fluorides, chlorides, bromides, and iodides.

halite Rock salt (NaCl).

halo The general name of a class of optical meteors which appear as colored or whitish rings and arcs about the sun or moon when seen through an ice crystal cloud or in a sky filled with ice crystals. Halos are due to the refraction of the light which passes through the crystals, and, in the case of the whitish halos, to reflection from their surfaces. A colored halo may be distinguished from a corona by the fact that it is read nearest the sun or moon, whereas the corona has the red in the outer rings.

halomorphic soil Soil characterized by the accumulation of salts.

halophyte A plant that grows in a saline environment.

hammada Rocky uplands in a desert. Also spelled hamada.

hanging valley A tributary valley that enters the main valley at a considerable elevation above its floor.

hanging wall The rock immediately above a fault plane or mass of ore.

haploid An organism or cell with one rather than two sets of chromosomes.

harbor A portion of the coast which provides vessels with protection from winds and currents.

hard frost See: black (or hard) frost

hardness 1. The capacity of a substance to scratch or be scratched. 2. The quality attributed to water containing dissolved calcium and magnesium salts, the presence of which inhibits the formation of soapsuds.

hardpan A shallow layer of earth material that has become relatively hard and impermeable, usually through deposition of minerals.

hardwood tree A broadleafed tree. Some may have very soft wood.

harmattan The northeast trade wind of the Guinea coast. In the dry season it blows off the Sahara desert and, contrasted with the warm, humid winds of the wet season, is pleasant. Because of its supposedly healthful qualities it is sometimes called "the doctor."

harmonic tremor A slight, continuous vibration of the earth occurring during volcanic eruptions.

haul (v.) 1. In nautical parlance, to change the course of a ship, especially so as to sail closer to the wind. 2. Applied to wind, to change direction.

Hausa A Negro people of northern and central Nigeria.

haze Fine dust or salt particles dispersed through the atmosphere, so small that they cannot be distinguished individually by the unaided eye, but diminishing horizontal visibility, giving the atmosphere a characteristic hazy and opalescent appearance, and casting a uniform bluish or yellowish hue over the landscape and subduing its colors.

haze line In the lower atmosphere, the boundary between the relatively clean and bright air above a temperature inversion and the dust-filled air beneath.

head 1. In hydraulics, the pressure exerted by flowing water. 2. The hottest and most active front of a forest fire.

headland A steep cliff jutting out into the sea.

head of navigation The farthest point reached by vessels.

head race 1. A channel that conducts water to a water wheel. 2. A forebay.

headward erosion Erosion that occurs in the upstream end of the valley of a stream, causing it to lengthen its course in that direction.

heartwood The central portion of a tree trunk, entirely dead and without function; usually darker and more durable than the outer portion, or sapwood.

heat A form of energy, most commonly manifested to the senses by a rise in the temperature of a material body to which heat is being transferred. Heat is the kinetic energy of random motion of the molecules; a mass whose molecules vibrate faster that those of another is the hotter. As a substance grows cooler, the molecular vibration grows less, until finally, with no vibration at all, absolute zero of temperature is reached, and the body has no heat.

heat balance The equilibrium that exists, on the average, between the radiation received from the sun and that emitted by the earth. That such equilibrium exists is proven by the fact that the earth's temperature is constant, for if it were emitting less radiation than it receives, its temperature would rise, and vice versa. On the average, regions of the earth nearer the equator than 35° latitude receive more energy from the sun than they are able to reradiate, while latitudes higher than 35° receive less. The excess heat is carried from the low latitudes to the higher ones by atmospheric and oceanic circulation and is re-radiated there. Thus there exists, in general, a meridional transport of warm air poleward, chiefly at high levels in the atmosphere, and a similar flow of cold air at the surface toward the equator. The heat balance forms the basis of the general circulation of the atmosphere. See: general circulation, radiation

heat budget See: energy budget

heat capacity The quantity of heat, in calories, required to raise the temperature of a body one degree. See: thermal capacity

heat equator The line connecting the places on the earth's surface having the highest mean annual temperatures.

heat exhaustion A condition of low body temperature and low internal heat production caused by failure of the body to respond normally to excessively warm weather.

heath An extensive tract of open, uncultivated ground, usually flat.

heat island An area with heat measurably greater than that of its surroundings. Urban areas form heat islands.

heat laws See: thermodynamics, fusion

heat lightning Illumination generally seen near the horizon toward or during the evening, often in a clear sky. It is usually explained as the reflection

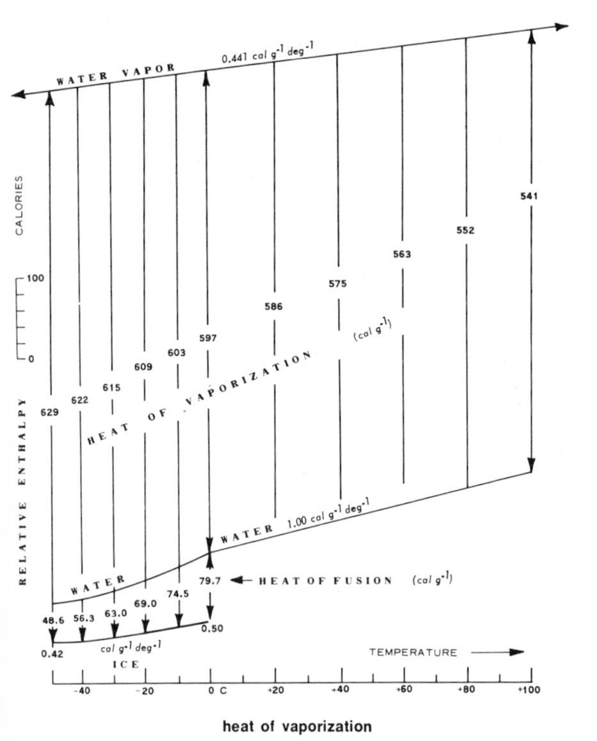

heat of vaporization

from hazy air of the lightning flashes of storms below the horizon and too distant to be audible. It is called heat lightning because it is characteristic of hot weather when local thunderstorms occur.

heat of fusion (or freezing) The amount of heat required to convert the unit mass of a solid to its liquid state at constant temperature, or the amount of heat given up by one gram of a substance in passing from the liquid to the solid state while the temperature remains constant. For the same substance, the two amounts are the same. The heat of fusion of ice (or latent heat) is 79.7 calories per gram. See: latent heat, heat of vaporization or condensation

heat of sublimation The amount of heat used when a solid changes directly to its gaseous state in doing external work (that of expansion) and internal work (that required to increase the molecular velocity). It is customarily measured in calories per gram, and would equal the sum of the heats of fusion and of vaporization if all these processes took place at the same temperature.

heat of vaporization or condensation The amount of heat given up by the unit mass of a substance when it passes from the vapor to the liquid state, or the amount of heat absorbed by the unit mass of a substance when it passes from the liquid to the vapor state, both at constant temperature. The amount is the same in both cases for the same substance at the same temperature. The heat of vaporization of water at 100°C is about 540 calories per gram. It is greater at lower temperatures. See: heat of fusion; heat of sublimation; latent heat

heat sink Anything that absorbs heat; usually part of the environment, such as the air, a river, or outer space.

heat wave A period of abnormally high temperature lasting more than a day. Also known as a hot or warm wave.

heaving See: frost heaving

Heaviside layer The Kennelly-Heaviside layer or E layer. See: E layer

heavy rain Rain falling in excess of 0.30 inch per hour (0.03 inch in 6 minutes).

heavy soil A term formerly used for clayey or fine-textured soil.

heavy water Water containing significantly more than the natural proportion (one in 6500) of heavy hydrogen (deuterium) atoms to ordinary hydrogen atoms. Heavy water is used as a moderator in some reactors because it slows down neutrons effectively and also has a low cross section for absorption of neutrons.

hectare See: Appendix II

hedgerow A hedge consisting of a line of shrubs or small trees.

Heidelberg man A man of the Middle Pleistocene period whose jawbone was found in 1907 near Heidelberg, Germany.

height Commonly, vertical distance, expressed as so many linear units (e.g., feet or meters) above sea level or some other specified point.

heliocentric Having or representing the sun as a center.

Helios The sun god of the Greeks, represented as driving a four-horsed chariot through the heavens.

heliotropic wind The diurnal component of the wind velocity, leading to a diurnal shift in the direction of the wind with the sun. The average direction of the wind changes slightly during the day, over both plains and mountaintops, in the direction of the hottest section of the earth. Thus the wind tends to travel east during the forenoon, south (in the northern hemisphere) during the early afternoon, and west during the late afternoon and early evening.

helium A rare gas contained in the atmosphere.

hematite One of the important iron ores usually found in secondary deposits associated with sedimentary rocks. It gives a red color to soils and rocks.

hemisphere One half of the earth's surface, formed when a plane bisects its center.

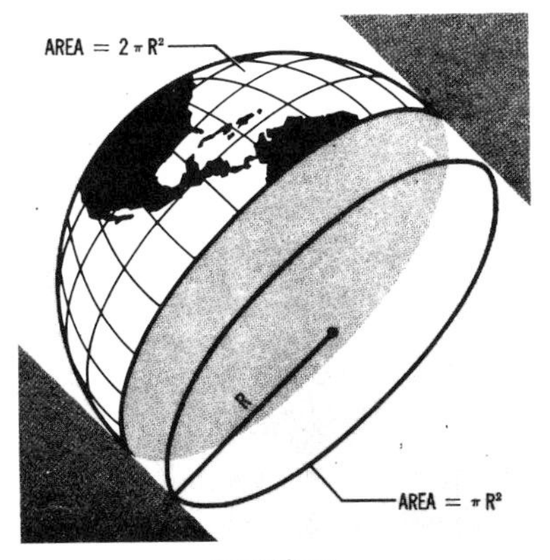

hemisphere

hemoglobin The respiratory pigment of all vertebrate animals, present as the coloring matter in red blood cells. Hemoglobin is a large complex molecule composed of a protein, globin, and an iron-containing pigment, hematin. Hemoglobin, along with related compounds, gives meat its characteristic fresh color and, through degradation, develops off-colors during storage.

hemostatic Serving to arrest the flow of blood.

hemp A tall Asiatic herb utilized for its fiber. It is also a source of drugs such as hashish and marijuana.

herb A soft, succulent seed plant that does not develop woody tissue.

herbicide A chemical or chemical compound used to kill herbs.

herbivore An animal that feeds on herbs and grasses.

heredity 1. The transmission of physical characteristics of an organism to its offspring. 2. The characteristics thus transmitted.

Hereford A beef cow bred originally in the British Isles. It is now the breed most commonly raised in the United States.

herring An important food fish found in the North Atlantic and Pacific oceans.

heterotrophic (adj.) Obtaining nourishment from organic matter.

heterozygous Containing genes from both members of a pair of gene-bearing parents.

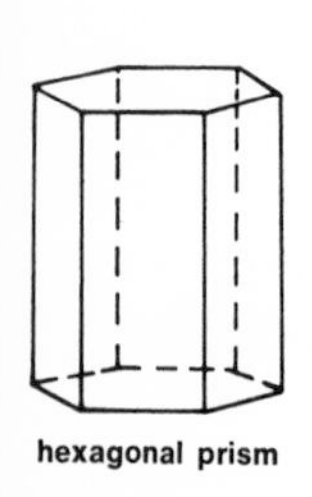

hexagonal prism

hexagonal prism A six-sided polygon.

hibernal (adj.) Pertaining to winter. Its contrasting term is estival, which pertains to summer.

hibernation A state of dormancy, or winter sleep, of many animals. With some, it is very deep, so that they can hardly be aroused, while with others it is less deep and even intermittent.

high (adj.) 1. A descriptive term in the International System of Cloud Classification. "High clouds" is the name for Family A, comprising cirrus, cirrocumulus, and cirrostratus clouds. See: cloud classification. 2. A modifying term in connection with wind, temperature, humidity, and pressure, generally with no exact quantitative significance. (n.) Synonymous with anticyclone.

high latitudes The areas of the earth north of the Arctic Circle and south of the Antarctic Circle, between 66½° and 90° north or south of the equator.

high pressure area Synonymous with high barometric area, high, or anticyclone.

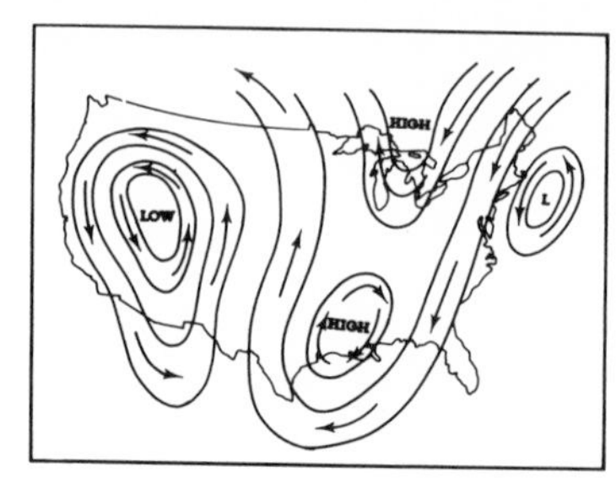

high pressure area

high tide Maximum height reached by each successive tide.

hill A small portion of the earth's surface elevated above its surroundings. It is smaller than a mountain and generally less than 1000 feet above the surrounding terrain.

hinterland 1. The region around a port that supplies most of the exports and utilizes the bulk of the imports. 2. The remote or less-developed part of a country.

histamine A colorless crystalline organic compound found in ergot and formed by bacterial decomposition of histidine and by at least one enzyme. It is formed when tissue injury occurs and when antigen antibody reactions take place, thus accounting for the shock that results from severe injuries and for the symptoms of allergy.

histogram A bar graph of a frequency distribution in which one dimension of each bar represents a given range of frequencies and the other dimension represents the number of frequencies within that range.

histology The study of the minute structure of plant and animal tissue.

historic flood Any flood for which the stage or flow can be estimated or is recorded. See: flood, recorded flood

historic flow The flow recorded at a gaging station.

histosol A soil characterized by high organic content throughout; e.g., bog soils.

hoarfrost Atmospheric moisture deposited through sublimation upon terrestrial objects in the form of ice crystals, by the same process which causes dew but at a time when the temperature of the objects is at or below freezing.

hogback A long, narrow ridge marked by a steep escarpment and beds that slope abruptly away.

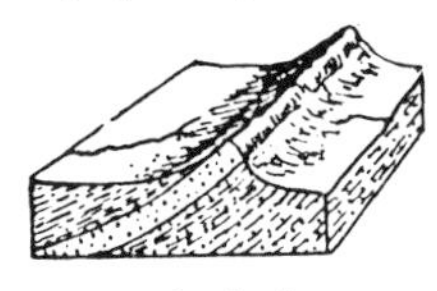

hogback

Holocene See: Appendix I

Holstein-Friesian A dairy cow, black with white markings, bred originally in the Netherlands. It is renowned as the largest milk producer of any breed.

homeostasis A state of equilibrium among the interdependent systems and subsystems of an organism.

Homestead Acts A series of U.S. federal laws granting land in units of 160, 320, or 640 acres to individuals who complied with provisions of the acts. The Homestead Act of 1862 was the first and most significant.

homocline Rock layers in which the beds dip uniformly in one direction.

homogeneous atmosphere A hypothetical atmosphere in which the density is the same throughout and the pressure at the surface is the same as that of the actual atmosphere. Such an atmosphere would have an approximate altitude of 8,000 meters.

homolographic projection 1. A map projection that preserves the actual proportion of the parts; i.e., an equal-area projection. 2. Generally, Mollweide's projection, in which the globe is shown within an elliptical projection.

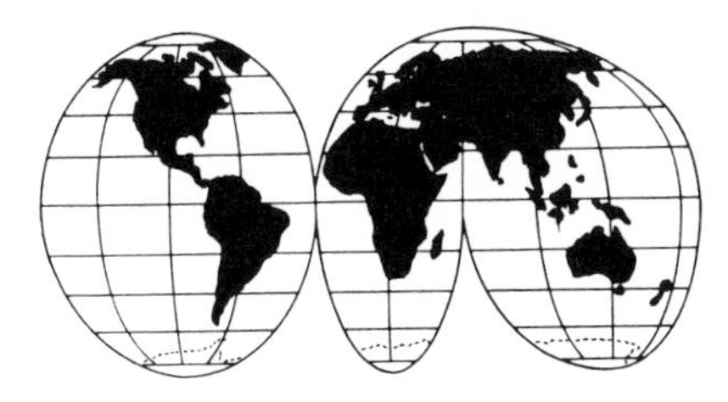

homolographic projection

homozygote An organism that has identical pairs of genes with respect to any given pair of inherited characteristics and will thus breed true to these characteristics.

hook A spit that is curved at one end.

hook gage A pointed, U-shaped hook attached to a staff or vernier scale and used in the accurate measurement of the elevation of a water surface. The hook is submerged, and then raised, usually by means of a screw, until the point just makes a pimple on the water surface.

hookworm A parasitic nematode.

Hopi A tribe of American Indians living on the Colorado Plateau.

horizon 1. In astronomy, the great circle on the celestial sphere halfway between zenith and nadir. 2. In geography, the line where earth and sky appear to meet.

horizon (soil) A layer of soil having distinctive characteristics. The C-horizon is relatively unmodified rock material; the A-horizon is topsoil and includes organic material; and the B-horizon is the zone of eluviation and modification.

hormone An organic cell product, transported from one part of an organism to another, which produces specific effects within the cells in another part of the organism.

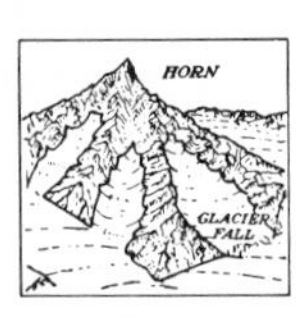

horn

horn A high, steep-sided peak formed by the intersecting walls of three or more cirques.

hornblende See: amphibole

horny endosperm See: endosperm

horse latitudes Belts of prevailing high pressure, located in the mean at 35° north and south latitude, which migrate north and south with the sun and are characterized by light variable winds and clear fine weather due to the descending air.

horst An elevated portion of the earth's crust lying between two more or less parallel faults.

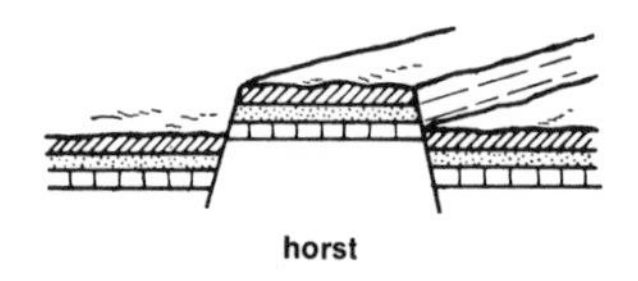
horst

horticulture Cultivation of flowers, fruit, or vegetables.

hot-logging A logging operation in which logs go from the stump to the mill without pause.

hot wave See: heat wave

howling fifties The belt of westerly winds between 50° and 60° south latitude; so named because of the strength of the winds and their effect on ship's rigging.

hot spring or thermal spring A flow of hot water occurring wherever hot rock material lies near the surface.

huayco An avalanche of water, rock, and mud produced when an earthquake causes the collapse of a natural dam containing the waters of the lakes of the high Andes.

huerta In Spain, land that is double- or triple-cropped each year.

human biology The study of relations between man and his environment from a biological viewpoint.

human ecology The study of relations between man and his environment from a geographical or sociological viewpoint.

humic acids Alkali-soluble end products of the decomposition of organic matter in soil and in composts. The term sometimes is used interchangeably for humus.

humid climate A climate with enough precipitation to support a forest vegetation, although there are exceptions where the plant cover includes no trees, as in the Arctic or high mountains. The lower limit of precipitation may be as little as 15 inches in cool regions.

humidity The state of the atmosphere with respect to its water vapor content. It may be measured in many different ways. See: absolute humidity; dew point; mixing ratio; relative humidity; saturation deficit; specific humidity vapor pressure; water vapor

humification A process or condition of decay in which plant or animal remains are so thoroughly decomposed that their initial structures or shapes can no longer be recognized.

humus Decomposed or partly decomposed organic matter in soil.

hurricane 1. A tropical cyclone, especially in the West Indian region. 2. The designation of the highest wind force on the Beaufort wind scale, applied to any wind exceeding 75 miles an hour (or 35 meters per second) in speed.

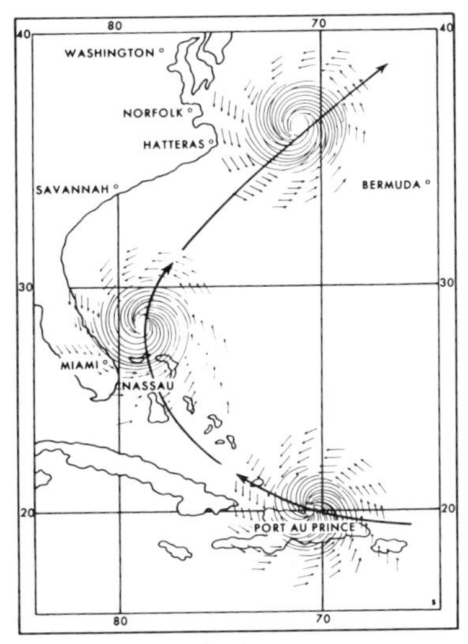

hurricane

hybrid An organism descended from parents belonging to different varieties, species, or races.

hydraulic head The height of the free surface of a body of water above a given point beneath that surface.

hydraulic mining

hydraulic mining Mining accomplished with the use of water under pressure to break loosely consolidated rock or gravel and wash it into a sluice box.

hydraulics The study of water in motion.

hydrocarbon A compound containing only carbon and hydrogen (e.g. gasoline). Hydrocarbons are a major component of photochemical smog.

hydroclimatology The branch of the atmospheric sciences that deals with water.

hydrodynamics The branch of mechanics that treats of the motions of fluids.

hydroelectricity Electricity produced by falling water.

hydrogen A chemical element; the lightest known gas. It is colorless, odorless, and inflammable.

hydrogen sulfide A toxic air contaminant with a smell like rotten eggs.

hydrogenation The act of combining with hydrogen, one of the chemical elements, as in the hydrogenation of an oil. Liquid oils or fats of soft consistency are hardened by the addition of hydrogen to yield hydrogenated products of greatly increased consistency which require a higher temperature to melt or liquefy. Thus an oil, such as cottonseed, may be modified to give it the characteristics of margarine or shortening.

hydrogen ion concentration See: pH

hydrograph A graph showing, for a given point on a stream or conduit, the stage, velocity of flow, available power, or other function of discharge with respect to time.

hydrographic survey An instrumental survey to measure and determine characteristics of oceans, lakes, streams, and other bodies of water within an area.

hydrography The study of the surface waters of the earth.

hydrologic benchmark A hydrologic unit such as a basin or a groundwater body that has been designated, because of its expected freedom from the effects of man, as a benchmark or standard. Data from such basins may provide a standard with which data from less independent basins can be compared so that changes wrought by man's interference can be distinguished from changes caused by variations in the natural regimen.

hydrologic budget A record of the inflow to, outflow from, and storage in a hydrologic unit, such as a drainage basin, aquifer, soil zone, lake reservoir, or irrigation project.

hydrologic cycle A series of water phenomena or phases, namely: (a) precipitation from clouds falling on sea and land; (b) flow of water from the land to sea as run-off or seepage; (c) evaporation from falling moisture (in all forms) and from all moist surfaces, forming the water vapor of the atmosphere; and (d) condensation of water vapor, producing clouds and thus completing the cycle.

hydrologic equation The water inventory equation (inflow = outflow + Δ

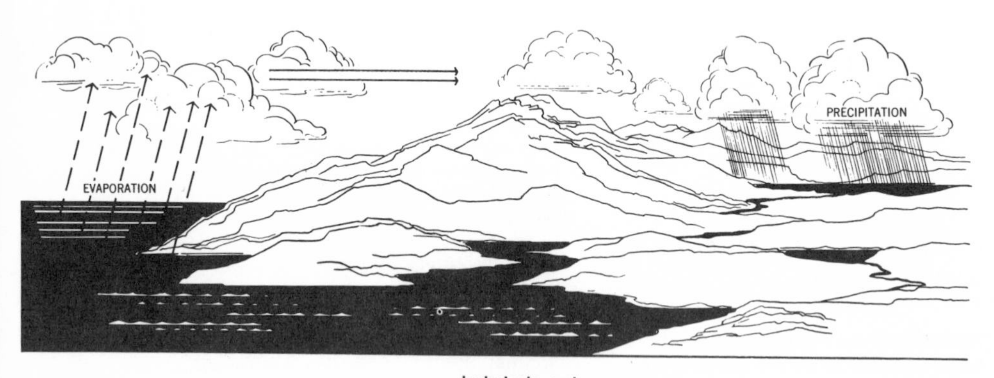

hydrologic cycle

storage), which expresses the basic principle that during a given time interval the total inflow to an area must equal the total outflow plus the net change in storage.

hydrology The science that treats of the phenomena of water in all its states; of the distribution and occurrence of water in the earth's atmosphere, on the earth's surface, and in the soil and rock strata; and of the relation of these phenomena to the life and activities of man.

hydrolysis A chemical process of decomposition by the addition of the elements of water. Water may cause the salt of a weak acid or weak base to break down in part into the acid and the hydroxide of the metal with which it is combined.

hydrometeor Any product from condensation of atmospheric water vapor, whether formed in the free atmosphere or at the earth's surface.

hydrometeorology A branch of meteorology treating of the water in the atmosphere and of precipitation and its aftereffects, such as run-off, floods, etc.

hydrometer An instrument used to determine the specific gravity of liquids.

hydronium ion The predominant form of occurrence of hydrogen ions in solution, each hydrogen ion being associated with a single water molecule, H_3O^+.

hydrophyte A plant that grows naturally in water or in saturated soil. See: mesophyte; phreatophyte; xerophyte

hydroponics The growing of plants in water in which nutrients essential for plant growth are dissolved.

hydrosphere The water on or surrounding the surface of the earth, as distinguished from the solid part of the earth, which is called the lithosphere.

hydrostatic equation The equation that describes the dynamic state of the atmosphere when completely at rest relative to the surface of the earth.

hydrostatic pressure The pressure of a liquid. The pressure acting at any point in a liquid below its surface is dependent on the weight of the liquid above it—that is, the height of the liquid column times its density.

hydrous (adj.) Containing water.

hydroxyl A chemical group consisting of one hydrogen and one oxygen atom. If a hydroxyl group is attached to an aliphatic organic radical, the compound formed is an alcohol; if attached to an aromatic organic radical, it is a phenol. Hydroxyls also occur in other types of organic and inorganic compounds.

hyetal (adj.) Of or pertaining to rain.

hygrograph An instrument that makes a continuous record of the relative humidity; most hygrographs have human hair elements.

hygrology The science that deals with humidity.

hygrometer An instrument that indicates either the relative humidity or, in the case of the dew point hygrometer, the dew point of the air.

hygrometry The branch of physics that treats of the measurement of the humidity of the atmosphere and other gases.

hygrophyte See: hydrophyte

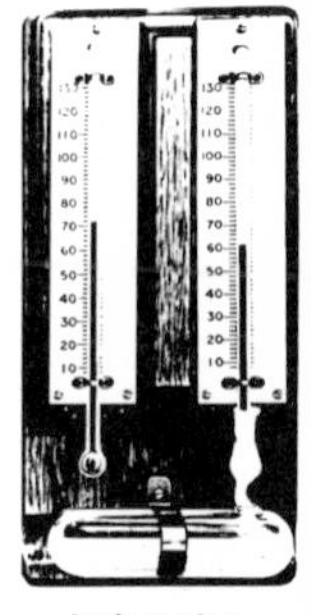

hydrometer

hygroscopic (adj.) Capable of taking up moisture from the air.

hygroscopic coefficient The amount of moisture in a dry soil when it is in equilibrium with some standard relative humidity near a saturated atmosphere (about 98 percent), expressed in terms of percentage on the basis of oven-dry soil.

hygroscopic nuclei Condensation nuclei, principally salts, upon which water condenses at relatively low relative humidities (75 percent for sodium chloride).

hygroscopic water Water that is so tightly held by molecular attraction to soil particles that it cannot be removed except as a gas.

hygrothermograph A self-recording instrument combining the registration of both relative humidity and temperature on one record sheet.

hyperemia A superabundance or congestion of blood in any part of the body. Blushing is a hyperemia of the skin.

hyperplasia The abnormal multiplication or increase in number of tissue elements.

hypothesis A tentative statement intended to explain certain observations. It is used to guide an investigator in his search for the true explanation of a problem.

hypsographic curve A diagrammatic curve often used to represent the shape of the earth's surface. Elevations and depressions are represented by distances above or below a horizontal line indicating sea level.

hypsography The observation or description of topographic relief.

hypsometer An instrument for measuring the boiling points of liquids, especially in order to estimate elevations above sea level.

hypsometric formula A formula for reducing barometric readings to sea level or for determining heights by the barometer.

hypsometric map A map that shows relief by using colors to represent areas at different elevation.

hypsometry The science of measuring heights above sea level or any other fixed plane.

hythergraph See: climagram

i

ice The solid state of water.

ice age A period of geologic history during which considerable portions of the earth were covered with glacial ice. There have been many ice ages in the geological history of the earth; during some, the ice sheets were situated in the polar regions, but in others they were in equatorial regions. In the last ice age, the Pleistocene, about one-fifth of the earth's surface was glaciated at one time or another.

iceberg A mass of land ice that has broken away from its parent formation on the coast, and either floats in the sea or is stranded in the shallows.

ice cap A perennial cover of ice and snow over an extensive area of land or sea. There are several ice caps in the world, the most important being those on the Antarctic Continent, on Greenland, and in the Polar Sea, which lies to the north of North America and eastern Siberia and extends beyond the Pole.

ice-cap climate The climate peculiar to an ice-cap region; in general, the mean annual temperatures vary from about –10°F at the North Pole to about –50°F in the Antarctic, and freezing weather prevails all the time. The precipitation is light, especially in the interior of the ice-cap regions, where there are practically no storms.

ice crystals The macroscopic forms in which ice occurs. They form cirus clouds and act as sublimation nuclei in clouds in the middle and high latitudes.

ice-crystal theory See: Bergeron-Findeisen theory

icefall A part of a glacier flowing down a very steep slope, resulting in a zone of crevasses and jagged pinnacles.

ice field A uniform, unbroken ice sheet covering a large area.

ice floe A mass of floating ice detached from the main polar ice.

ice fog A hydrometeor which consists of a fog formed of ice spicules, usually in clear, cold, windless weather. Most frequent in the higher latitudes, it also is known as frost in the air, frozen fog, and pogonip.

Icelandic low The semipermanent low-pressure area centered over the northern part of the Atlantic Ocean, generally in the vicinity of Iceland.

ice needle A thin shaft of ice, also called an ice spicule, which seems to float in the air when rendered visible by sunshine. Cirrus clouds are mostly composed of ice needles. See: ice crystals

ice nucleus Any particle that serves as a nucleus in the formation of ice crystals in the atmosphere. See: freezing nucleus; sublimation nucleus

ice point The temperature (32°F, 0°C, 273.1°K) at which water freezes under normal atmospheric pressure (1,013 millibars).

Ice Pole The center around which ice in the northern latitudes is distributed in a series of concentric belts. The Ice Pole, or pole of inaccessibility, as it is sometimes called, lies in the vicinity of latitude 83° to 85°N. and longitude 170° to 180°W.

ice storm A storm characterized by the falling of rain from a relatively warm layer of air aloft to the surface of the earth when the latter is at subfreezing temperatures, and the consequent freezing of the rain as a film of ice on terrestrial objects. Also called freezing rain, silver thaw, glaze, and, mistakenly, sleet storm.

ichthyology The study of fishes.

icing (of aircraft) The formation of ice on an aircraft in flight, often to the extreme peril of its occupants. It is caused by the freezing of liquid water droplets that strike the leading edges and exposed surfaces.

igneous rock Rock formed by the solidification of magma (hot molten material).

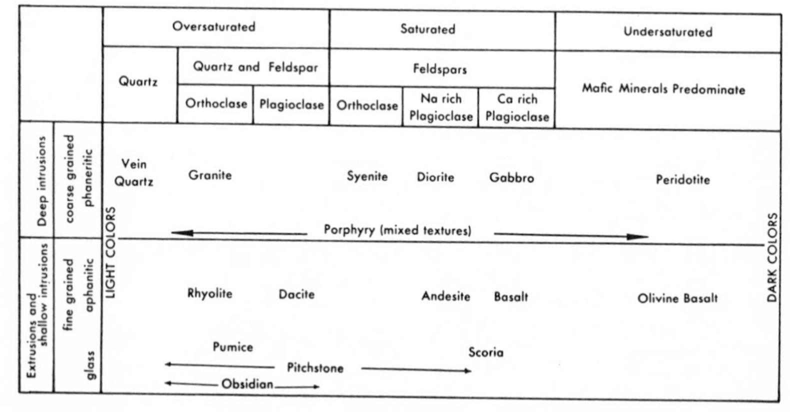

igneous rock

illuviation An accumulation of material in a soil horizon through the deposition of suspended mineral and organic matter originating from horizons above. Since at least part of the fine clay in the B horizons (or subsoils) of many soils has moved into them from the A horizons above, these are called illuvial horizons.

immature soil A soil lacking clear individual horizons because of the relatively short time for soil-building forces to act upon the parent material since its deposition or exposure.

impermeable (adj.) Having a texture that does not allow water to move through perceptibly.

impervious (adj.) A term applied to a material through which water cannot pass or passes with great difficulty. See: aquiclude; aquifuge

inbreeding The interbreeding of closely related individuals.

Incas Indians who lived along the western coast of South America before the coming of white men.

incense cedar An evergreen belonging to the pine family, named for its fragrance. It is used chiefly for poles, fence posts, shingles, etc.

inceptisols Soils formed in humid climates ranging from polar to tropical, generally on forested land. Their soil profiles are not well developed.

inch-degrees The product of inches of rainfall times temperature in degrees above freezing (Fahrenheit scale), used as a measure of the snow-melting capacity of rainfall.

incised meander See: entrenched (or incised) meander

inclination The dip of a fault, vein, stratum, or other more or less flat body.

inclined staff gauge A gauge for measuring stream depth. It is placed on the slope of a stream bank and graduated so that the scale reads directly in vertical depth.

increasers Nutritious, sod-forming grasses that increase in quantity under moderate grazing but decrease under heavy grazing pressure.

index An indicator, usually numerically expressed, of the relation of one phenomenon to another. For an example, see zonal index.

index of wetness The precipitation for a given year expressed as ratio to the mean annual precipitation.

Indian summer A spell of warm, quiet, hazy weather occurring in October or even in early November. In some years there may be only a few days of such weather or none at all, while in other years there may be one or more extended periods.

indigenous (adj.) Native to a particular region or country.

indirect flood damage Expenditures made as the result of a flood for items other than repair, such as relief and rescue work, moving silt and debris, etc.

industrial consumption The quantity of water consumed in a municipality or district for mechanical, trade, and manufacturing purposes, in a given period—generally one day.

Industrial Revolution The change from the use of animate to inanimate forms of energy to do man's work. It is generally said to have begun in 1770 with the development of the steam engine by Watt.

inert Not having or showing active properties; especially not having any chemical action.

infanticide The killing of infants.

infant mortality rate The ratio of the number of deaths of infants under one year of age to the number of live births.

inference Something assumed to be true as a result of reasoning from known facts and assumptions.

infiltration The movement of water from the surface into the soil. Infiltration is equal to the total precipitation less the losses due to interception by vegetation, retention in the depressions upon the land surfaces, evaporation from all moist surfaces, and surface runoff.

infiltration capacity The maximum rate at which soil, when in a given condition, can absorb falling rain or melting snow.

infiltration-capacity curve A graph showing the time variation of infiltration capacity. A standard infiltration-capacity curve shows the time variation of the infiltration rate that would occur if the supply were continually in excess of infiltration capacity.

infiltration index An average rate of infiltration, in inches per hour, equal to the average rate of rainfall, so that the excess volume of rainfall equals the total direct runoff. Also designated as the ϕ or W index.

infiltration rate 1. The rate at which infiltration takes place, expressed in depth of water per unit of time, usually in inches per hour. 2. The rate, usually expressed in cubic feet per second or millions of gallons per day per mile of waterway, at which ground water enters an infiltration ditch or gallery, drain, sewer, or other underground conduit.

infrared (adj.) Pertaining to heat rays that have wavelengths longer than light rays but not longer than about 0.03 cm.

ingestation The taking of food into the body for digestion.

inherited characteristics Characteristics of an organism which are attributable to genetic transmission.

initial detention The volume of water on the ground, either in depressions or in transit, at the time active runoff begins.

initial loss In hydrology, rainfall preceding the beginning of surface runoff. Unless otherwise specified, it includes interception, surface wetting, and infiltration.

inlet A small indentation in a coastline.

inlier A mass of older stratified rocks surrounded by newer strata. See: outlier

inorganic (adj.) A term referring to substances occurring as minerals in nature or obtainable from minerals by chemical means. The term characterizes all such matter except the compounds of carbon, but includes carbonates.

inorganic nitrogen Nitrogen existing in combination with mineral elements, and not in animal or vegetable form.

Examples are ammonium sulfate and sodium nitrate, while proteins contain nitrogen in organic combination.

inselberg A lone relict mountain occurring in an arid region.

insolation The rate at which radiant energy is incident directly from the sun per unit horizontal area at any place on or above the surface of the earth. Its value depends upon: (a) the solar constant, (b) the distance of the point from the sun, (c) the inclination of the sun's rays to the horizontal plane at the point under consideration, and (d) the transparency of the atmosphere.

instability A state in which the vertical distribution of temperature is such that an air particle, if given either an upward or a downward impulse, will tend to move away with increasing speed from its original level. (In the case of unsaturated air, the lapse rate for instability will be greater than the dry adiabatic lapse rate; in that of saturated air, greater than the saturated adiabatic lapse rate.) The principal kinds of instability are absolute, conditional, convective, and latent.

instability shower Rainfall, brief and intermittent, caused by steepening of the lapse rate in any way, such as the rapid warming of the lower layers of a cold current as it moves over a relatively warm surface.

instinct A natural aptitude, impulse, or capacity.

instrument shelter In the United States, a structure with louvered sides and a double roof for ventilation, in which thermometers and other instruments may be exposed.

intangible flood damage Flood damage done as a result of the disruption of business, dangers to health, shock, and loss of life. In general the estimation of such damage involves the computation of all costs not directly measurable.

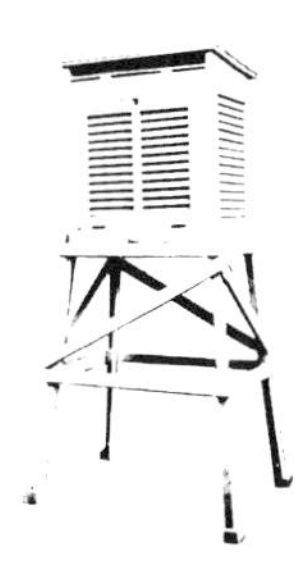
instrument shelter

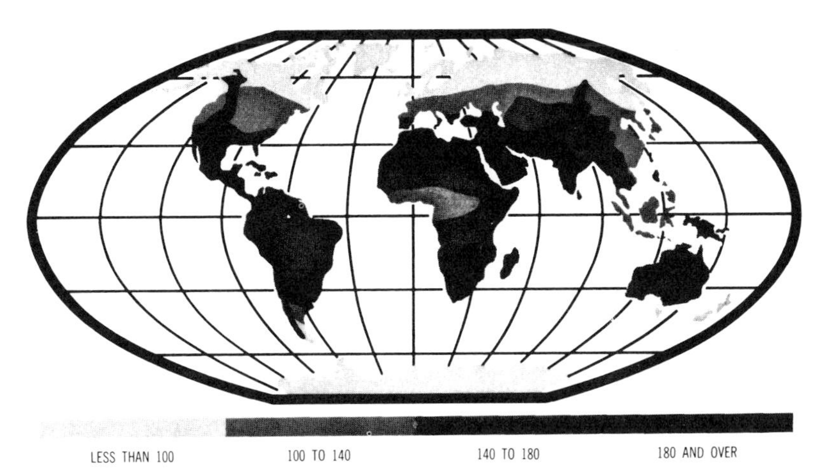

insolation

integrated logging A method of logging designed to make the best use of all timber products. It removes in one cutting all timber that should be cut and distributes the various timber products to the industries that can use them to best advantage.

intensive cultivation A system of agriculture utilizing large quantities of capital or labor or both to achieve high yields per acre.

intercalary planting The practice of planting two different crops on the same fields, such as citrus under date palms.

interception The process by which trees, underbrush, standing crops, and other objects prevent a certain amount of rainfall from reaching the soil.

interface The contact surface separating two different substances.

interfluve The ridge between two adjacent river valleys.

interglacial period An interval between successive advances of the ice sheets during a given ice age, as, for example, the Pleistocene period. Also known as an interglacial stage.

interior drainage Drainage into basins without outlets.

interlobate moraine See: medial moraine

intermediate zone The subsurface water zone below the plant root zone and above the capillary fringe.

intermittent (or seasonal) stream One that flows only at certain times of the year when it receives water from springs or from some surface source such as melting snow in mountainous areas.

International Biological Program A United Nations program to study the major world ecosystems.

international candle The international unit of luminous intensity, being the light emitted by five square millimeters of platinum at the temperature of solidification.

International Date Line The line approximating the meridian 180° W. or E., where the date changes one day as it is crossed. A traveller crossing the date line westward adds a day; one crossing it eastward subtracts a day.

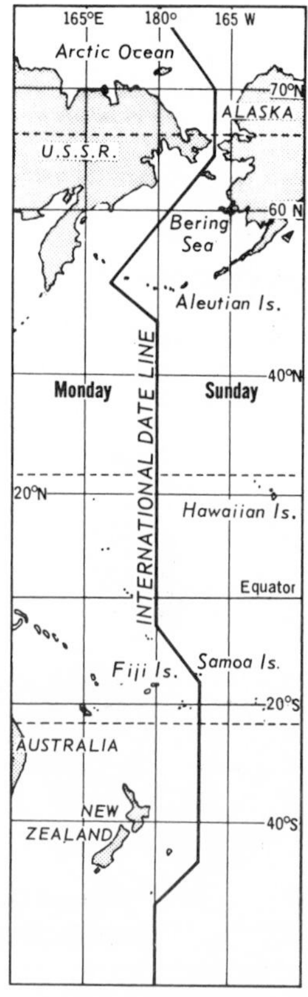

International Date Line

International Geophysical Year (IGY) A period of a year and a half in 1957 and 1958 when 67 countries of the world pooled their resources to study the physical characteristics of the sun and the earth.

International Hydrological Decade (IHD) A program sponsored by UNESCO to develop research and education programs in the field of water resources during the 1965-74 period.

international index numbers The numbers agreed upon by an international committee to designate meteorological stations.

intertidal zone An area covered with water at high tide and uncovered at low tide.

intertilled crop A crop having or requiring cultivation during growth.

intertropic convergence zone The region of the frontal discontinuity that develops when an outbreak of cold air penetrates the doldrums and interacts with the equatorial air. Along this zone the most severe weather of the tropics is to be found: overcast skies, strong squalls, heavy rains, and severe local thunderstorms.

intrazonal soil A soil differing (because of the effects of local factors such as relief, parent material, or age) from the soil that would be expected from the prevailing climate and vegetation.

intrusion A body of igneous rock that has penetrated older rock.

intrusive rocks Rocks that have solidified at some distance beneath the surface of the earth.

invaders Native or exotic weeds or woody plants of low forage value that replace valuable grasses as rangeland deteriorates under heavy grazing pressure.

inverse square law The law that states that the intensity of a force (electrostatic, magnetic, or gravitational) between two objects decreases as the square of the distance between their centers.

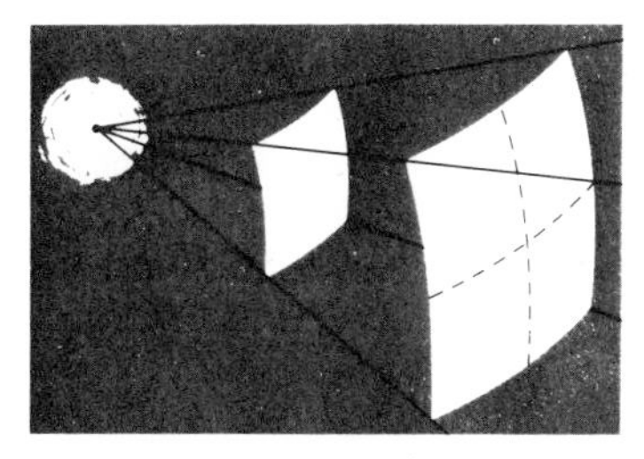

inverse square law

inversion The condition that exists in the atmosphere when the temperature increases rather than decreases with height through a layer of air.

inversion of rainfall The phenomenon, observed in mountainous regions, in which the amount of rainfall increases up to a certain level on the mountain and then decreases.

iodine A grayish-black crystalline solid used as a medicine and needed as a trace element by many organisms. Iodine-131 is one radioactive element thought to be harmful to man.

ion An atom or molecule that has a smaller or greater number of electrons than in its normal electrically neutral state.

ion exchange A chemical process involving the reversible interchange of various ions between a solution and a solid material, usually a plastic or a resin. It is used to separate and purify chemicals, such as fission products, rare earths, etc., in solutions.

ionization The process of adding one or more electrons to, or removing one or more electrons from, atoms or molecules, thereby creating ions. High temperatures, electrical discharges, or nuclear radiations can cause ionization.

ionosphere The region of the earth's atmosphere in which a relatively great degree of ionization exists and causes reflection of radio waves. It extends from about 50 to about 250 miles above the earth.

iridescent clouds Cirrostratus or cirrocumulus clouds that exhibit brilliant spots or borders of colors, usually red and green.

irradiation Exposure to any form of radiation.

irrigated area The gross farm area upon which water is applied for the production of crops, with no reduction for access roads, canals, or farm buildings.

irrigation The controlled application of water to arable lands to supply water requirements not satisfied by rainfall.

irrigation efficiency The percentage of water applied that can be accounted for by moisture increase in the soil to be used.

irrigation requirement The quantity of water, exclusive of precipitation, that is required for crop production. The figure should include surface-evaporation and other unavoidable wastes.

irrigation return flow Applied water that is not consumptively used and returns to a surface or ground-water supply. Under conditions of water-rights litigation, the definition may be restricted to measurable water returning to the stream from which it was diverted. See: return flow

irruption A marked increase in the size of a population, causing environmental problems.

isabnormal line A line on a map or chart connecting places having the same anomaly, usually of temperature. Also called an isanomalous line.

isallobar A line connecting points on a weather map having the same barometric tendency with regard to both magnitude and direction (positive or negative) of change.

isallotherm A line connecting points having equal temperature variations within a given period of time.

isanomalous line See: isabnormal line

isanthesic line A line drawn on a map to connect places at which the blossoming of any plant occurs simultaneously.

isarithm See: isogram

isentropic (adj.) Taking place without change of entropy. An adiabatic line or an adiabat is also an isentropic line. See: isentropic graph

isentropic analysis An analysis, by means of data obtained from aerological soundings, of the physical and dynamic processes taking place in the free atmosphere, on the basis of the location and configuration of the various isentropic surfaces and the distribution of air properties and air movement on these surfaces.

isentropic graph The graph representing the variables in a transformation during which entropy remains constant.

isentropic surface A surface in the atmosphere on which potential temperature is a constant. It is not a level surface, but has domes and troughs, since it is higher in a cold air mass than in a warm one.

island A body of land entirely surrounded by water.

island arc A curved chain of islands, such as the Aleutian Islands off the Alaskan coast.

iso A combining form from the Greek, meaning "equal": a familiar word with this prefix is isotherm. Before a vowel, iso- becomes is-: an example is isallotherm.

isobar On a weather map, a line connecting places having the same barometric pressure. Usually the barometric pressures at all points are reduced to mean sea level.

isobath A line on a map connecting all points that are the same vertical distance above the upper or lower surface of a water-bearing formation or aquifer.

isogonal (adj.) See: agonic line

isogonic line A line on a map connecting places having equal magnetic declination.

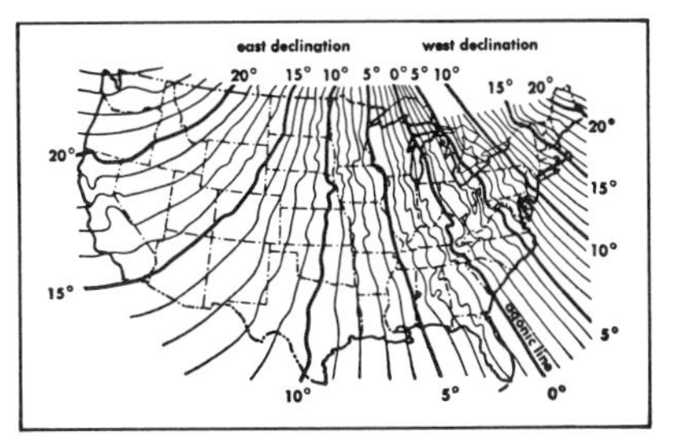

isogonic line

isogradient A line connecting points on a chart having the same horizontal gradient of barometric pressure, temperature, or other element.

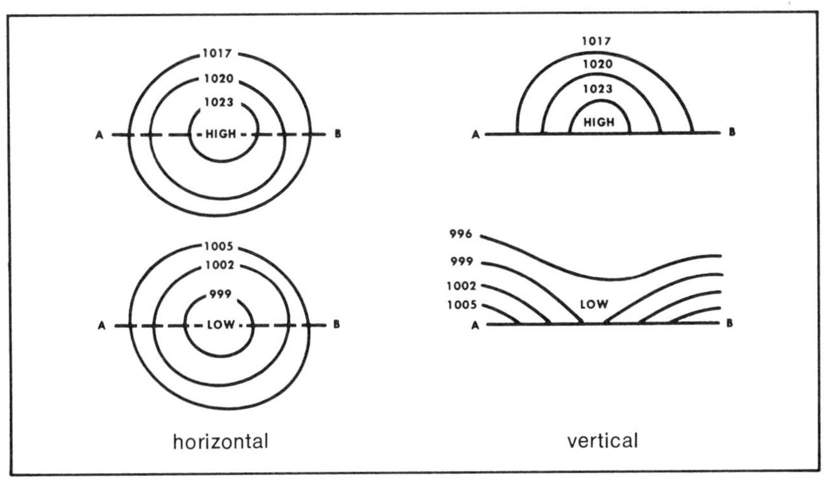

isobar

isogram A line on a map or chart connecting points of equal value. See: isopleth

isohel A line on a map or chart connecting points having the same amount of sunshine during any specified period.

isohyet A line on a map connecting points having the same amount of precipitation for any specified period.

isoline See: isogram

isomer 1. A line on a map indicating an equal proportion. 2. A molecule that has the same kind and number of atoms as another molecule but a different structural arrangement.

isomorphic (adj.) Different in ancestry but having the same form or appearance.

isoneph A line drawn on a map through points having the same degree of cloudiness.

isopleth See: isogram

isoseismal (adj.) Referring to points at which an earthquake shock is equally intense.

isostasy The balance of large sections of the earth's crust, a condition in which the elements tending toward elevation are in balance with those which tend to depress. The effect is as if the crust floated on a denser underlying layer.

isotherm A line on a weather map connecting points having the same temperature.

isothermal layer Any layer in the atmosphere through which the temperature does not change appreciably with height.

isotope Atoms of the same element but with different numbers of neutrons in the nucleus and different atomic weights are called isotopes of that element. Isotopes have the same number of protons and are essentially identical in their chemical properties and can be distinguished or separated only on the basis of slight differences in those physical properties that depend on the mass of the atom.

isthmus A narrow strip of land joining two large land areas.

jack mackeral A small fish, weighing from three to five pounds, of commercial importance in West Coast fisheries.

jack pine The two-needle North American pine. It is relatively small and grows farther north than any other pine.

January thaw A period of mild weather, popularly supposed to recur each January in the United States, especially in New England and the middle Atlantic states.

Java man One of the predecessors of modern man (Pithecanthropus), whose fossil remains were discovered in Java in 1891-92.

Jersey A dairy cow bred originally in the British Isles. Its milk has the highest butterfat content of any breed.

jet stream A rapidly moving stream or streams of air taking sinuous paths through the atmosphere at elevations between 25,000 and 40,000 feet. In general, they flow from west to east, are associated with the polar front, and may have speeds of several hundred miles per hour.

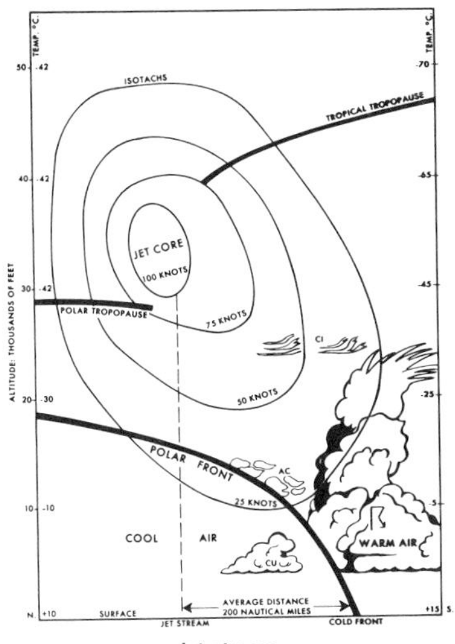

jet stream

jetty A structure extending into the body of water on estuaries or open sea-coasts that is designed to prevent shoaling of a channel by littoral materials by directing or confining the flow.

joint A fracture in rock along which no movement has occurred. See: fault

jointing 1. The system or combination of joints in a rock. 2. The process by which joints are formed.

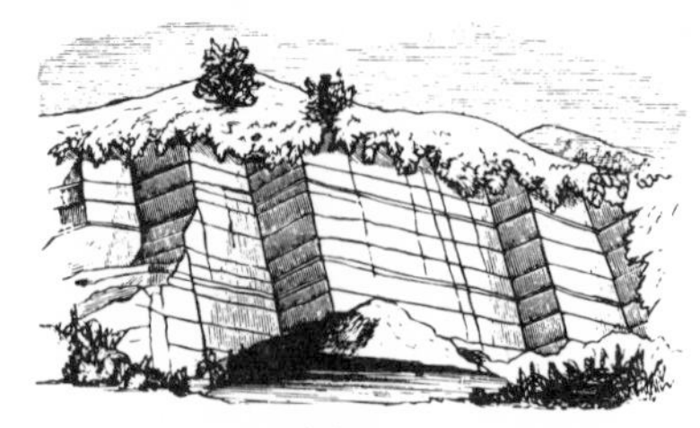

jointing

joule A practical unit of work; equal to 10^7 ergs.

Joule's law A thermodynamic law that states that all the work done in compressing a gas at constant temperature is converted into heat, and, conversely, that when the gas expands at constant temperature, the quantity of heat absorbed is equal to the quantity of work done.

jungle A tropical forest with dense undergrowth.

juniper Any of the conifers belonging to the genus Juniperus. It is found widely in the semiarid parts of the western United States.

Jupiter 1. The chief Roman god; god of the elements. 2. The largest of the planets of our solar system.

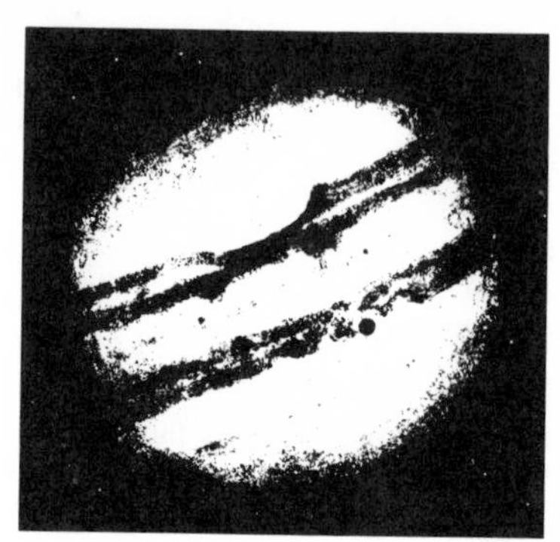

Jupiter

Jurassic See: Appendix I

juvenile water Water formed chemically within the earth and brought to the surface in intrusive rock.

Kaikias Greek name of the northeast or north-northeast wind or its personification.

kaléma A very heavy surf breaking on the Guinea coast during the winter, even when there is no wind.

kame A mound of gravel deposited by a stream flowing along a glacier.

kaolin A whitish clay formed by the decomposition of the feldspar in granite. It is used in the making of china.

kaolin minerals A group of nonswelling clay minerals in which one layer or sheet of silicon and oxygen alternates with a sheet made up of aluminum, oxygen, and hydrogen.

karaburan A "black storm" in central Asia, in which violent east- northeast winds begin early in spring and continue by day till the end of summer. Blowing with gale force, they carry with them clouds of desert dust which darken the air.

karakul sheep A breed of sheep raised in the dry lands of central Asia for its skins.

karst region (karstland) A limestone region in which most or all drainage is underground.

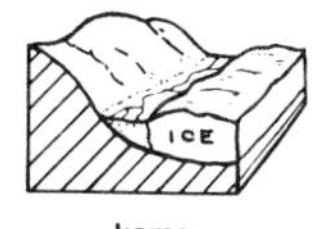

kame

katabatic wind A wind flowing down an incline; caused by the cooling of surface air, which then, impelled by gravity, flows downward. Also called mountain wind, canyon wind, and gravity wind.

kayak A type of canoe developed by the Eskimos, consisting of a light frame of wood or bone covered with tightly stretched seal or walrus skin.

kelp Large, brown seaweeds; a source of iodine.

Kelvin scale of temperature An ideal,

absolute temperature scale proposed by Kelvin (1848). On this scale, the freezing point of water is 273.13°K, and the boiling point is 373.13°K, under one atmosphere of pressure.

Kennelly-Heaviside layer See: E layer

Kepler's laws of planetary motion 1. Law of ellipses. The orbit of a planet is an ellipse with the sun located at one focus. 2. Law of equal areas. The line joining any planet to the sun sweeps out equal areas in equal times. 3. The harmonic law. The square of the orbital period of a planet is proportional to the cube of its mean distance from the sun. ($P^2 = D^3$, where P is the orbital period of a planet in earth years, and D is the mean distance of the planet from the sun in units of the earth-sun distance.)

kernite An important ore of boron.

kettle hole

kettle hole A bowl-shaped depression, usually 20 to 50 feet deep and several hundred feet in diameter, formed in glacial drift when a buried block of ice melts and the surface layers collapse.

key A relatively small, low island of sand or coral lying near the coast. Sometimes spelled kay or cay.

key day A day the weather of which is popularly supposed to be a sign of the weather to come, although the idea is without basis in fact. Sometimes called a control day. Examples are January 1; Ground Hog Day (February 2), and St. Swithin's Day (July 15).

khamsins In Egypt, dry winds blowing from the south. They are most common in the spring, when the air is thick with drifting dust and sand.

kinetic energy Energy due to motion.

kinetic theory of gases The theory that gases are made up of molecules which are minute, perfectly elastic particles, ceaselessly moving about at high velocities and colliding with each other and with the walls of the containing vessel. The pressure exerted by a gas is due to the combined effect of the impacts of the moving molecules upon the walls of the containing vessel. The magnitude of the pressure is dependent upon the kinetic energy of the molecules and their number.

Kirchoff's law A radiation law that states that at a given temperature the ratio between the absorptive and emissive power for a given wave length is the same for all bodies. See: radiation laws; black body

knik wind In the vicinity of Palmer (Matanuska Valley), Alaska, the local name for a strong southeast wind, most frequent in the winter, though it may occur at any time during the year.

knob and kettle topography A region of small conical hills and small depressions. See: kame; kettle hole

knot A unit of speed equal to one nautical mile per hour.

koembang A foehn type of wind in Java.

kona storm A storm over the Hawaiian Islands, characterized by strong southerly or southwesterly winds and heavy rains. The kona rains occur over the normally leeward sides of the mountain slopes, i.e., the southerly, and are occasioned by cyclonic storms whose centers pass north of Hawaii. The kona storms occur, on an average, about five times a year; they are accompanied by heavy rains amounting to an inch or more, and the wind force is often destructive, but on the whole they are beneficial.

kraal A native village of Africa, usually built in the form of a circle and surrounded by a protective fence.

Krakatoa An island in the Strait of Sunda. In 1883, an explosive volcanic eruption occurred there, one of the greatest on record. The volcanic dust produced observable color effects in the sky for more than three years.

Kuroshio (Japan Current) A warm-water current flowing northward along the coast of Asia and eastward in the North Pacific.

l

labile (adj.) Subject to change, as when heated or exposed to conditions that lead to alteration in chemical or physical structure.

laccolith A large mass of igneous material that is forced upward from the earth's interior. It does not reach the surface but bows the surface layers upward and hardens as a dome-shaped rock mass.

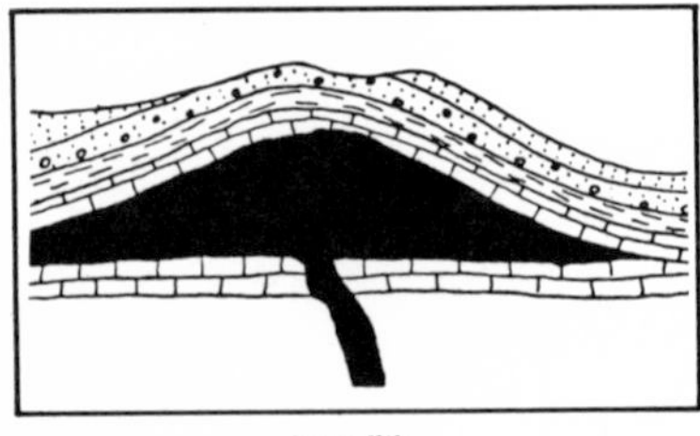

laccolith

lactic acid A fairly strong organic acid that exists in sour milk. When pure it is a colorless liquid, miscible with water, alcohol, or ether in all proportions. It is a metabolic intermediate in many living forms and is produced in quantity by many microorganisms.

lacustrine deposits Materials deposited from lake water. Many nearly level soils have developed from such deposits from lakes that have long since disappeared.

lagoon A shallow stretch of water that is separated from the sea by a narrow strip of land.

laissez faire The practice or doctrine of noninterference with others, especially with reference to individual conduct or freedom of action.

lake An extensive body of water enclosed by land. The name is sometimes loosely applied to the widened part of a river or to a body of water lying along a coast even when it is connected with the sea.

Lake Bonneville A fresh-water lake that occupied much of western Utah during and at the end of the last ice age.

lake breeze A wind, generally light, produced by the differential heating of the waters of a lake and the surrounding land, in much the same manner as a sea breeze.

Lake Lahontan A fresh-water lake formed in western Nevada at the close of the Pleistocene as the glaciers in the western mountains melted.

Lambert's azimuthal projection An equal-area or homolographic projection that is also azimuthal. It is commonly used to show hemispheres.

Lambert's azimuthal projection

laminar flow The flow of a fluid in which successive flow particles follow similar path lines.

lamination Layering or bedding less than two-fifths of an inch thick in sedimentary rock.

lamprey An eel-like marine fish that enters fresh water to spawn and often becomes landlocked. It has ruined the commercial lake trout and whitefish industries of the Great Lakes.

land breeze See: sea breeze

land-capability classification A grouping of kinds of soil into special units, subclasses, and classes according to their capability for intensive use and the treatments required for sustained use.

land classification Any system for identifying and categorizing land, generally, used by planning agencies in mapping use of land.

landes Low-lying, sandy plains in southwest France.

land hemisphere That half of the globe in which most of the land masses lie.

land hemisphere

landscape The sum total of the characteristics that distinguish a certain kind of area on the earth's surface and give it a distinguishing pattern in contrast to other kinds of areas. See: natural landscape; cultural landscape

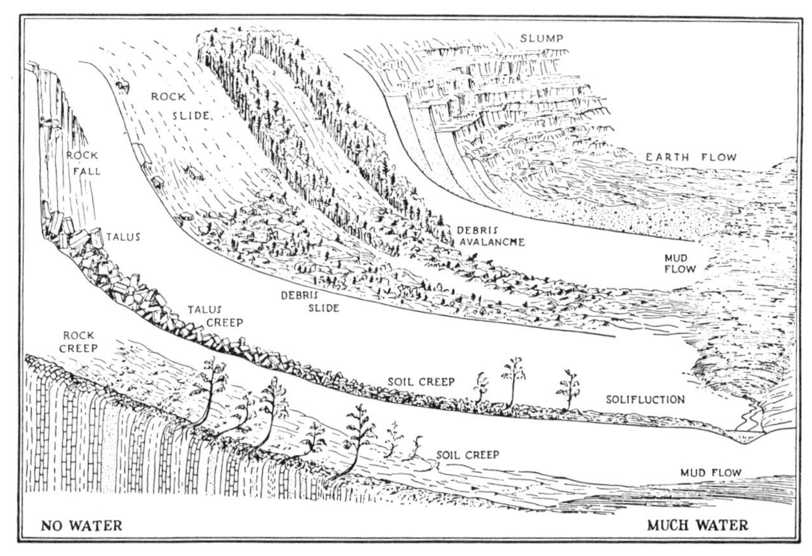

landslide

landslide or landslip The downward movement of large masses of rocks and earth.

landspout Former term for a tornado, by analogy with its marine counterpart, the waterspout.

land-use planning The development of plans for the uses of land that, over long periods, are intended to best serve the general welfare, together with the formulation of ways and means for achieving such uses.

langley A unit of energy per unit area equivalent to a gram calorie per square centimeter.

lapilli Small rock fragments ejected from a volcano. They are also called cinders.

Laplace's formula An equation expressing the relation between atmospheric pressure and height above sea level, used to reduce barometric readings to sea level or to any other desired level.

Lapp A member of a people inhabiting the northern treeless lands of Europe extending from Norway to the Soviet Union.

lapse rate 1. In general, the rate of change in the value of any meteorological element with elevation. 2. The rate of decrease of temperature with elevation. Thus the lapse rate of temperature is synonymous with the vertical temperature gradient.

large-scale convection Vertical motion in the atmosphere over extensive areas, as in a hurricane.

latent heat The heat absorbed by a substance, without change in temperature, while it is passing from a liquid to a vapor or from a solid to a liquid state, and released in the reverse

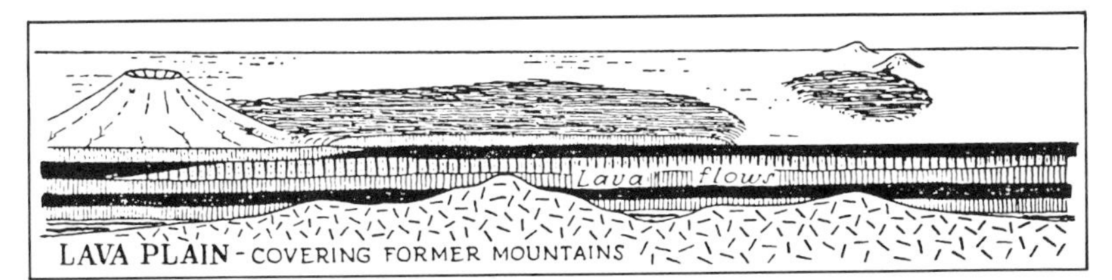

lava plain

change of state. See: heat of fusion; heat of vaporization or condensation

latent instability A type of conditional instability that can be released by vertical displacement of air.

lateral moraine The pile of rock debris left along the sides of a glacier.

laterite A red soil that develops in moist tropical or subtropical regions where drainage is good. It contains high concentrations of iron oxides and aluminum hydroxide.

latex 1. The milky liquid from certain plants (for example, the India rubber tree) which coagulates on exposure to air. 2. The synthetic counterpart of this liquid, used to make synthetic rubber or plastic.

latifundium A large estate or farm, particularly in Spain or Argentina.

latitude Distance north or south of the equator, generally expressed in degrees. See: longitude

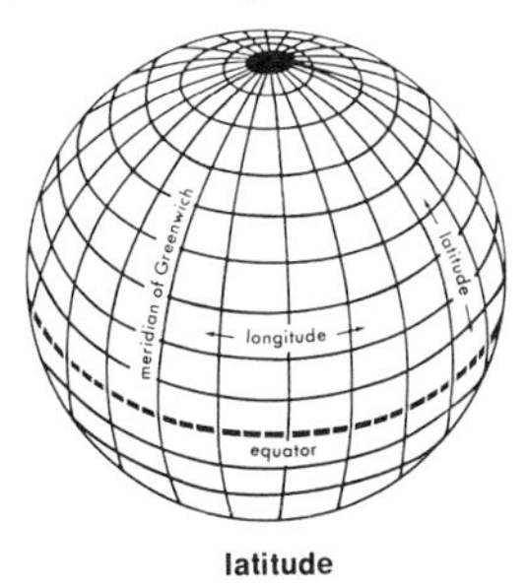

latitude

lava Molten rock material on the earth's surface or the solidified material that is derived from it.

lava plain A level, gently undulating surface created by the flow of lava across the land in an area of extensive volcanic activity.

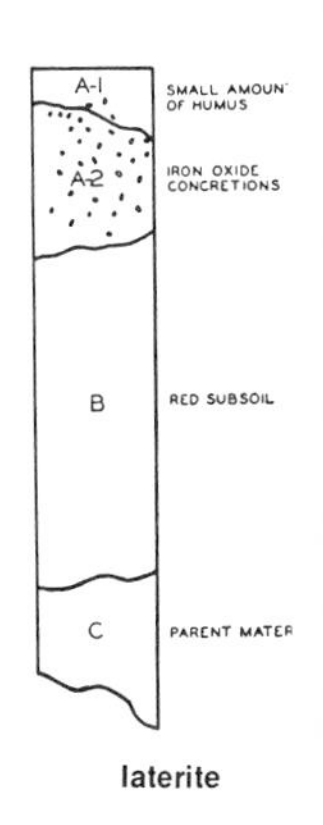

laterite

law of reflection A law of physics that states that when a ray of light is reflected from a surface the angle of incidence is equal to the angle of reflection and the incident and reflected rays are both in the same plane.

law of the conservation of energy A law of physics that states that energy can be neither created nor destroyed, or, in other words, that the total amount of energy in the universe is a constant. Energy may, however, be transferred from one system to another and transformed from one kind into another.

law of the minimum The law that states that the number of organisms that can be supported in a given ecosystem is limited by that element present in the smallest quantity.

leach (v.) To remove soluble constituents from soils or other material by infiltrating or percolating water.

Leap Year A year of 366 days, occurring every fourth year. It was devised to allow for the fact that the duration of the earth's revolution around the sun is about 365¼ days.

Le Chatelier's principle In physics, the principle that if some energy stress is brought to bear upon a system in equilibrium, a change will occur so that the equilibrium is displaced in a direction that tends to relieve the effect of the stress.

leeward The side of an object protected from the wind.

legume A plant, such as the bean, alfalfa, or vetch, which bears nitrogen-fixing bacteria on its roots.

lemming A rodent inhabiting polar regions. Lemmings are noted for their destructive migrations, in which millions of them march across the tundra and into the sea.

lenticular cloud A stationary cloud, resembling a huge lens, formed above a hill or mountain by the condensation of an ascending current of moist air, and sometimes over plains or rolling surfaces by the same process.

lespedeza Perennial bush clover found throughout Asia, Australia, and North America and used as forage by livestock.

leste A hot, dry, easterly wind of the Madeira and Canary Islands. It is of the sirocco type, blowing from the African desert and is most frequent during July, August, and September. It usually lasts three days but may vary in duration from one to seven days.

levante Spanish name for the east wind.

levanter An east wind in the Strait of Gibraltar, accompanied by cloudy, foggy weather much feared by mariners in sailing-ship days. It is also accompanied by rain, especially in the winter.

levantera A persistent east wind of the Adriatic, usually accompanied by cloudy weather.

levanto A hot southeasterly wind that blows over the Canary Islands.

leveche In Spain, a local name for the sirocco which is chiefly experienced along the southeast coast from Valencia to Malaga, and which generally blows from a direction between southeast and southwest.

level borders irrigation See: basin irrigation

levee The natural bank of a river formed during flooding by the deposition of sediment. Also the man-made banks built to confine the flow of a river.

level of free convection That level at which an air mass that is being lifted will continue to rise because it has become less dense than the surrounding air.

liana or liane A climbing plant found in equatorial forests.

libration of the moon An apparent slow oscillatory motion of the moon which causes the areas near its edge to appear and disappear alternately from view.

lichen A plant that is composed of a fungus in symbiotic union with an alga. It grows in leaflike or crustlike form on rocks, trees, etc.

lichen

life cycle The continuous sequence of changes that occur as an organism develops.

life zone A biogeographical region delimiting the distribution of plants and animals. Life zones exist in a horizontal dimension and also in a vertical dimension, extending up the slopes of mountains.

lifting condensation level See: condensation level

light (adj.) 1. Applied to the wind to indicate, in general, a relatively low velocity. For example, in the Beaufort wind scale a light air is a wind of one to three miles per hour, and a light breeze is a wind of four to seven miles per hour. 2. One of the degrees of intensity of various hydrometeors and lithometeors, which are usually reported as light, moderate, or heavy according to certain arbitrary standards of rate of fall, visibility, etc.

lightning A discharge of electricity from one part of a thundercloud to another part, from cloud to cloud, or between cloud and earth.

lightning storm A thunderstorm with or without rain reaching the ground. The term is used synonymously with the terms electric storm and thunderstorm, particularly in the western portions of the United States in discussion of the causes of forest fires.

light rain Rain falling with an intensity between a trace and 0.10 inch per hour (0.01 inch in 6 minutes). See: precipitation

light soil A term formerly used for sandy or coarse-textured soils.

light year A unit of measurement equivalent to the distance that a beam of light will travel in a year (5.88×10^{12} miles).

lignin An organic substance that incrusts the cellulose framework of plant cell walls. It is made up of modified phenyl propane units. It is dissolved only with difficulty and is more inert chemically and biologically than other plant constituents. Lignin increases with age in plants.

lignite An imperfect coal of woody appearance. It is the second stage in the transformation of vegetable matter into coal.

lime Generally the term lime, or agricultural lime, is applied to ground limestone (calcium carbonate), hydrated lime (calcium hydroxide), or burned lime (calcium oxide), with or without mixtures of magnesium carbonate, magnesium hydroxide, or magnesium oxide, and materials such as basic slag, used as amendments to reduce the acidity of acid soils.

lime requirement The amount of standard ground limestone required to bring a 6.6-inch layer of an acre (about two million pounds in mineral soils) of acid soil to some specific lesser degree of acidity (usually slightly or very slightly acid).

limestone A sedimentary rock composed mostly of calcium carbonate.

limiting factor Any environmental factor limiting the range or number of organisms living in an area.

limnology That branch of hydrology pertaining to the study of lakes.

limonite Brown hydrous iron oxide.

linear accelerator A device for speeding up protons and deuterons as projectiles for disintegrating the nucleus of atoms. Superimposed high-potential fields cause the projectile to gain speed in a linear path up to about 40 million volts.

line squall A squall or series of squalls occurring along a squall line and usually marked by a sudden change of wind direction, generally from the southeast or south to west, northwest and north. A line squall is frequently accompanied by heavy rain, snow, or hail, thunder and lightning, a sudden fall of temperature, and a rise of pressure and relative humidity.

linkage An association of two or more genes located on the same chromosome; this tends to cause the characters determined by these genes to be inherited as an inseparable unit.

Linnaean (adj.) Conforming to the principles of binomial nomenclature advocated by Linnaeus, a Swedish botanist (1707-1778).

lint The fibers of the cotton plant, used in making cotton thread and cloth.

linters The short fibers remaining on cottonseed after ginning. The word is also applied to the machines that remove this short fiber from the seed (also called delinters). Linters are too short for textile use, but they are used in making cotton batting, upholstery, and mattress stuffing, and, after purification, as a raw material for chemical conversion to cellulose derivatives.

liquified natural gas (LNG) A low-emission fuel which can be used to power automobiles that have been modified. Major drawbacks include the necessity for a special insulated gas tank and the fact that this type of fuel is not readily available.

liquified petroleum gas (LPG) A low-emission fuel which can be used to power automobiles. It requires a special storage tank and is not as commonly available as gasoline. LPG-powered vehicles have approximately one-half the hydrocarbon emissions and about one-fourth the carbon monoxide emissions of the same vehicle operating on gasoline.

lister furrows A small trough in the soil left by a special plow used just before planting.

lithification The process by which newly deposited sediment is eventually converted into hard rock.

lithology The science dealing with the physical character of rock.

lithometeor The generic term for a class of atmospheric phenomena, of which dry haze and smoke are the most common examples. A lithometeor is composed of solid dust or sand particles, or the ashy products of combustion.

lithosol A soil having little or no evidence of soil development and consisting mainly of a partly weathered mass of rock fragments or of nearly barren rock.

lithosphere That part of the earth composed predominantly of rock as opposed to the hydrosphere.

littoral (adj.) Pertaining to the shore.

littoral transport The movement of material along a shore by waves and currents.

live oak The name given to evergreen oaks.

llama A genus of animals found only in South America. They are used as beasts of burden and their wool is used for fabric.

llama

llanos The savanna of the Orinoco basin and the Guiana highlands in South America.

loam The textural class name for soil having a moderate amount of sand, silt, and clay. Loam soils contain 7 to 27 percent of clay, 28 to 50 percent of silt, and less than 52 percent of sand.

loblolly pine A timber tree of the southeastern United States.

lobster A marine crustacean of considerable importance as a seafood. It is most abundant in New England waters.

local time The time at any place on the earth's surface calculated by the position of the sun; thus, noon local time occurs when the sun reaches its highest position in the sky.

lock In a river or canal, an artificial enclosure with gates at each end. The level of the water in a lock is controlled by pumps, so that boats can be raised or lowered to the level of the next section of the waterway.

locust The grasshopper. The locust is an insect pest of significance especially in the Middle East.

locust tree An ornamental tree bearing elongated seed pods, some of which may be eaten.

lode A vein or fissure in the earth filled with mineral ore.

lode mining The mining of minerals occurring in veins or lodes, generally underground.

lodgepole pine A timber tree of significance in the western United States.

loess A deposit of fine silt or dust that has been transported by the wind to its present location.

logging The process of removing the trees from a forested area.

longitude Distance east or west of the prime meridian, generally expressed in degrees. See: latitude

longitudinal profile A cross section of anything that extends lengthwise along its greatest dimension.

longitudinal valley A valley lying between two parallel mountain ranges.

longleaf pine A tree valuable as a source of resin and lumber. It grows in the southern United States.

long-range forecast A weather forecast encompassing a period longer than five days.

longshore drift The flow of water along the shore of the ocean, carrying sand and small pebbles with it.

long-wave radiation See: terrestrial radiation

looming An optical phenomenon in which objects below and beyond the horizon appear in the view of the observer, and even the horizon itself is extended. This is caused by an abnormal decrease of density with elevation, which occurs when the temperature of the surface layers of air is extremely low, causing abnormal refraction of light coming from distant objects to the eye of the observer.

low A region of the atmosphere where the barometric pressure is below normal, usually surrounded by closed isobars with the point of minimum pressure in the center. The term low is not strictly synonymous with cyclone since lows often appear on the weather map without a well-defined cyclonic wind circulation.

lowland The low-lying land of a region, found generally in river valleys and along coasts.

low latitudes The region of the earth between the North and South Tropics, between 0° and 23½° north or south of the equator.

loxodrome A line of constant compass direction. See: rhumb line

lumen 1. A unit of light intensity, being the light emitted by a uniform source of one international candle. 2. In zoology, the passageway of a tubular organ such as a gland.

luminous cloud 1. A cloud illuminated by an interior electrical discharge. See: sheet lightning. 2. A name sometimes given to a nacreous or noctilucent cloud.

lunar eclipse The total or partial obscuring of the moon when the earth's shadow falls on it.

lunar month The period of time required for the moon to revolve around the earth (29.5 days).

luster The light reflected by a mineral. Among the several kinds of luster are metallic, adamantine, vitreous, resinous, and pearly.

lux 1. A practical metric unit of illumination, equal to one lumen per square meter. 2. The illumination of a surface at a uniform distance of one meter from a symmetrical point source of one candle.

lysimeter 1. A device to measure the quantity or rate of downward water movement through a block of soil usually undisturbed. 2. A device to collect such percolated water for analysis as to quality.

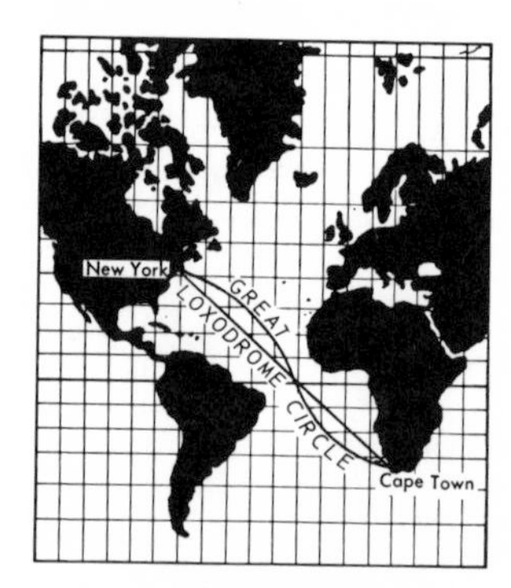

loxodrome

macchia See: maquis

mackerel Important commercial fish of both the Atlantic and Pacific oceans.

mackerel sky A formation of rounded and isolated cirrocumulus clouds resembling the pattern of scales on the back of a mackerel.

macroclimate The general climate over a comparatively large area considered as a whole, as distinguished from the detailed variations within very small areas.

macrometeorology 1. The study of the weather over a period of weeks or months, for the purpose of examining general large-scale weather changes such as, for instance, those pertaining to the general circulation. 2. The study of meteorological phenomena as they occur over the whole earth or a considerable area of it.

macropore A large or noncapillary pore: the pores, or voids, in a soil from which water usually drains by gravity.

macroscale A continental or subcontinental scale.

maestro A northwesterly wind in the central Mediterranean area.

magma Molten or otherwise fluid rock occurring naturally within the earth. Igneous rocks are believed to have been formed by solidification of magma.

magnetic declination See: declination

magnetic equator The line on the surface of the earth where the magnetic needle remains horizontal or does not dip: that is, where the magnetic lines of force are horizontal. Also called the aclinic line.

magnetic field A region of space at each point of which a definite magnetic force exists.

magnetic lines of force Lines in a magnetic field that are everywhere in the direction of the magnetic force: i.e., the magnetic force at any point is tangent to the line through that point.

magnetic pole The magnetic north or south to which a compass points.

magnetic storm A large disturbance of the earth's magnetic field, associated with great sunspot activity. See: sunspot

magnetism (terrestrial) The magnetic properties of the earth, causing it to act like a huge bipolar magnet.

magnetite Magnetic iron ore.

magnetometer An instrument used in measuring intensities of magnetic fields, particularly of very weak fields, such as those found on the earth's surface.

magnitude 1. Size. 2. Brightness (of a star).

mahogany A tropical lumber and veneer tree of Africa, Central and South America, and the Philippines.

maize rain In East Africa, a heavy rain occurring between February and May.

malaria A disease produced by a parasitic protozoa that multiplies in and destroys red blood cells. It is marked by waves of cold, fever, and sweating.

malleability The capability of being shaped or stretched by hammering without breaking or rolling. Gold is an unusually malleable substance.

mallee Australian scrub growth, mostly of eucalyptus bushes.

malnutrition Faulty nutrition due to insufficient food or wrong kinds of food.

maloka In the tropical rain forests of South America, a long building housing all of the Indians of a village.

Malthusian theory The theory that population increases geometrically while food supply increases arithmetically, so that, unless the population increase is checked, poverty, war, and vice are inevitable.

malting The process of germinating grain, generally barley, to develop the enzyme diastase.

mammal A vertebrate that nourishes its young with milk.

mammatus (adj.) Any cloud whose lower surface resembles the shape of pouches or breasts.

mammoth (adj.) Enormous (n.) An extinct hairy elephant of the Pleistocene epoch.

mango A tropical fruit tree and its oblong, yellowish-red fruit.

mangrove A tropical maritime tree or shrub of the genus Rhizophora. They grow in a swampy coastal environment and are characterized by enormous prop roots extending in all directions away from the trunk.

manioc A tropical plant, also called the cassava. Its roots look somewhat like sweet potatoes and are used for food.

manometer An instrument for measuring the pressure of gases.

mantle A layer of loose rock fragments overlying solid rock.

manure Generally, the refuse from stables and barnyards, including both animal excreta and straw or other litter. In some other countries the term manure is used more broadly and includes both farmyard or animal manure and chemical manures, for which the term fertilizer is nearly always used in the United States.

Maoris The native people of New Zealand, who settled the land over 1,000 years ago.

map The representation on a flat surface of specific features of the earth's surface.

map legend The part of a map containing an explanation of the symbols used.

map projection A systematic representation of the parallels and meridians of the earth on a plane surface, by means of any geometrical construction that sets up a correspondence between the points of the earth's surface and the points of a region in a plane.

map scale The ratio of distance on a map to distance on earth; thus, a scale of one inch to one mile may also be expressed as 1:63,360. A map scale may also be expressed graphically.

maquis In parts of the Mediterranean region, low scrub consisting of shrubs and trees able to survive drought.

marble Coarse-grained, crystalline metamorphic rock derived from limestone.

march The progression or successive occurrence of any of the meteorological elements through the day, month, or year.

mares' tails A popular name given to well-defined cirrus clouds thickening into cirrostratus and then gradually lowering into watery altostratus. The clouds resemble mares' tails and may prove the precursors of a storm.

Mariana trench An ocean trench southeast of Guam.

marine-built terrace A coastal terrace elevated above sea level and built by the deposition of material eroded by wave action.

marine climate The climatic type controlled by an oceanic environment, found on the windward shores of continents and on islands, and characterized by small diurnal and annual temperature ranges and high relative humidity.

marine-cut terrace A plain cut by wave action and since elevated above sea level.

marine rocks Sedimentary rocks formed as a result of deposition in an ocean.

maritime (adj.) Of or related to the ocean. The term is used to identify air masses and climates with particular characteristics due to ocean influences.

market gardening Intensive cultivation of vegetables for sale as fresh food.

marl An earthy deposit consisting mainly of calcium carbonate commonly mixed with clay or other impurities. It is formed chiefly at the margins of fresh-water lakes. It is commonly used for liming acid soils.

marsh Wet, low-lying land, generally under water.

Masai A Negro people of the east African grasslands.

mass The quantity of matter in a body. Often used as a synonym for weight, which, strictly speaking, is the force exerted by a body under the influence of gravity. See: atomic mass unit; atomic weight

mass curve A graph of the cumulative values of a hydrologic quantity (such as precipitation or runoff), generally plotted as ordinate, against the time or date as abscissa.

mass-energy equation The statement developed by Albert Einstein, that "the mass of a body is a measure of its energy content." The statement was subsequently verified experimentally by measurements of mass and energy in nuclear reactions. The equation, usually given as: $E = mc^2$, shows that when the energy of a body changes by an amount, E, (no matter what form the energy takes) the mass, m, of the body will change by an amount equal to E/c^2. (The factor c^2, the square of the speed of light in a vacuum, may be regarded as the conversion factor relating units of mass and energy.) This equation predicted the possibility of releasing enormous amounts of energy (in the atomic bomb) by the conversion of mass to energy. It is also called the Einstein equation.

massif A mountainous mass, consisting of a group of connected peaks, in which isolated mountains rarely occur.

mass number The total number of neutrons and protons in the nucleus of an atom. Different isotopes of the same element have different mass numbers.

mass wasting The slow downward movement of rock debris.

mast The accumulated fruits or nuts found on the forest floor. The major contributors are such species of trees as oaks, beeches, chestnuts, and some pines. Mast is used as a food for hogs, deer, turkeys, and other wildlife.

Matanuska wind In the vicinity of Palmer, Alaska, the local name (taken from the Matanuska River) for a strong, gusty northeast wind that occasionally occurs during the winter.

matter The substance of which a physical object is composed. All materials in the universe have the same inner nature, that is, they are composed of atoms, arranged in different (and often complex) ways; the specific atoms and the specific arrangements identify the various materials. See: atom; element

matterhorn Any sharp, hornlike mountain peak resembling the Swiss peak of that name.

mature river valley A valley characterized by low stream gradients, many stream tributaries, and a broad, flat floodplain.

mature soil Any soil with well-developed soil horizons having characteristics produced by the natural processes of soil formation and existing in close equilibrium with its present environment.

maximum The highest value observed during a specified period.

maximum thermometer A thermometer that automatically registers the highest temperature occurring since its last setting. See: minimum thermometer

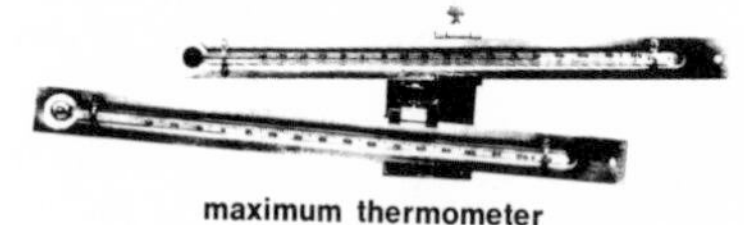

maximum thermometer

mean 1. The sum of the magnitudes of all items of a set divided by the number of items. 2. In meteorology, the average value of any element such as temperature, pressure, or humidity.

mean annual flood The average of all annual flood stages or discharges of record. It may be estimated by regionalization, correlation, or some other process that is believed to furnish a better estimate of the long-term average than is provided by observed data.

mean annual precipitation The average over a period of years of the annual amount of precipitation.

mean annual runoff The average value of all annual runoff amounts, usually estimated from the period of record or during a specified base period from a specified area. See: base period; runoff

mean annual temperature The average of the monthly means for the year.

mean daily temperature The average of the daily maximum and minimum temperatures.

mean depth The average depth of water in a stream channel or conduit. It is equal to the cross-sectional area divided by the surface width.

mean monthly temperature The average of the mean monthly maximum and minimum temperatures.

mean sea level (M.S.L.) The average level of the sea for all stages of the tide.

mean solar time The system of time commonly used; derived by calculating average values from the varying daily values which occur throughout the year. See: apparent solar time, equation of time

meander The winding of a stream channel.

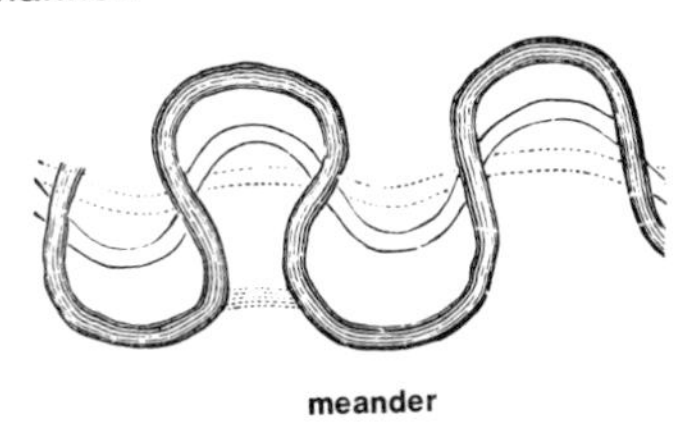
meander

meander belt Area between lines drawn tangential to the extreme limits of fully developed meanders.

mechanical weathering The processes which work to break rock down by physical means: e.g., differential expansion of rock crystals due to heating and cooling, root growth, etc.

medial moraine Drift laid down between the lobes of two glaciers.

median In a series, the value of the middle term if the number of terms is odd, or the average of the two middle terms if the number of terms is even.

medical climatology The study of the effects of weather on the health of human beings and animals and of the effects experienced when they are transferred to a climate to which they are not native.

medical geography That subfield of geography that studies the factors involved in the geographical variation of disease. See: epidemiology

Mediterranean climate A marine or littoral climate characterized by: (a) mild winters of light to moderate rainfall, because of westerly winds and traveling cyclones; (b) warm to hot summers, tempered by sea breezes, with a considerable period rainless or nearly so; (c) abundant sunshine during both winter and summer; and (d) a natural vegetation consisting of broad-leaved evergreens and drought-resistant trees and shrubs.

megalopolis An urbanized area formed by the merging of a number of large metropolitan areas. The term was applied originally by Gottmann to the urbanized area of the northeastern United States.

megathermal zone In Köppen's classification, the region of the earth enjoying a winterless climate, where megatherms (tropical plants requiring continuous high temperatures and abundant rainfall) grow.

meiosis A two-stage division of cells in which each of the resulting four new cells contains one-half the number of chromosomes present in the original cell.

meltemi Turkish name for the etesian winds.

melting point The temperature at which fusion takes place. All crystalline solids have definite melting points under specified pressures.

Mendel's laws Laws of heredity discovered by Gregor Mendel (1822-1884), according to which physical characteristics such as height and color depend on the presence of genes as determining factors. Second and later generations of crossbreeds exhibit these characteristics in all possible combinations.

menhaden A fish found along the Atlantic coast and used commercially for oil and fertilizer.

meniscus The curved upper surface of a liquid observed in a tube when the latter is partially plunged into the liquid. It may be either concave or convex.

Mercalli scale A scale used to rate the intensity of an earthquake. The ratings are generally expressed in Roman numerals. See: earthquake intensity

Mercator's projection A mathematically derived map projection with straight meridians and parallels, developed originally to be used as an aid to navigation.

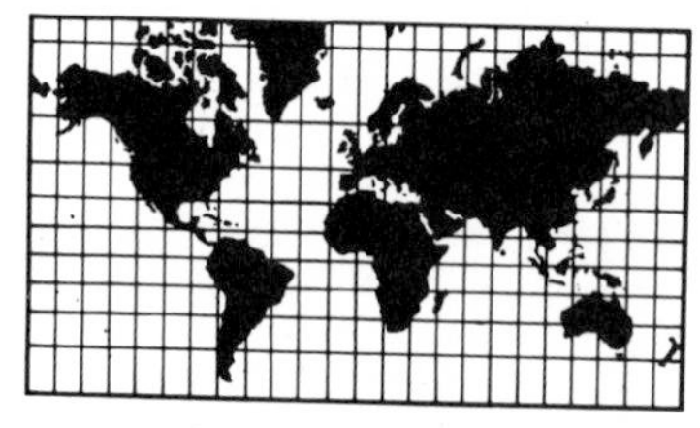

Mercator's projection

mercurial barometer A device invented by Torricelli in 1643 to measure atmospheric pressure.

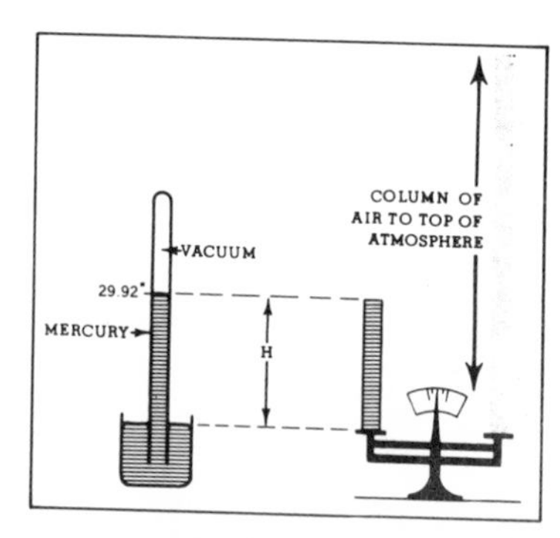

mercurial barometer

mercury A metallic liquid resembling silver in appearance, used in barometers as a weight and in many thermometers as the expanding and contracting medium. Its melting point is –38.87°C. or –37.97°F., and its boiling point is 356.90°C. or 674.42°F. It is a chemical element. Its symbol is Hg and its atomic number is 80.

meridian On the globe, one-half a great circle extending from pole to pole. It is used to measure longitude.

mesa A tableland, or flat-topped hill with at least one steep side.

meseta An extensive plateau comprising the interior of Spain.

mesoclimate The climate of a portion of a region that may not be representative of the climate of the region as a whole.

mesometeorology The study of atmospheric phenomena such as tornadoes and thunderstorms which occur between meteorological stations or beyond the range of normal observation from a single point, i.e., on a scale larger than that of micrometeorology, but smaller than the cyclonic (synoptic) scale.

mesopause The top of the mesosphere. It is the coldest part of the atmosphere (about –95°C.).

mesophyte A plant that grows under medium conditions of atmospheric moisture supply, as distinguished from a zerophyte (one which grows under dry or desert conditions) or hydrophyte (wet conditions). See: hydrophyte, phreatophyte, xerophyte

mesoscale In meteorology and climatology, pertaining to middle values or areas of middle size (e.g., cities and counties) investigated by scientists.

mesosphere The atmospheric zone lying between 50 and 85 kilometers above the stratosphere. In this zone the temperature decreases with height.

mesothermal climate A type of climate marked by moderate temperature and rainfall and ample sunshine.

Mesozoic era See: Appendix I

mesquite A spiny, deep-rooted tree or shrub of the American Southwest, bearing bean-like pods.

mestizo The offspring of a European and an American Indian. In Brazil, the offspring of a European and a Negro.

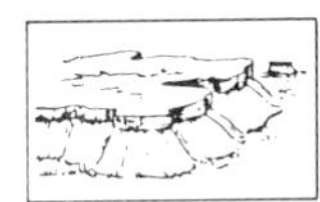

mesa

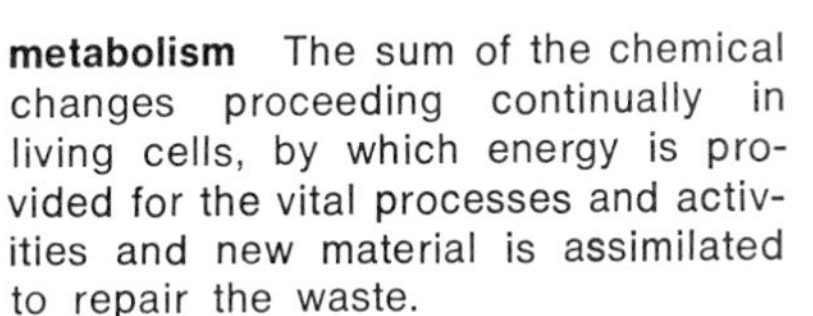

metabolism The sum of the chemical changes proceeding continually in living cells, by which energy is provided for the vital processes and activities and new material is assimilated to repair the waste.

metamorphic rock A rock that has been greatly altered from its previous condition through the combined action of heat and pressure. For example, marble is a metamorphic rock produced from limestone, gneiss is produced from granite, and slate is produced from shale.

metamorphism The change in the composition of rock due to heat and pressure.

metaplasm That part of a cell consisting of lifeless matter.

meteor Originally a general term for any atmospheric phenomenon, but now more commonly restricted to astronomical meteors, which are sometimes called shooting or falling stars; a fiery streak in the sky produced by a meteoroid passing through the earth's atmosphere.

meteoric water Water derived from precipitation.

meteorite A piece of solid matter that has fallen to the earth's surface from outer space.

meteorograph 1. An instrument that

automatically records on a single sheet the measurements of two or more meteorological elements, such as air pressure, temperature, humidity, precipitation, etc. 2. A self-recording meteorological instrument carried aloft by a plane, balloon, or other means, to furnish data on conditions in the upper air.

meteoroid A small, solid body moving through outer space or through the earth's atmosphere. See: meteor

meteorological acoustics A branch of meteorology concerned with sounds of a distinctly meteorological origin, such as the rumbling of thunder, as well as with the effects of various meteorological conditions on the travel, distribution and audibility of all sounds, whatever their sources.

meteorological elements 1. Six quantities — air temperature, barometric pressure, wind velocity (direction and speed), humidity, cloudiness (type and amount of clouds), and precipitation— that specify the state of the weather at any given time and place. 2. In addition to the above, any of the various other meteorological phenomena that distinguish the particular condition of the atmosphere, such as sunshine, visibility, radiation, halos, thunder, mirages, lightning, etc. See: climatic elements

meteorological equator Synonymous with heat equator.

meteorological symbol A letter, number, diagrammatic sign, or character used as a kind of shorthand to indicate, in weather records or on weather maps, the state of the weather or the occurrence of any hydrometeor or other meteorological phenomenon.

meteorology Originally, knowledge and lore of the weather and all other phenomena of the atmosphere. Now the term is often restricted to the branch of physics dealing only with phenomena directly related to the weather.

meter A primary standard of length, originally defined as a ten-millionth part of the length from the equator to the North Pole of the meridian passing through Paris, France, but now the length at the temperature of 0°C. of the platinum-iridium bar deposited at the International Bureau of Weights and Measures at Sèvres, near Paris, France.

methane The colorless, odorless, flammable gas CH_4. It is obtained from natural gas and is the chief constituent of marsh gas and the firedamp of coal mines.

methoxyl A chemical grouping composed of a carbon atom linked to an oxygen atom and three hydrogen atoms. The conventional symbol is OCH_3.

metric system A system of measurement based on the meter (39.3701 inches) and the kilogram (2.2 pounds) in which all other units are derived by division or multiplication of the base units by factors of ten.

metrology The science of weights and measures.

mica Any of a group of mineral silicates that tend to split into thin sheets.

micaceous Composed of or similar to mica.

micro- A prefix meaning: 1. Very small. 2. A one-millionth part of something. 3. That which makes use of a microscope, such as microbiology.

microbar A one-millionth part of a bar, taken in meteorology as equal to the pressure of one dyne per square centimeter.

microbarograph A barograph designed to record variations of atmospheric pressure smaller than those that can be detected by the barographs in general use.

microbe A microorganism, especially a pathogenic bacterium.

microclimate The climate of a very small area of the earth's surface, e.g., a single forest or even a corn field, over which small variations exist from place to place.

microclimatology The detailed study of the climate of a very small area.

microcosm A world in miniature.

micrometeorology The study of the variations in meteorological conditions over very small areas such as hillsides, forests, river basins, and individual cities. See: macrometeorology

micronutrient A nutrient that a plant needs in only small, trace, or minute amounts.

microorganism A form of life either too small to be seen with the unaided eye or barely discernible.

microphyte Any microscopic plant.

micropore The small spaces in soil through which water moves by capillary action and in which it is held against the pull of gravity.

microrelief Relief which exhibits small-scale differences such as small mounts, swales, or pits up to three feet in depth or elevation. Microrelief is significant to soil-forming processes, preparation of the soil for cultivation, and plant growth.

microthermal climate A type of climate found on the equatorward margins of the polar regions and distinguished by long, cold winters and short summers.

microwave An electromagnetic wave with a length of from about one-quarter inch to six inches. Its main use is in radar and television.

midden A refuse heap, especially of a primitive culture.

middle clouds Family B of the international cloud classification, comprising the altocumulus and the altostratus clouds, with their subgenera and varieties.

mid-latitudes The area lying between the tropics and the Arctic and Antarctic circles.

mid-latitude mixed forest The forest of temperate lands, consisting of both conifers and deciduous species.

midnight sun The sun as seen in high latitudes in summer when it appears above the horizon throughout the 24 hours.

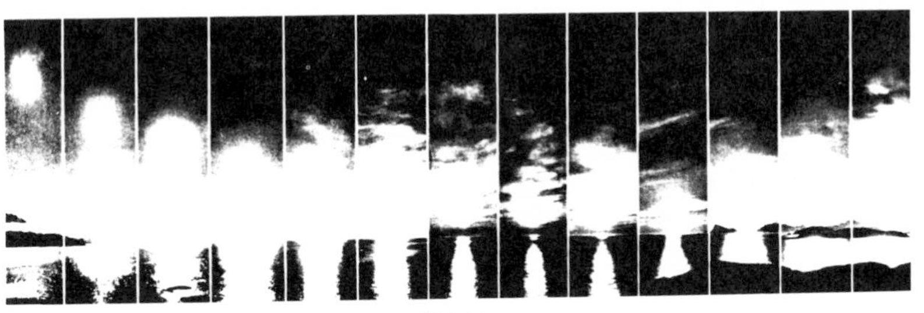

midnight sun

mid-oceanic ridge The great median arch extending the length of an ocean basin and roughly paralleling the continental margins. Such ridges exist in the Arctic, Atlantic, and Indian oceans.

mile (statute) A unit of measurement equal to 1,760 yards or 5,280 feet.

Milky Way The galaxy of which our solar system is a part, seen as a luminous tract stretching across the heavens.

millet A grass producing very small kernels that are used as grain.

millibar A subunit of pressure equal to one one-thousandth of a bar. In meteorology it is equal to a force of 1,000 dynes per square centimeter. The values of atmospheric pressure are now usually expressed in millibars. 1,013 millibars is standard atmospheric pressure.

mineral An inorganic substance that possesses a definite chemical composition.

mineralogy The study of minerals.

mineral spring A hot spring containing appreciable amounts of mineral matter, often thought to have medicinal values.

miner's inch The rate of discharge of water through an orifice one inch square under a specific head of pressure.

minimum The lowest observed value of the temperature, pressure, or other weather element during any given period.

minimum thermometer A thermometer that automatically registers the lowest temperature that occurs after it has been set. See: maximum thermometer

minuano A cold southwesterly wind of southern Brazil, occurring in the winter months (June to September) and named for the Minuano Indians, who inhabit the region from which it blows.

minute 1. A unit of time equal to one-sixtieth of an hour. 2. A unit of measurement of latitude or longitude equal to one-sixtieth of a degree.

Miocene Epoch See: Appendix I

mirage An optical illusion due to the refraction of light as it passes through nonhomogeneous layers of the atmosphere. Distant objects are seen in an unnatural position, sometimes elevated, sometimes depressed, and often inverted. The phenomenon occurs most frequently in hot climates over surfaces that are warmed by insolation, such as sandy plains.

Mississippian Period See: Appendix I

mist 1. A thin fog of relatively large particles of water. 2. A very fine rain,

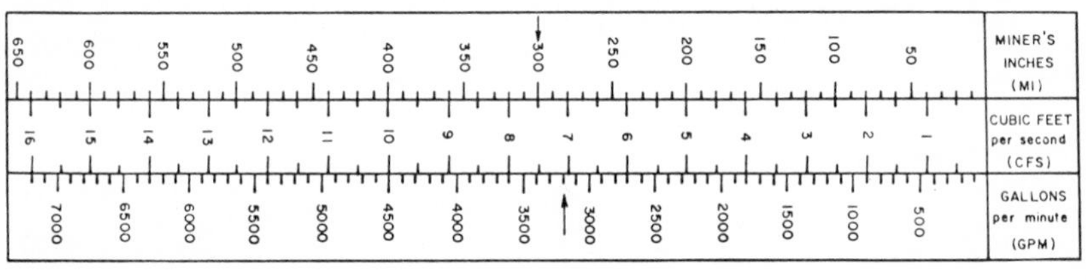

miner's inch

lighter than a drizzle and similar to a damp fog with falling particles of water.

mistbow A fogbow.

mistral A cold, dry wind blowing from the north over the north-west coast of the Mediterranean Sea, particularly over the Gulf of Lions. It blows when there is a low-pressure area centered over the Gulf of Genoa and a high-pressure area centered to the north-west, and often with force enough to overturn railroad trains.

mitosis The division of the nucleus of a cell, sometimes applied to the division of the cell itself.

mixed forest A forest in which conifers and broadleafed deciduous trees intermingle.

mixing ratio The mass of water vapor per unit mass of perfectly dry air in a humid mixture. The mixing ratio is for all practical purposes numerically equal to the specific humidity.

moa Any of several flightless birds of New Zealand, now extinct.

mode The value that occurs most frequently in a set of observed values of a quantity and that may therefore be taken as typical.

model A representation of anything. It may be visual, mathematical, or intellectual, and is generally used to present an object or concept clearly so as to bridge the gap between the observational and theoretical.

moderate rain Rain falling with an intensity between 0.11 inch per hour (0.01+ inch in 6 minutes) and 0.30 inch per hour (0.03 inch in 6 minutes). See: precipitation

modified flow That streamflow that would occur if the works of man in the stream channels and in the drainage basin had been consistent throughout the period of record.

moho discontinuity The interface between the earth's crust and the mantle, lying about 10 kilometers below the oceans and 35 kilometers below the continents.

mohole A deep borehole penetrating the earth's mantle.

moist adiabat See: wet adiabat

moist adiabatic lapse rate See: wet adiabatic lapse rate

moisture equivalent The ratio of the weight of water that the soil, after saturation, will retain against a centrifugal force 1,000 times the force of gravity, to the weight of the soil when dry. The ratio is stated as a percentage.

moisture stress See: moisture tension

moisture tension The force at which water is held by soil, usually expressed as the equivalent of a unit column of water in centimeters. 1,000 centimeters equal 1 atmosphere equivalent tension.

mold The common name applied to a fungus that in nature appears on damp or decaying matter. Certain molds produce the appearance and flavor of such cheeses as Roquefort and Camembert. Others are used for the production of various organic acids, enzymes, and the antibiotic drug penicillin.

mold rain See: bai-u

mole 1. A mass numerically equal to the molecular weight of the substance. The gram-mole or gram-molecule is the mass in grams numerically equal to the molecular weight. For example, a gram-mole of oxygen is 16 grams; of carbon dioxide, 44 grams. 2. A breakwater of large stones to protect harbors from storms.

molecular diffusion The spread of atmospheric constituents or properties as a consequence of random molecular motion, which take place as a result of thermal agitation. This agitation varies directly with temperature and inversely with pressure.

molecular theory A scientific theory that all matter is composed of electrical energy organized into atoms and molecules. These molecules are in constant motion, and their collision produces temperature. The relative amount of heat energy in an object is measured by temperature scales. When radiant energy is absorbed by a molecule, the molecular energy content is increased: molecular activity speeds up and a higher temperature results.

molecular weight The weight of any molecule; and, therefore, the sum of the weights of the atoms of which it is composed.

molecule The smallest particle of an element or compound that retains the chemical properties of that substance. Molecules are made up of atoms.

mollisol The soil group that includes most of the soils formerly known as chernozem, prairie, chestnut, and brown and characterized by a thick, dark surface horizon with moderate to strong structure and saturated with bivalent cations.

mollusk A member of a group of invertebrate animals including clams, snails, slugs, squids, oysters, and many others. Sometimes spelled mollusc.

Mollweide's projection A type of equal-area or homolographic projection in which all parallels and the central meridian are straight lines.

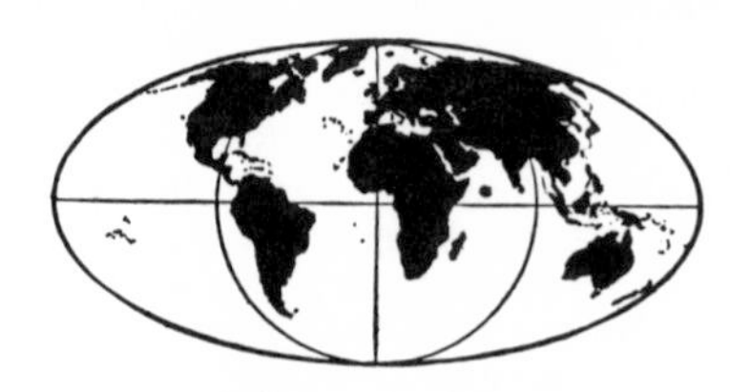

Mollweide's projection

molybdenum A mineral used in making hardened forms of steel.

monadnock An isolated hill or mountain of resistant rock surmounting a peneplain.

monatomic (adj.) Having one atom per molecule.

mongoloid (adj.) Pertaining to a racial division of mankind marked by yellowish complexion, prominent cheekbones, epicanthic folds about the eyes, straight black hair, and broad face.

monocline In geology, a steplike bend in otherwise nearly horizontal strata.

monocotyledon An angiosperm with only one cotyledon in each seed.

monoculture The use of land to grow only one type of crop; typical of plantation agriculture.

monogamous (adj.) Marrying only once.

monovalent cation An ion having a single positive charge, a deficiency of one electron from the neutral state.

monsoon A seasonal wind blowing from the continental interior to the ocean in winter, and in the opposite direction in summer. These winds result from the temperature differences arising between land and ocean; and, since their path is long, their direction is greatly influenced by the earth's rotation. The term monsoon is of Arabic origin and means season. It is used often in that sense in India, where monsoon refers to the rain occurring from June to September.

monsoon air An air mass formed over India and the Bay of Bengal by a direct and gradual transition of air from the Siberian polar continent to the equatorial Indian Ocean.

monsoon fog A coastal advection fog, which usually occurs during periods when a marked temperature contrast exists between adjacent land and water surfaces.

Montaña A region of the eastern slopes of the Andes having heavy precipitation and dense forests.

montmorillonite A finely platy, alumino-silicate clay mineral that expands and contracts with the absorption and loss of water. It has a high cation-exchange capacity and is plastic and sticky when moist.

moon The natural satellite of any planet.

moor A tract of open, peaty, and poorly drained wasteland overgrown with heather, located in high latitudes or altitudes.

mor Raw humus; a type of forest humus layer of unincorporated organic material, usually matted or compacted or both.

moraine A pile of drift carried or deposited by a glacier either at its side (lateral moraine) or at its lower end (terminal moraine).

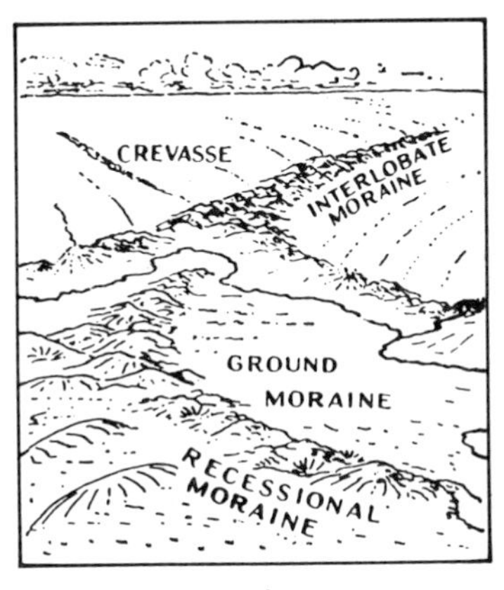

moraine

morphology The form and structure of an object or organism.

mortality rate See: age-specific death rate; crude death rate

mother-of-pearl clouds Nacreous clouds.

mottled (adj.) Irregularly marked with spots of color. A common cause of the mottling of soil horizons is imperfect or impeded drainage, although there are other causes, such as soil development from unevenly weathered rock. Different minerals in soil may also cause mottling.

mountain A mass of land considerably higher than its surroundings. Its summit area is smaller than its base and is at least 1000 feet above the surrounding terrain.

mountain breeze A breeze that blows down mountain slopes because of the gravity flow of cooled air. It is of the same type as canyon, gravity, and katabatic winds.

mountain climate The climate on mountains, controlled primarily by the altitude factor, and distinct from that of valleys in many respects. It is characterized by low pressure due to the elevation, by strong sunshine rich in ultraviolet rays (owing to the thin and clear air), and by uniformly low temperatures and humidity. The precipitation depends greatly on exposure, the windward slopes being the wettest. The temperature, while lower in general than that of the lowlands, is hot on sunny slopes at midday. The winds, in the absence of storms, flow up the slopes during the day and down at night.

mountain sickness An ailment experienced by persons subjected to air pressures much less than those to which they are accustomed. The same symptoms may be experienced in a balloon or airplane. These symptoms are: suffocation, exhaustion, nausea, headache, sleeplessness, bleeding at nose and lips, lapse of memory, and inability to think. It may occur with or without exertion, and is usually felt by people accustomed to living at or near sea level when they ascend to about 10,000 feet elevation.

mountain wave A wave created by the flow of air over a mountain, and used by glider pilots for prolonged flights. Evidence of waves is seen in the lenticular and roll clouds appearing along the crests of mountains.

muck Highly decomposed organic soil material developed from peat. Generally, muck has a higher mineral or ash content than peat and is decomposed to the point that the original plant parts cannot be identified.

mud flat A muddy area of ground on or near the shore, usually submerged at high tide.

mudflow A flow of rock debris lubricated by a large amount of water.

mud volcano A cone-shaped hill, generally less than 250 feet high, formed of mud brought to the surface by gas.

mulatto The offspring of white and Negro ancestors.

mulberry tree A small tree raised mainly in the Far East. Its leaves are used for food by silkworms.

mulch A natural or artificially applied layer of plant residues or other materials on the surface of the soil. Mulches are generally used to help conserve moisture, control temperature, prevent surface compaction or crusting, reduce runoff and erosion, improve soil structure, or control weeds. Common mulching materials include compost, sawdust, wood chips, and straw. Sometimes paper, fine brush, or small stones are used.

mulga Scrub growth in western and central Australia consisting largely of a species of acacia also known as mulga.

mull A humus-rich layer of forested soils consisting of mixed organic and mineral matter. A mull blends into the upper mineral layers without an abrupt change in soil characteristics.

multiple use The use of a piece of land in several different ways; thus, in national forests the timber may be harvested and the land grazed or used for recreational purposes.

municipal use of water The various uses to which water is put in developed urban areas, including domestic use, industrial use, street sprinkling, fire protection, etc.

muskeg A swampy area in taiga or tundra regions.

musk ox A buffalo-like mammal originally found in the polar lands of Europe, Asia, and North America but now confined to limited areas in Canada and Greenland. It is killed by Eskimo hunters for meat and fur.

mutagens Any substance increasing the normal mutation rate of a species.

mutant An organism produced by mutation and thus having qualities differing from those of its ancestors.

mutation The process by which a new type of organism, differing in certain respects from its parents, is created. Mutation is due to genetic changes in the offspring.

mycelia The threadlike bodies of simple organisms, such as the common bread mold.

mycology The study of fungi.

mycorhiza (mycorrhiza) The morphological association, usually symbiotic, of fungi and roots of seed plants. The feeding roots are enshrouded and partially penetrated by fine filaments of fungi. Such roots commonly are more branched and lose their root hairs.

myxomatosis A highly infectious viral disease of rabbits.

nacreous cloud A luminous, iridescent cloud occurring at an elevation of about 75,000 feet and made visible by reflected and diffracted light approximately 25 minutes before sunrise or after sunset. It is also called a mother-of-pearl cloud.

nadir A point in the heavens directly opposite the zenith.

narwhal An arctic whale about 20 feet long, hunted by the Eskimos.

natality rate See: crude birth rate; fertility ratio

national forest A tract of land, valuable for its vegetative cover, that has been set aside and managed by the federal government.

national monument An area or site of scenic or historical value set aside by presidential action.

national park A unique and scenic area set aside by an Act of Congress and managed by the National Park Service.

natural area An area little affected by man.

natural biota The native plants and animals of an area.

natural bridge A rock arch created by erosion—generally by running water in an area of sedimentary rock.

natural bridge

natural flow The natural rate of water movement past a specified point on a natural stream. In a true natural flow there have been no effects resulting from stream diversion, storage, import, export, return flow, or changes in consumptive use caused by man. Natural flow rarely occurs in a developed country.

natural gas Combustible gas produced underground in much the same way that petroleum is produced. About 80 percent is methane and most is used for fuel.

natural hazard Any natural event that can result in death or endanger the works of man.

natural increase The increase in numbers in a population due to the excess of births over deaths.

natural levee See: levee

natural region An area that is relatively homogeneous in regard to one or more elements of the natural environment, i.e., vegetation, landforms, soil, climate.

natural resource An area, material, or organism useful to man. Today the prevalent view is that everything on earth represents a resource.

natural selection The theory that in each generation certain individuals possess traits that permit them to survive while others perish; eventually these traits are the ones that persist among the dominant members of an ecosystem.

natural vegetation The vegetation native to an area prior to modification by man.

nautical mile A unit of distance equal to approximately one minute of arc along a meridian (6,080 feet).

NAWAPA (North American Water and Power Alliance) An alliance proposed for the total development of the water resources of the North American continent.

Neanderthal man Primitive man occupying Europe and parts of Asia during the Upper Pleistocene.

neap tide A tide of small amplitude occurring when the gravitational pull of the sun opposes that of the moon.

Nearctic Realm Wallace's biotic region encompassing most of Mexico, the United States and Canada.

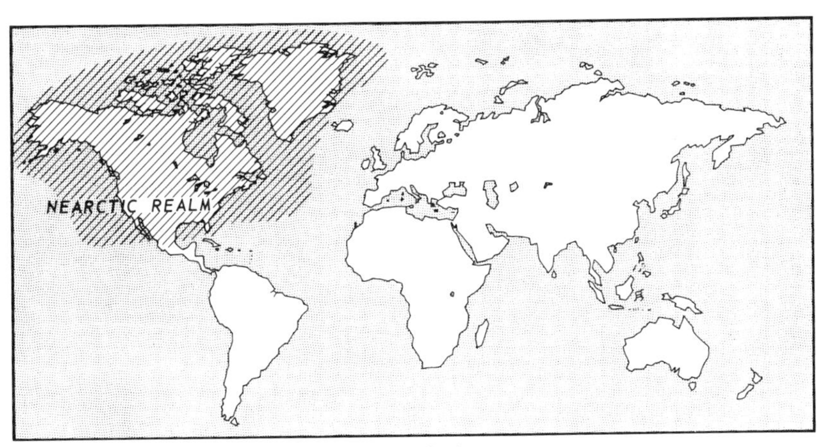

Nearctic Realm

nebula

nebula A luminous or dark mass composed chiefly of dust and gases.

nebular hypothesis The suggestion that our solar system was once a nebula, which then cooled, shrank, and rotated faster as the planets formed. The earth, originally molten, cooled; the atmosphere and hydrosphere were formed; and the earth's solid crust came into existence.

necrosis The localized or general death of plant tissue caused by low temperatures or fungi.

needle ice See: frazil

Negrito A member of a small-sized Negro people of Asia and Africa.

Negroid (adj.) Having the characteristics of the Negro race: dark skin, broad flat nose, everted lips, and wooly hair.

nematode An unsegmented worm. Many nematodes are agricultural pests.

neolithic (adj.) Pertaining to the last period of the Stone Age, characterized by the use of highly finished stone implements, the domestication of animals, and simple agriculture.

neon One of the rare gaseous constituents of the atmosphere. It is an inert gas comprising 0.0018 percent of the whole atmosphere by volume. Its atomic number is 10 and its weight 20.2. Its symbol is Ne.

Neotropical Realm Wallace's biotic region encompassing Central and South America.

nephology The study of clouds.

nephometer An instrument designed to measure the amount of cloudiness.

nephoscope An instrument for determining the direction of cloud movement.

net consumptive use See: consumptive use

net radiation The difference between incoming and outgoing radiation.

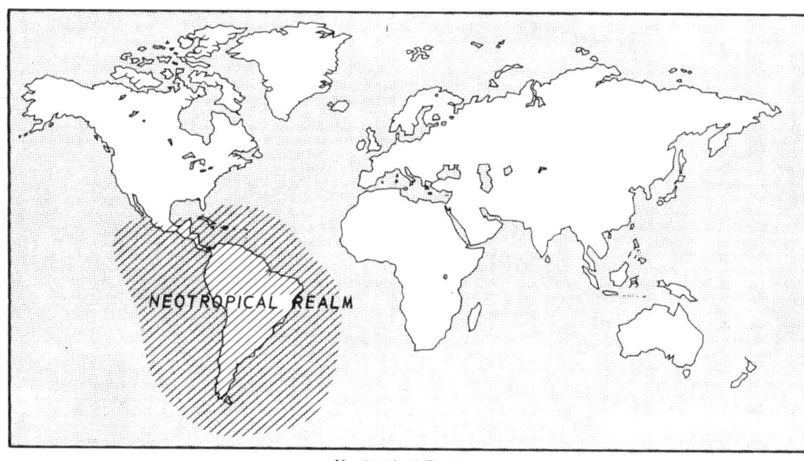

Neotropical Realm

net reservoir evaporation The difference between the total evaporation from a reservoir water surface and the evapotranspiration from the reservoir area under prereservoir conditions, with identical precipitation considered for both conditions.

neutralization Adjustment of hydrogen- or hydroxyl-ion concentration toward the point (pH 7.0) at which the number of hydroxyl ions is exactly equal to the number of hydrogen ions. If the number of hydrogen ions is the greater, as at pH values less than 7.0, it is necessary to add a source of hydroxyl ions (a base, e.g., sodium hydroxide). If the number of hydroxyl ions exceeds the number of hydrogen ions, a source of hydrogen ions (an acid) must be added to equalize the numbers of the two ions.

neutral soil A soil that is significantly neither acid nor alkaline. Strictly, a neutral soil has pH of 7.0; in practice, a neutral soil has a pH between 6.6 and 7.3.

neutrino An elementary particle with no mass or electrical charge, moving at the speed of light and capable of penetrating matter for vast distances without being stopped.

neutron One of the fundamental particles making up the nucleus of an atom. It has no electrical charge but has a slight magnetic field, and its mass is slightly greater than that of a proton.

nevada Spanish name for a cold wind descending from a mountain glacier or snowfield.

nevado A katabatic mountain wind of Ecuador.

névé See: firn

newton (N) A unit of force which when applied for one second will give to a one-kilogram mass a speed of one meter per second. A newton is equivalent to approximately two-tenths of a pound of force.

Newton's laws of motion These three laws are as follows: 1. Inertia. A body remains at rest or continues to move in a straight line at constant velocity unless acted upon by some outside force. 2. Acceleration. The acceleration of a body varies inversely as the mass and directly as the force applied to it. ($a = F/m$) 3. Reaction. For every action there is an equal and opposite reaction.

Newton's universal law of gravitation The law that states that any two bodies attract each other with a force that is directly proportional to the product of their masses and inversely proportional to the square of the distance between them.

nieve penitente A field of pinnacled snow, found on high mountains in many parts of the world.

nimbostratus A low, amorphous, and rainy cloud layer, dark gray in color.

nimbus A rain cloud.

nitrate 1. A fertilizer containing potassium nitrate or sodium nitrate. 2. Any compound containing nitrogen oxides.

nitrification The formation, by microorganisms, of nitrates and nitrites from ammonia or ammonium compounds, as in soils.

nitrifying bacteria Bacteria taking nitrogen from the atmosphere and the soil and converting it to a form useful to plants.

nitrogen One of the principal components of the earth's atmosphere. It accounts for 79 percent of the volume and mass of the air.

nitrogen cycle The movement of nitrogen from the atmosphere into organic life, then into the soil, and eventually back to the atmosphere again.

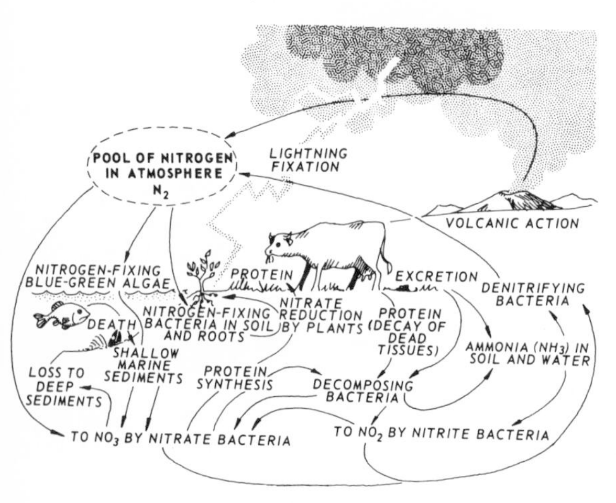

nitrogen cycle

nitrogen fixation Generally, the conversion of free nitrogen combined with other elements; specifically in soils, the assimilation of free nitrogen from the soil air by soil organisms and the formation of nitrogen compounds that eventually become available to plants. The nitrogen-fixing organisms associated with legumes are called symbiotic; those not definitely associated with the higher plants are nonsymbiotic.

nitrogen-fixing organisms See: nitrifying bacteria

nitrogen oxides Combinations of nitrogen and oxygen forming nitrites and nitrates, some of which are useful to man as fertilizers while others (the products of petroleum combustion in automobiles) are noxious gases detrimental to health.

nivation Erosion caused by the action of snow.

noctilucent cloud A luminous cloud about 82 kilometers above sea level. It resembles a cirrus cloud, is silvery or bluish-white, and is rendered luminous by the sun.

nocturnal Of or pertaining to the night.

nocturnal cooling The decrease in air temperature during the night or at any time when the outgoing radiation of heat from the earth exceeds the incoming solar radiation. The loss of radiant energy lowers the temperature of the earth's surface. The air temperature is thereafter reduced by conduction.

nocturnal radiation Synonymous with terrestrial radiation, which is the more accurate term. Probably the term nocturnal has been used because the phenomenon is more apparent and more easily measured at night when incoming solar radiation is absent.

noise pollution See: acoustic pollution

nomadism The frequent changing of habitat by primitive peoples.

nomograph A diagram for the graphical solution (with a straightedge) of problems that involve formulae in two or more variables.

nonconformity In geology, an unconformity in which older rocks either dip at a sharp angle or are of plutonic origin.

nonconsumptive use (water) The use of water in a way that does not reduce the supply. Examples include use for hunting, fishing, boating, water-skiing, and swimming.

nonmetallic minerals Minerals not containing metals. They may be inorganic, like coal or petroleum, or composed of such elements as carbon or nitrogen.

nonrenewable resource Resources that are consumed when used, such as fuels.

noosphere That portion of the earth upon which man has imposed his will so that the landscape completely reflects the hand of man.

normal year In hydrology, a year during which precipitation or streamflow approximates the average for a long period of record.

norte A norther, especially in the Gulf of Mexico or Central America.

North Frigid Zone See: Arctic Zone

northeaster 1. A wind or gale from the northeast. 2. A moderate to strong wind blowing from the northeast over the New England and middle Atlantic states. 3. A steady northeast gale, which sometimes blows for three days with violent squalls, off New Zealand and the coast of New South Wales.

northeast trades See: trade winds

norther In the southern part of the United States, the Gulf of Mexico, the western Caribbean, and Central America, a cold strong wind from a northerly quarter in winter. It is caused by the southward movement of a polar anticyclone in the wake of an intense cold front.

northern lights See: aurora

northwester (or nor'wester) 1. A wind or gale from the northwest. 2. A dry, warm, foehn wind, blowing east of the mountains of the South Island of New Zealand.

Notos Greek name for the south wind or its personification.

NRM wind scale A wind scale adapted by the U.S. Forest Service for use in the forested areas of the Northern Rocky Mountains (NRM).

nuclear energy The energy liberated by a nuclear reaction (fission or fusion) or by radioactive decay. See: fission; fusion

nuclear fission The breaking up of the nucleus of an atom into two or more new nuclei, accompanied by the release of much energy. Nuclear reactors and atomic bombs release energy through this process.

nuclear fusion The combination of two or more nuclei to form a single nucleus. When two nuclei are fused, enormous amounts of energy are released. The hydrogen bomb, the sun, and other stars release energy through this process.

nuclear power plant Any device, machine, or assembly that converts nuclear energy into some form of useful power, such as mechanical or electrical power. In a nuclear electric power plant, heat produced by a reactor is generally used to make steam to drive a turbine that in turn drives an electric generator.

nuclear reaction A reaction involving a change in an atomic nucleus, such as fission, fusion, neutron capture, or radioactive decay, as distinct from a chemical reaction, which is limited to changes in the electron structure surrounding the nucleus.

nuclear reactor A device in which a fission chain reaction can be initiated, maintained, and controlled. Its essential component is a core with fissionabl fuel. It usually has a moderator, a reflector, shielding, coolant, and control mechanisms. Sometimes called an atomic "furnace", it is the basic machine of nuclear energy. See: fission

nucleation Any process by which the phase change of a substance to a more condensed state is initiated on particular nuclei within the less condensed state.

nucleic acid Any of the group of complex acids occurring in all living cells, in particular as a component of cell-nucleus proteins. See: DNA, RNA

nucleon A proton or neutron.

nucleus 1. In meteorology, a particle of any nature whatsoever on which condensation of atmospheric moisture occurs. Some of these particles consist of hygroscopic gases, such as oxides of sulphur and nitrogen, and some of inorganic and organic dust particles. 2. In physics, that part of an atom that is supposed to be the seat of its effective mass and to control the motions of its orbital electrons. 3. In biology, a mass of protoplasm encased in a membrane and found in most living cells, and forming an essential element is their growth, function, and genetic transmission.

nuclide An atomic species in which all atoms have the same atomic number and mass number.

nullah A normally dry watercourse in India.

numerical model A concept or process or groups of concepts and processes expressed by mathematical formulae.

nunatak An isolated mountain peak projecting above a snow or ice field.

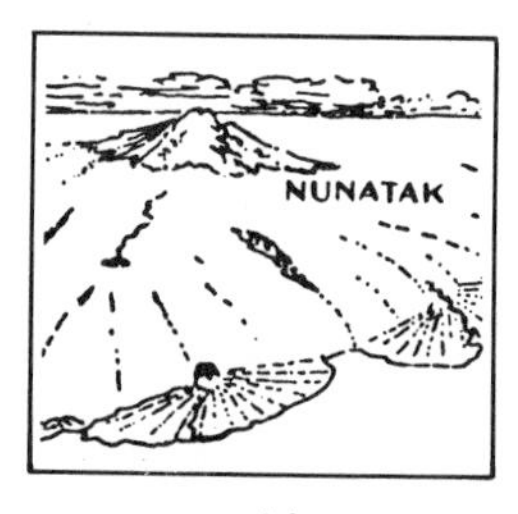

nunatak

nutation The periodic oscillation of the earth's axis.

nutrient Food providing nourishment for growth and development.

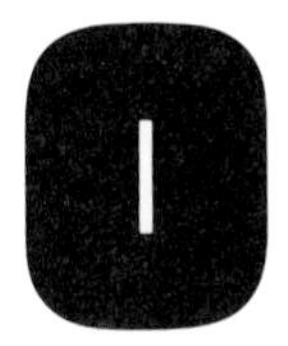

oak A sturdy long-lived tree, belonging to two genera found in nearly all parts of the northern hemisphere and in limited areas of South America. It is valued as a shade tree and for its acorns and wood.

oasis Any productive, watered area in the desert.

observation well A nonpumping well used for observing the elevation of the water table or the piezometric surface.

obsidian A hard, dark gray or black rock formed when lava cools so quickly that crystals do not have time to form. When broken, sharp edges are produced. Obsidian was used by primitive people for weapons and tools.

occluded cyclone See: occlusion

occluded front The front formed when the cold front overtakes the warm front in an extratropical cyclone.

occlusion The process in which the warm sector of an extratropical cyclone is gradually restricted in size and ultimately lifted entirely from the earth's surface as the cold front overtakes the warm front; the warm sector then exists as a trough of warm air aloft until the cyclone dissolves.

ocean Any of the large bodies of salt water surrounding all major land masses.

ocean current The movement of surface waters of the ocean produced mainly by the planetary winds. In general, warm currents flow poleward and cold currents equatorward.

oceanity The degree to which a climate is affected by marine influences.

oceanography The study of oceans and marine life.

octahedron An eight-sided polygon.

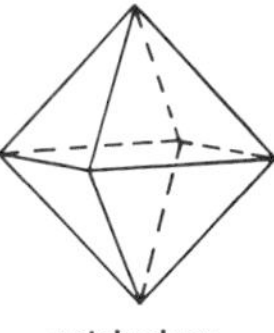

octahedron

octane rating In engines, a method of indicating a fuel's resistance to knock. Premium fuels have an octane rating of approximately 97 and should be used in high-compression engines. Regular grade fuels have an octane rating of approximately 94 and are suitable for automotive engines with compression ratios of about 9:1.

octoroon The offspring of a quadroon and a white person.

oe A whirlwind off the Faeroe Islands.

offshore bar See: barrier reef

off shore drilling The practice of drilling for oil on the continental platforms. It is common in the Gulf of Mexico and along the coast of Southern California.

oil field An area in which oil occurs in pools, in which it has been trapped by a geologic structure.

oil shale Rock impregnated with petroleum. Large quantities are located in Colorado. It is regarded as a potential source of petroleum products.

off shore drilling

Oleoresin A natural product containing oil and resin.

Oligocene See: Appendix I

oligotrophic (adj.) Refers to lakes with a low accumulation of dissolved nutrient salts and usually with considerable dissolved oxygen in the bottom

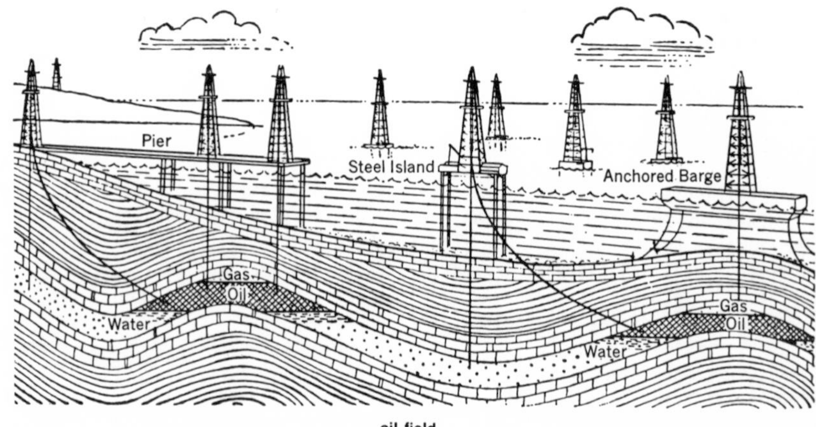

oil field

waters (owing to a low organic content).

omnivore An organism that feeds on both animal and plant substances.

ontogeny All the phases in the development of an individual organism.

ooze A mud deposit that covers a deep region of the ocean floor.

open-pit mining The practice of obtaining ore by removing the layers of rock covering it and then shoveling it out of the ground.

open-pit mining

open space Within a metropolitan area any land on which buildings are not erected.

optical meteor Any phenomenon of the atmosphere explained by the laws of light, such as a mirage.

optimum growing conditions For any organism, the ideal light, temperature, and moisture conditions for growth.

orbit The path of a heavenly body around a given point in space. The orbit of the earth is around the sun.

orchard heater Any device for heating an orchard to protect the plants from frost.

order The largest category in soil classification. The three orders are zonal soils, intrazonal soils, and azonal soils.

ordinate The y-coordinate, representing the distance of a point from the x-axis, measured parallel to the y-axis. See: Cartesian coordinates

ordnance survey An accurate and detailed geographical survey (British).

Ordovician Period See: Appendix I

ore A naturally occurring group of minerals from which metal can be extracted.

organ In an organism, a group of cells or tissues that performs a specific function.

organic acid A compound that contains primarily carbon, hydrogen, and oxygen and tastes sour. Examples are acetic acid (CH_3COOH), in vinegar and lactic acid ($CH_3CHOHCOOH$), in sour milk.

organic compound A compound containing carbon.

organism A living thing organized in a definite pattern with a definite function and consisting of an individual

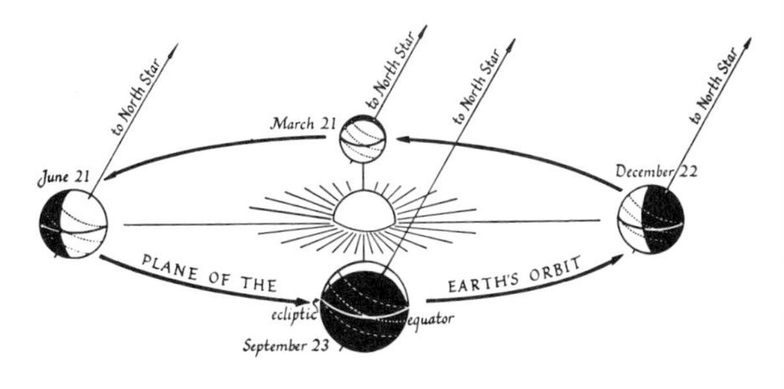

orbit

cell or complexity of cells. It undergoes dynamic changes and contains a colloidal medium—protoplasm, which is the physical and chemical basis of all life.

organic soil A general term applied to a soil or to a soil horizon that consists primarily of organic matter, such as peat soils, muck soils, and peaty soil layers.

organophosphate An insecticide that inactivates the enzyme responsible for breaking down acetylcholine, a substance transmitting nerve impulses. New versions of the nerve gases developed by the Germans in World War II are sold under different names: parathion, malathion, azodrin, TEPP, phosdrin, etc.

Oriental Realm Wallace's biotic province encompassing India, China, and Southeast Asia, including many of the islands of Indonesia.

origin The point of intersection of coordinate axes.

orogenesis The process by which mountains are built.

orogeny A mountain-building period.

orographic lifting The lifting of an air mass that occurs as it flows across an orographic, or mountain barrier.

orthographic lifting

orographic precipitation Precipitation that results from the lifting of moist air over a topographic barrier such as a mountain range. It may occur some distance upwind and a short distance downwind as well as on the barrier feature.

orographic rainfall Rainfall resulting when moist air is forced to rise by mountain ranges or other land formations lying athwart the path of the wind.

orography The branch of physical geography dealing with mountains.

orthogenesis The theory that evolution is caused and controlled by laws of growth that function regardless of environmental conditions.

orthographic projection A map projection in which the globe is viewed as if from an infinite distance, so that only one hemisphere is depicted.

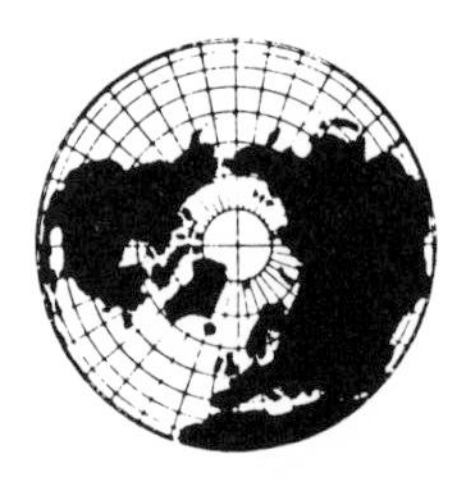

orthographic projection

orthomorphic map or projection See: conformal map

orthorhombic crystal system A system of crystals having three axes of unequal length, each of which is perpendicular to the plane of the other two.

oscillation Any periodic recurrence of a phenomenon.

oscillator An instrument that produces vibrations.

osmosis The process of liquid diffusion through a membrane. Osmosis tends to equalize the liquid concentrations on either side of the membrane.

outcrop An exposure of bedrock through the covering soil or alluvium.

outlier A mass of younger rocks surrounded by older rocks. See: inlier

outwash plain The plain formed beyond the leading edge of a glacier by the meltwater carrying glacial drift away from the glacier.

ovary A female reproductive organ that produces eggs in animals and megaspores in plants.

overburden Regolith or loose rock material overlying a mineral deposit near the earth's surface.

overcast (adj.) The state of the sky when more than nine-tenths of the visible canopy is covered with clouds.

overdraft That quantity of water pumped in excess of the safe yield.

overfishing The depletion of a fishery resource to such an extent that a stable population cannot be maintained through normal reproduction rates.

overfold A geological fold in which the compressional forces overturn the folded rock layers.

overgrazing Excessive grazing, causing the deterioration of native grass cover, which is replaced by invaders of low nutrient value.

overland flow The flow of rainwater or melted snow over the land surface toward stream channels. After it enters a stream, it becomes runoff.

overseeding Cloud seeding in which an excess of nucleating material is released. Occasionally this is accidental, but purposeful overseeding is done to minimize hail damage by reducing hailstone size.

overthrust fault A fault in which the fault plane is low-angled or nearly horizontal so that one set of strata slides across another.

ovule 1. A rudimentary seed occurring in the ovary. 2. The body that contains the embryo sac which develops into the seed after it is fertilized.

ovum An egg cell.

ox-bow lake or cutoff A lake formed when a meandering river cuts a new course through the neck of a meander, isolating part of the old stream course and creating a lake.

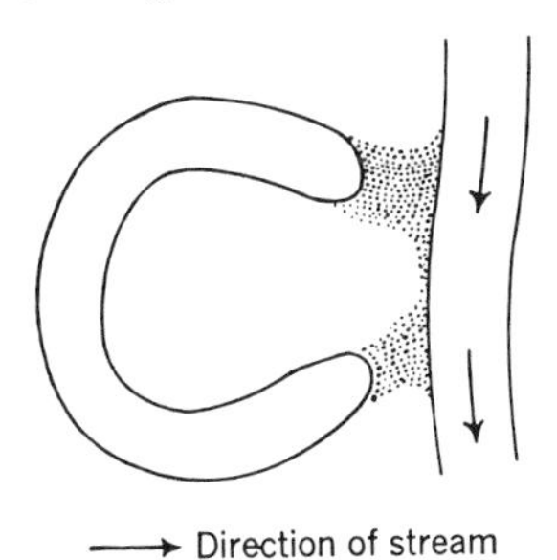

⟶ Direction of stream

ox-bow lake

oxidation 1. The process of combining with oxygen. Rusting is a familiar example of this process. 2. The removal of electrons from an ion or atom.

oxide A compound of any element with oxygen alone.

oxides of nitrogen A molecular structure consisting of atoms of nitrogen and oxygen. This pollutant is a principal component of photochemical smog and accounts for the characteristic reddish brown color. Oxides of nitrogen result from any type of burning but are produced in high concentrations under conditions of high combustion temperatures and pressure.

oxisol A soil of the humid tropical forest on an old land surface. See: lateritic

oxygen A chemical element. Its symbol is O, atomic number 8, and atomic weight 16.000. It is an odorless and colorless gas, fifth of the earth's atmosphere in uncombined molecular form. Chemically combined with other elements, it makes up about one-half of the earth's crust.

oxygen cycle The cycle in which oxygen is taken in by animals and given off in the form of carbon dioxide, which is then taken in by plants and given off in the form of oxygen.

oxygen demand See: biochemical oxygen demand (B.O.D.)

oxygen saturation capacity The maximum quantity of dissolved oxygen that a liquid exposed to the atmosphere can contain at a given temperature and pressure.

oyster One of a family of bivalve mollusks. Oysters have been cultivated for several thousand years for food and pearls. They are found widely in protected areas in warm and temperate waters.

ozone A form of molecular oxygen, each molecule consisting of three atoms; it is colorless but has a characteristic odor. Ozone is produced in the high atmosphere principally by the action on oxygen of the ultraviolet radiation from the sun.

ozone layer A layer in the atmosphere that absorbs ultraviolet radiation from the sun. It lies 25 to 35 kilometers above the surface of the earth.

pack A body of drift ice consisting of several pieces, the complete extent of which cannot be seen; the term open pack is used when the pieces do not touch, and close pack is used when they are pressed together.

pack ice A group of ice blocks larger than ice floes, formed when an ice-field is broken by wind and waves.

pahoehoe The Hawaiian term for basaltic lava flows which have billowy or ropy surfaces.

painter A dirty fog frequently experienced on the coast of Peru. The brownish deposit from it is sometimes called Peruvian paint.

Palearctic Realm Wallace's biotic region including most of Europe and Asia.

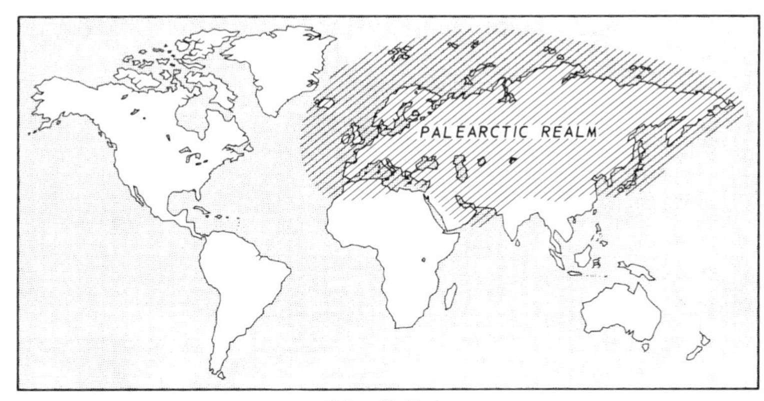

Palearctic Realm

paleobotony See: paleontology

Paleocene See: Appendix I

palaeoclimatology The science that treats of the climates of the world throughout the geological ages. Its data are the distribution of glacier deposits, nature of plant and animal fossils, topography and geography of former periods, and character of sedimentary rocks.

paleogeography The study of water and land distribution during geologic time.

paleolithic (adj.) Pertaining to early human culture, characterized by rough or chipped stone implements.

paleontology The science dealing with the fossil remains of animals and plants. In strict usage, paleontology is the study of animal fossils, while paleobotany is the study of plant fossils.

Paleozoic See: Appendix I

palm A tropical evergreen, often called the most useful of all trees because it provides man with food, drink, fiber, oil, fuel, and timber.

paloverde A thorny shrub of the American desert. It is covered with yellow flowers in the spring.

palynology The study of pollen and spores.

pampas The mid-latitude grasslands of Argentina, Uruguay, and southern Brazil.

pampero A wind of gale force blowing from the southwest across the pampas of Argentina and Uruguay. It is comparable with the norther of the United States.

pan A layer or soil horizon within a soil that is firmly compacted or is very rich in clay. Examples include hardpans, fragipans, claypans, and traffic pans.

Papagayo A violent north wind, similar to the Tehuantepecer, occurring in the Gulf of Papagayo.

parallax The apparent change in position of an object when seen from two different points.

parallel A circle extending around the globe in an east-west direction and used in the measurement of latitude.

parallel drainage A stream drainage pattern in which the tributaries or adjacent streams flow nearly parallel to one another over a fairly extensive area. It may be produced by parallel folded or faulted rock strata or by any other set of parallel topographic features.

parallelism The relationship of the earth's axis at any two given points in its orbit.

parameter A quantity to which an investigator may assign arbitrary values. Also, a variable with a fixed value.

paramo An alpine region of the Andes, covered for the most part by lichens and mosses.

parasite An organism deriving sustenance from another.

parent A radionuclide that upon radioactive decay or disintegration yields a specific nuclide (the daughter), either directly or as a later member of a radioactive series. See: daughter; radioactive series

parent material Unconsolidated rock material from which soils are formed.

parhelion One of two mock suns or sundogs. Parhelia are two bright colored spots appearing as images of the sun at 22° to the right and left of it on the parhelic circle at the same altitude as the sun.

parsec A unit of measurement used in astronomy equivalent to 3.26 light years or 1.92 x 10^{13} miles.

partial potential temperature The temperature to which the dry air component would come if reduced from its partial pressure to 1,000 millibars.

partial pressure In any mixture of gases, the pressure which each gas exerts independent of the other gases.

Pascal's law of fluid pressure The law that states that pressure exerted at any point upon a confined fluid is transmitted undiminished in all directions.

pass A low gap through a mountain barrier.

pastoralism The breeding and rearing of herbivorous animals.

pathogen Any organism or substance causing disease.

pathology The branch of medicine concerned with diseases, their causes, and their effects on the human body.

peak A mountain or the top of a mountain with a small summit area.

peak discharge A period of maximum streamflow when the base flow is augmented and surface runoff is greatest.

pearl lightning See: beaded lightning

pea soup A popular name for a very dense fog.

peat A fibrous material; the first stage in the transformation of vegetable matter into coal.

peat bog A marshy area in which vegetable material is decomposing and becoming peat.

pectin A complex carbohydrate, present in the cell walls of fruits and other plant foods, which turns into gelatin when heated with sugar.

ped An individual natural soil aggregate such as a crumb, prism, or block, in contrast to a clod, which is a mass of soil produced by digging or another disturbance.

pedalfer A soil of humid regions, generally rich in iron and clay.

pediment The gently sloping, rock-cored area at the base of a mountain.

pedocal Soil of arid regions. It is generally rich in calcium and may have a calcium layer.

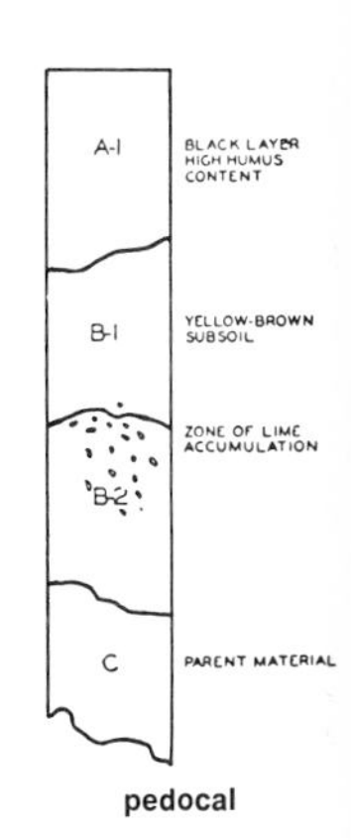

pedocal

pedogenesis The soil-forming process.

pedology The study of soils.

peeler 1. One who removes bark from timber cut in the spring months when bark "slips." 2. A log used in the manufacture of rotary-cut veneer.

pegmatite A coarsely crystalline granite or other rock occurring in veins or dikes.

Peking man (sinanthropus) A very early man living in China from 350,000 to 400,000 years ago. Fossils of his teeth, bones, and skull were found near Peking between 1927 and 1937.

pelagic (adj.) Referring to that part of the sea beyond the continental slope but not including oceanic deeps.

pelean cloud A group of hot, gas-charged lava fragments ejected horizontally from a volcano when upward movement has been obstructed.

pellagra A dietary disease characterized by skin lesions, gastrointestinal disturbance, and nervous symptoms.

peneplane or peneplain An area in which erosion has reduced the landscape to almost a plain.

penicillin A relatively nontoxic acid substance extracted from a green mold. It is a strong antibacterial agent.

peninsula An area of land surrounded by water except on one side.

Pennsylvanian Period See: Appendix I

pentad 1. In meteorology, a period of five days. 2. Any group of five.

penumbra The partly shaded region surrounding the umbra in an eclipse.

perched water table The water table of a relatively small ground-water body supported above the general ground-water body.

percolating waters Those waters that pass through the ground beneath the earth's surface without a definite channel and not shown to be supplied by a definite flowing stream.

percolation The movement, under hydrostatic pressure, of water through the interstices of rock or soil.

percolation rate The rate, usually expressed as a velocity, at which water moves through saturated granular material.

perennial stream A stream that flows at all times.

perennial yield The amount of usable water of a ground-water reservoir that can be withdrawn and consumed economically each year for an indefinite period of time.

pericarp The outer protective wall that covers the kernels of cereal grains.

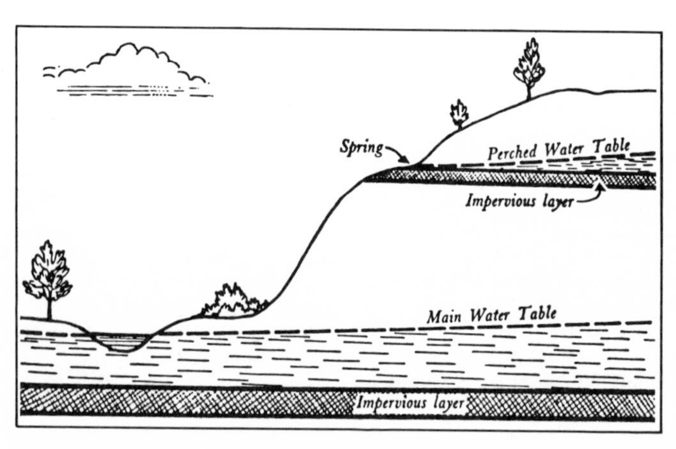

perched water table

pericyclone The ring of increasing pressure around a cyclone.

perigee The point in the orbit of a satellite or of a planet when it is nearest the earth. See: apogee

perihelion That point on the earth's orbit that is nearest the sun. It is now reached by the earth on about January 1, but the date varies irregularly from year to year and also has a slow secular change. See: aphelion

period The time taken by some stage of development or the regular recurrence of some event.

periodicity Quality or state of being periodical, or regularly recurrent; in plant physiology, the tendency of a plant to exhibit rhythmical changes in vital functions.

periodic law Elements when arranged in order of their atomic weights or atomic numbers show regular repetition in most of their properties.

permafrost Permanently frozen ground.

permanent pasture Pasture that occupies the soil for a long time, in contrast to rotation pasture, which occupies the soil for only a year or two in a rotation cycle with other crops. As used in the humid parts of the United States, the term permanent pasture is equivalent to the European long ley.

permeable (adj.) Capable of being penetrated through the pores, etc., as, for example, sandstone.

permeability In soil study, the quality of a soil horizon that enables water or air to move through it. It can be measured quantitatively in terms of rate of flow of water through a unit cross section in unit time under specified temperature and hydraulic conditions. Values for saturated soils usually are called hydraulic conductivity values. The permeability of a soil may be limited by the presence of one nearly impermeable horizon even though the others are permeable.

peroxyacyl nitrate (PAN) A chemical compound that is a component of photochemical smog and the primary plant toxicant of "smog" type injury; field levels of 0.01 to 0.05 ppm will injure sensitive plants. Causes eye irritations and other health effects in man.

perpetual resource See: renewable resource

personal equation A systematic error in observation due to the characteristics of the observer.

perturbation A change in the regular motions of a fluid or a body, produced by some change in the force acting upon it.

pervious (adj.) Permeable.

pesticide Any material used to kill rats, mice, bacteria, fungi, or other pests of man.

petiole A leafstalk; the slender stalk by which the blade of a leaf is attached to the stem.

petrified (adj.) Converted to rock. Fossils may be converted to stone as their tissues are gradually replaced by corresponding amounts of mineral material deposited from percolating solutions.

petrochemical (adj.) Pertaining to chemicals derived from a petroleum base.

petroglyph A carving or inscription on a rock: particularly one made in prehistoric times.

petrology The study of the composition, structure, and history of the rocks forming the earth's crust.

pH An abbreviation for potential hydrogen. It is a symbol, suggested by the Danish chemist Sorensen, for use with a numeral from 1 to 14 to denote the negative logarithm of the concentration of the hydrogen ion in gram atoms per liter, which concentration is an expression of true acidity or alkalinity.

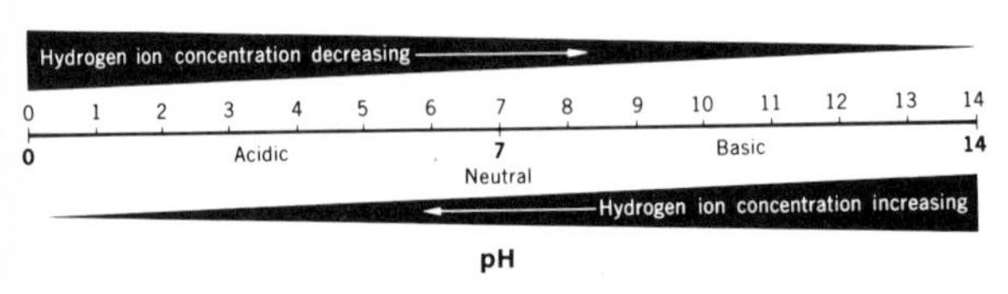

pH

phase In wave motion, an angular measure of the stage reached by the vibration in its progress through its cycle. 2. In soil study the subdivision of a soil type or other classificational soil unit having variations in characteristics not significant to the classification of the soil in its natural landscape but significant to the use and management of the soil. Examples of the variations recognized by phases of soil types include differences in slope, stoniness, and thickness because of accelerated erosion. 3. One of the stages or changes of appearance through which the moon passes during its revolution around the earth, known as new moon, full moon, etc.

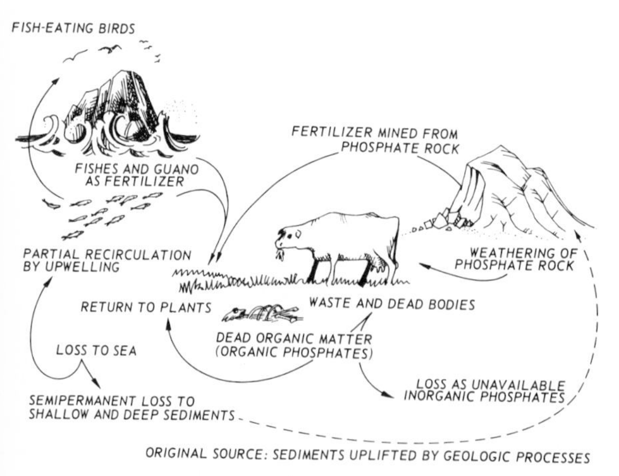

phase

phenol A crystalline compound produced by the distillation of many organic substances, especially coal tar. Commonly called carbonic acid and used as an antiseptic.

phenology The science that treats of the interrelations of climatological and meteorological conditions with biological phenomena, both plant and animal.

phenotype The physical characteristics of an individual resulting from the interaction of heredity and environment.

phloem In seed plants and ferns, tubular tissue that transports organic nutrients.

phosphate A phosphorous compound. They are used as fertilizers and as medicines.

phosphorus An element. See: Appendix II

phosphorous cycle The movement of phosphorus from rock minerals to the sea and back to rock minerals again.

photochemical (adj.) Pertaining to the chemical effects of light.

photochemical smog The predominant type of air pollution in Southern California. It is the result of sunlight reacting with hydrocarbons and oxides of nitrogen in the atmosphere. Ozone also is a component of photochemical smog.

photolysis 1. The first step in photosynthesis. 2. The breakdown of a substance in the presence of light.

photometer An instrument for measuring the intensity of luminosity of any source of light.

photometry The science of measuring light intensity.

photon The carrier of a quantum of electromagnetic energy. Photons have an effective momentum but no mass or electrical charge. See: radiation; quantum

photoperiodism The degree of response of an organism to the amounts of darkness and light that it receives.

photosynthesis The process by which sugar is manufactured in plant cells; it requires carbon dioxide, water, light, and chlorophyll.

phototropism Plant or animal movement in response to light.

phreatophyte A plant that habitually obtains its water supply from the zone of saturation, either directly or through the capillary fringe. See: hydrophyte; mesophyte; xerophyte

phylum A division of the Linnaean plant and animal classification system that includes one or more classes.

physical climate Solar climate, as modified by the oceans, mountains, and other surface features of the earth. These surface features react with the atmosphere and interfere with the uniform arrangement of the climate zones that would exist in a purely solar climate. The chief causes of this interference with the regular solar climatic zones are (a) the irregular distribution of land and water upon the earth's surface, (b) the aerial and oceanic currents, which are thereby compelled to follow certain definite paths, and (c) the differences in altitude of the land above sea level.

physical climatology The science that seeks the causes and effects of climate in climatic data, the facts of physical geography, and the laws of physics.

physical meteorology The branch of meteorology that seeks to explain all atmospheric phenomena by the accepted principles of physics.

physiognomy The external aspect or outward appearance of anything, particularly as regards form.

physiography See: geomorphology

physiology The study of the functions of cells, tissues, organs, and systems in living organisms.

phytogeography The study of the interrelation of plants with their environment.

phytoplankton Plant plankton.

pibal Pilot balloon observation.

piedmont The area at the foot of a mountain. It is generally an erosional form representing the roots of an older larger mountain mass.

piedmont glacier An extensive ice sheet on the lower slopes of a mountain, formed by the coalescence of valley glaciers.

piezometric surface An imaginary surface that everywhere coincides with the static level of water in an aquifier. It is the surface to which the water in such an aquifer or ground-water basin would rise if afforded an opportunity to do so.

pilot balloon A small rubber balloon that, when inflated with hydrogen and released and observed with a theodolite during its ascent, furnishes data from which may be calculated the direction and speed of the wind at all levels from the surface to the end of its flight.

pilot chart One of the charts published for each of the oceans by the Hydrographic Office of the Navy Department as a service for navigators.

pine A cone-bearing evergreen having bundles of two, three, or five needles. Pines grow throughout the northern hemisphere and have been planted in South Africa and Australia.

piñon pine A scrub pine growing on mountain slopes on the margins of the American desert.

pioneer community The first plant community to move into an area from which the original vegetation has been removed or in which the cover has been disturbed.

pirep An airplane pilot report presenting information on atmospheric conditions.

pistil The female reproductive organ of a flower.

pitch 1. A black sticky liquid derived from asphalt. 2. Angle of slope.

pitchblende A brown or black mineral, chiefly uranium oxide. It is a source of radium and uranium.

pith Soft tissue cells making up the central core of many plant stems and roots.

Pilot tube A tube with an end open square to a fluid stream. It is exposed with the open end pointing upstream to detect an impact pressure.

placer An accumulation of sand and gravel containing mineral deposits.

plague An epidemic disease with high mortality.

plain An extensive area of level or gently undulating land.

Planck's law An equation expressing the variation of the intensity of black-body radiation with wave length at a given temperature. See: radiation

plane of the ecliptic See: ecliptic

planet One of the spherical bodies revolving around a sun. There are several hundred, with nine much larger than the rest.

plane table A simple instrument used in surveying. It consists of a drawing board mounted on a stand and a ruler that can be accurately pointed at the object being observed.

planetary winds The general flow of air in the lower atmosphere consisting of the belt of calms near the equator, the trade winds, the westerlies, and the polar easterlies.

planetesimal hypothesis The theory that the planets of our solar system were formed by coalescence of large numbers of minute planets, or planetesimals, resulting from collisions and gravitational attraction.

planetoid A small piece of matter moving freely in space. See: asteroid

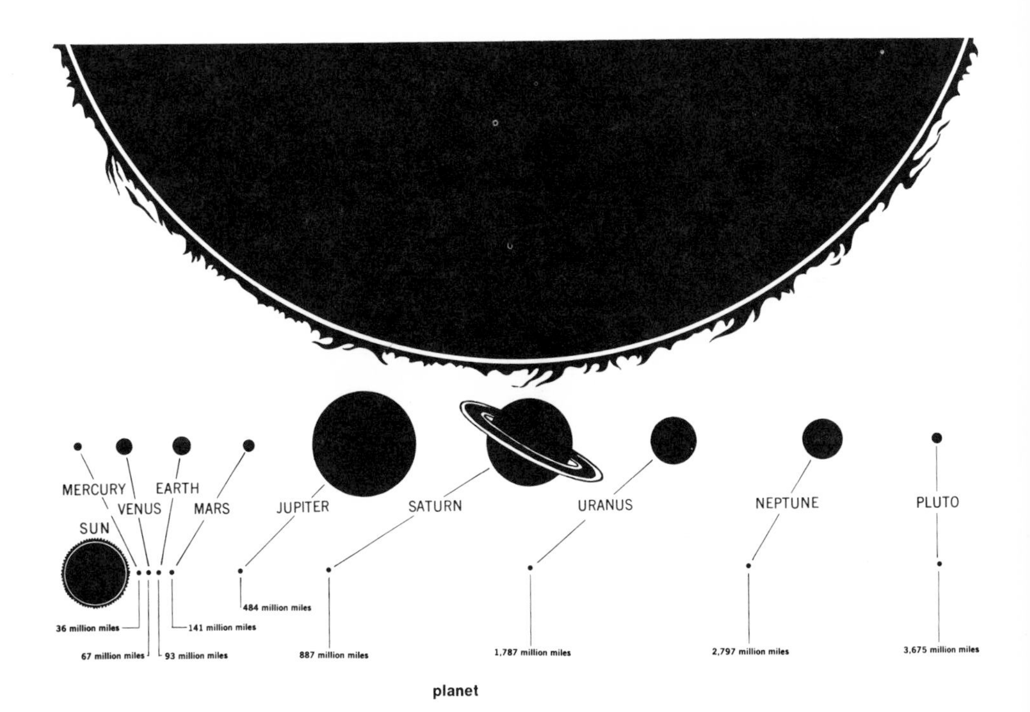

planet

planimeter An instrument for measuring areas on maps.

plankton See: zooplankton

planning, programming, budgeting system (PPBS) A systematic approach to the allocation of limited resources for the accomplishment of priority objectives.

planosol An intrazol group of soils with eluviated surface horizons underlain by claypans or fragipans, developed on nearly flat or gently sloping uplands in humid or sub-humid climates.

plantation An estate devoted to the production of one or more cash crops.

plant community See: community

plant succession See: succession

plasma 1. The fluid carrying blood cells. 2. An electrically neutral gaseous mixture of positive and negative ions.

plateau An extensive, level region elevated above the surrounding land.

platy soil structure Soil aggregates with thin vertical axes and long horizontal axes.

playa An area in an arid or semi-arid region that becomes temporarily filled with water after a rain or as a consequence of runoff.

Pleiocene See: Appendix I

Pleistocene See: Appendix I

Plowshare The Atomic Energy Commission program of research and development on peaceful uses of nuclear explosives. The possible uses include large-scale excavation, such as for canals and harbors, crushing ore bodies, and producing heavy transuranic isotopes. The term is based on a Biblical reference: Isaiah 2:4.

plucking A process by which a glacier may remove large pieces of projecting rock. Parts of the bedrock may be frozen into the glacier ice and broken away along joint planes as the glacier moves. Also called quarrying.

plume The air space containing smoke and other emissions from any point source of pollution.

plum rains Summer rains in Japan and eastern China. See: bai-u

plutonic rocks Igneous rocks that have solidified deep in the earth, where they have cooled slowly and crystallized; e.g., granite.

plutonium A heavy, radioactive, manmade, metallic element with atomic number 94. Its most important isotope is fissionable plutonium-239, produced by neutron irradiation of uranium-238. It is used for reactor fuel and in weapons.

pluvial (adj.) Relating to rain; but generally, in climatology, with reference to former periods of abundant rains, such as the pluvial periods of any region.

podzol A zonal group of soils having surface organic mats and thin, organic-mineral horizons above gray leached horizons that rest upon illuvial dark-brown horizons developed under coniferous or mixed forests or under heath vegetation in a cool-temperate, moist climate.

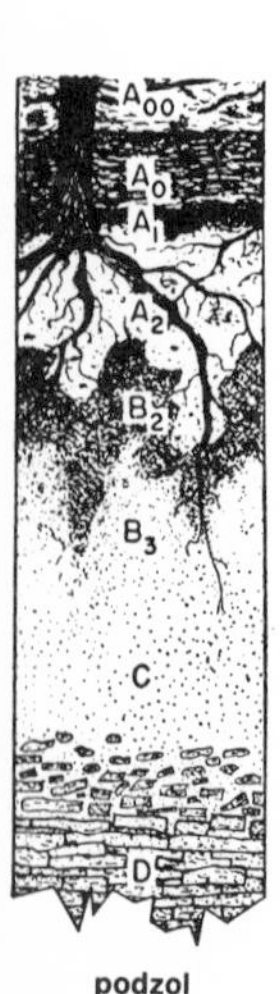

podzol

podzolization The process by which soils are depleted of bases, become more acid, and have developed leached surface layers from which clay has been removed.

pogonip An American Indian word applied to a frozen fog of fine ice needles occurring in the mountain valleys of the western United States.

point 1. In Australia, a unit used in measuring rainfall; equal to one one-hundredth of an inch. In the published tables the rainfall there is always given in points, not in inches and hundredths. 2. In France, a unit of length, equal to 1/12 ligne (0.0074 inch or 0.19 mm). 3. In England, a unit of length, equal to 1/6 line (0.0139 inch or 0.353 mm). 4. In nautical parlance, a measurement of direction, equal to 11¼ degrees, there being 32 points of division on the compass card. 5. A position or time of occurrence, as in boiling point and freezing point.

point precipitation Precipitation at a particular site, in contrast to the mean precipitation over an area.

Poisson's equation The equation expressing the relation between the pressure and the temperature of a perfect gas in an adiabatic process.

polar air Cold, dry air formed in the subpolar anticyclones, and divided into two types of air masses, polar continental (cP) and polar maritime (mP). See: polar continental air; polar maritime air

polar circle See: Arctic Circle; Antarctic Circle

polar continental air An air mass, designated cP in the Bergeron classification, formed in the northern regions of Europe, Asia, and North America, and characterized by: (a) low surface temperature; (b) stability in the lower layers—very often a temperature inversion; (c) very low moisture content, the value of the mixing ratio being seldom more than one gram per kilogram; and (d) shallow vertical extent, often less than 3 kilograms.

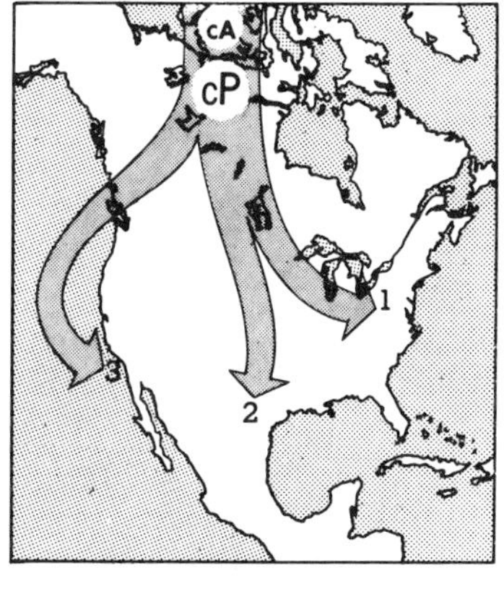

polar continental air

polar coordinates 1. In a plane, a system of curvilinear coordinates in which a point is located by its distance from the origin (or pole) and by the angle that a line joining the point and the origin makes with a fixed reference line known as the polar axis. 2. In three-dimensional space, polar coordinates are the same as spherical coordinates.

polar front The frontal zone between air masses of polar origin and those of tropical origin, located in the mean at a latitude of about 60°.

polaris The North Star, to which the axis of the earth points.

polarization A state of radiant energy in which the transverse electromagnetic vibrations take place in a regular manner: e.g., all in one plane, or in an ellipse or some other definite curve.

polar maritime air An air mass, mP in the Bergeron classification, that

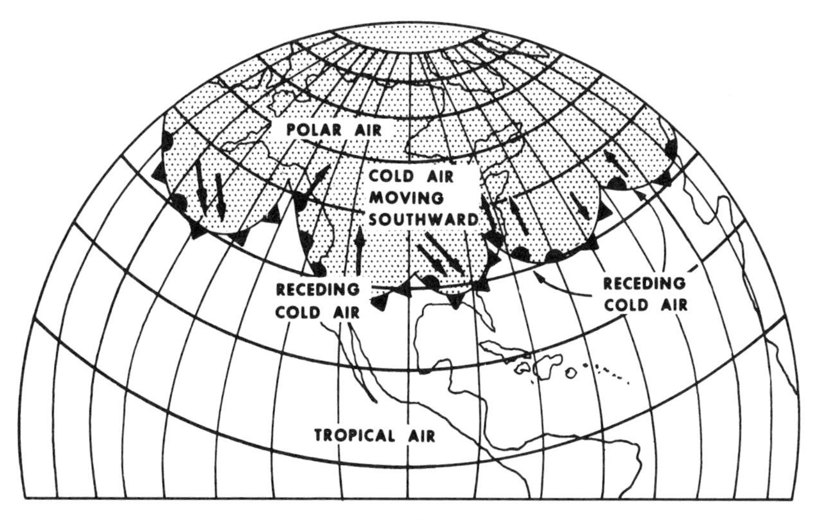

polar front

originates over the polar seas and has initial properties similar to those of polar continental air. During its passage equatorward over progressively warmer water, however, prolonged heating and moistening change the air mass from a cold, dry, extremely stable condition to one of marked conditional instability with a comparatively high moisture content in the lower strata.

polar projection Any map projection centered on the north or south pole.

polar projection

polar zones The zones lying between the Arctic and Antarctic circles and the poles.

polder Land reclaimed from the sea or a lake.

poles The points marking the ends of the earth's axis.

pole star See: Polaris

polje A large closed hollow in a karst region.

pollen The fertilizing dustlike powder produced by stamens; functionally the same as the male sperm in animal reproduction.

pollination The transfer of pollen from anther to pistil by wind, water, insects, or other means.

pollution The alteration of the physical, chemical, or biological properties of water or air, or a discharge of any substance into water or air that adversely affects the users.

polyconic projection A modified conic projection in which distances are true only along the central meridian and along each parallel. It is used in the United States for topographic maps.

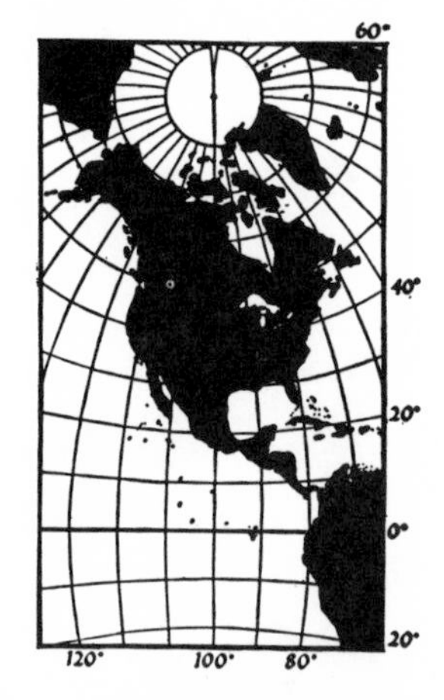

polyconic projection

polygamy The practice of having a number of wives.

polymer A chemical compound formed by combining two or more similar smaller molecules. The smaller molecules constitute the monomer. Polymers are called dimers, trimers, tetramers, and so forth, depending on how many molecules combine to form the new substance. The word polymer has come to be used chiefly to mean polymers of high molecular weight.

polymorphism The property of having or passing through many forms.

polynutrient fertilizer A fertilizer containing more than one major plant nutrient.

polyp A member of a group of small water animals, somewhat resembling a flower and usually having a hollow cylindrical body with a mouth at one end and tentacles at the other. A coral is a typical polyp.

pome Any fruit in which the seeds are enclosed in a core; e.g., an apple or pear.

pomology The science of growing trees for fruit, nuts, etc.

ponderosa pine A large, slow-growing, long-lived pine growing on the margins of dry environments in western North America. It is known also as yellow pine and is a valuable lumber tree.

ponor An open vertical hole in the ground created by solution in a karst landscape.

poorga See: purga

poplar (tulip tree) A broadleafed tree which is a member of the magnolia family. It is an ornamental and shade tree also used for lumber.

population The living organisms that inhabit a given area.

population density The denseness or sparseness of the population of a region. Population density is measured in terms of the average number of people living on each square mile of land.

population dynamics Population changes and the factors producing them.

population explosion Any rapid increase in the number of organisms in a given area.

population pyramid See: age pyramid

pore space The fraction of the bulk volume or total space within soils that is not occupied by solid particles.

pororoca The tidal bore of the Amazon River.

porosity In soil, the degree to which the soil mass is permeated with pores or cavities. Porosity can be generally expressed as a percentage of the whole volume of a soil horizon that is unoccupied by solid particles. In addition, the number, sizes, shapes, and distribution of the voids is important. Generally, the pore space of surface soil is less than one-half of the soil mass by volume, but in some soils it is more than half.

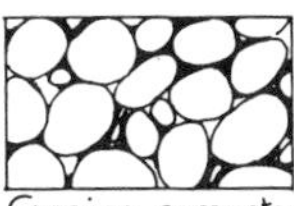

Grains cemented with mineral matter; little pore space

Porosity due to open solution cavities

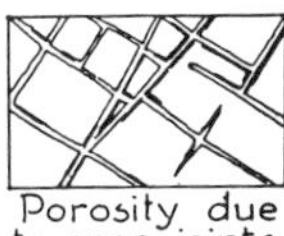

Porosity due to open joints and fractures

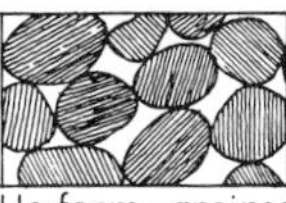

Uniform-grained deposit, having much pore space

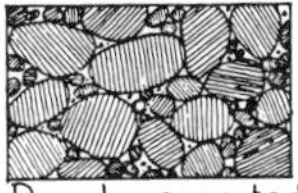

Poorly assorted deposit, having little pore space

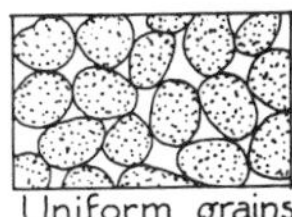

Uniform grains of porous material; much pore space

porosity

porphyritic (adj.) A type of rock texture in which large crystals are set in a finer groundmass.

porphyry A rock with porphyritic texture.

port Any city or other place where ships load and unload. More generally, a place where goods are exchanged.

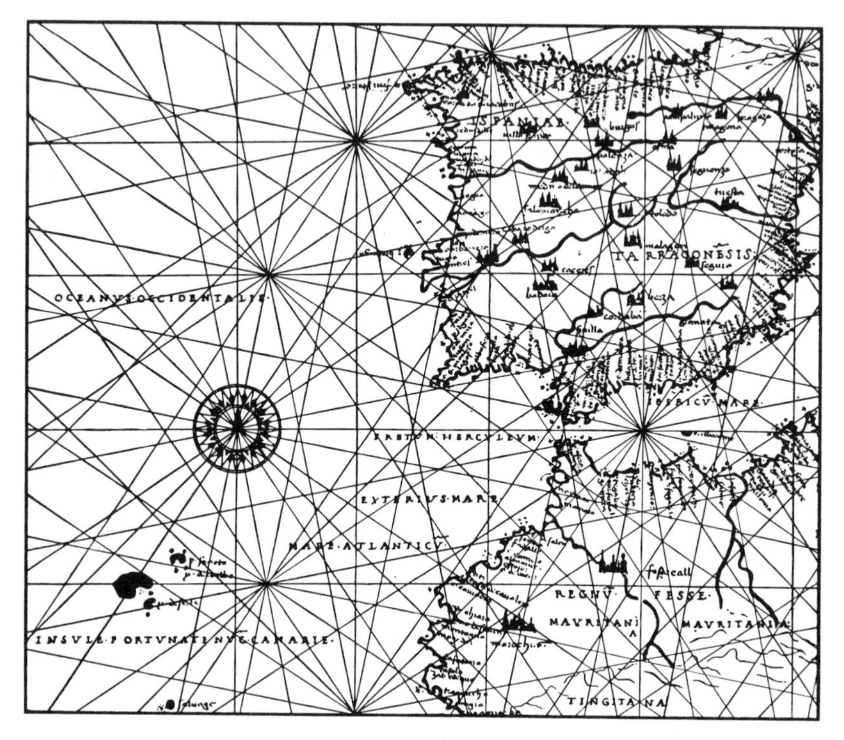

portolan chart

portolan chart A type of medieval sailing chart on which lines of constant compass direction (rhumb lines) were placed to aid navigators in sailing from one port to another.

potential energy The capacity of a body to do work due to its position or dynamical configuration. Water in a dam has potential energy due to its position, just as a bent spring possesses potential energy because of its configuration.

potential evapotranspiration Water loss that will occur if at no time there is a deficiency of water in the soil for use of vegetation.

potential gradient The difference of electric potential per unit distance vertically, between the earth and a point in the atmosphere. It varies greatly with location, season, hour, and weather conditions.

potential instability See: convective instability

potential temperature The temperature to which air would come were it reduced adiabatically to the standard pressure of 1,000 millibars. It is invariant for dry-adiabatic ascent and descent, provided no condensation or evaporation occurs.

pot-hole A hole worn in solid rock, usually near a waterfall or rapids where the current moves in eddies.

prairie A grassland region intermediate between forest and short-grass or steppe regions.

prairie soils A zonal group of soils having dark-colored surface horizons grading through brown soil material to lighter colored parent material at two to five feet, and formed under tall grasses in a temperate, humid climate.

Precambrian See: Appendix I

Precambrian shield An extensive, relatively flat erosional surface of some of the oldest exposed rock. These shields form the core areas of each continent. Around them other rock material has accumulated in the process of continent-building.

precession The slow change in the angle between the earth's axis and the plane of the ecliptic.

precipitable water Total water vapor contained in an atmospheric column of unit cross-sectional area, expressed in terms of liquid water of the same cross-sectional area.

precipitate A substance separated as a solid from a solution in consequence of some chemical or physical change, such as the action of a reagent, of cold, or of heat; usually a solid separated in a noncrystalline or minutely crystalline form. The precipitate may fall to the bottom, may be diffused through the solution, or may float at or near the top.

precipitation A general term for all forms of falling moisture. It includes rain, snow, hail, sleet, and their modifications. The more common term, rainfall, is also used in this general sense; for instance, it is commonly stated that the average annual rainfall at Washington, D.C., is 42.16 inches. This figure includes rain and the water equivalent of snow, hail, sleet, etc.

precipitation effectiveness A climatic element in Thornthwaite's classification of climates, which measures the ability of precipitation to promote plant growth, and grades precipitation effectiveness from a maximum in the tropical rain forest to a minimum approaching zero in the tropical desert.

predator Any animal that captures and eats other animals.

preemption Act or right of purchasing before others. For example, the Preemption Act of 1853 gave United States citizens the right to acquire up to 160 acres of the public domain.

prehensile Capable of grabbing or seizing an object.

pressure Force per unit area. In meteorology, most commonly expressed either in terms of the length in inches of the column of mercury sustained by the force, or in terms of millibars, though dynes per square centimeter and millimeters of mercury are sometimes used. See: atmospheric pressure

pressure altitude The altitude indicated by an altimeter when the barometric scale is adjusted to the standard sea-level atmosphere pressure of 29.92 inches.

pressure gradient See: gradient

pressure-height curve Any diagram showing the variation of pressure with altitude.

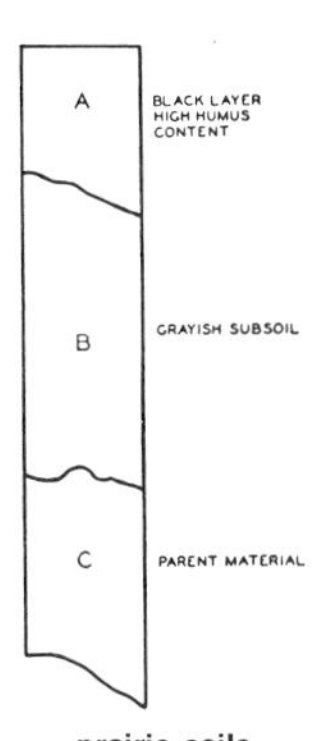

prairie soils

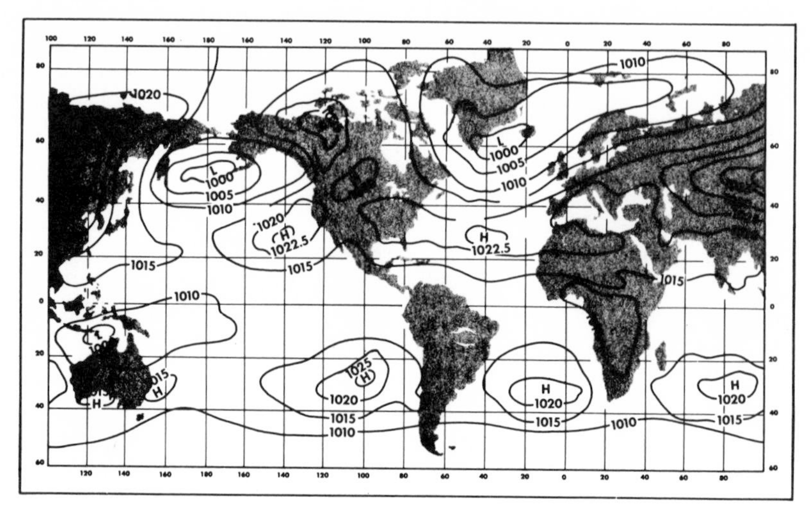

pressure system

pressure system An area of low or high pressure that appears at the earth's surface. See: cyclone; anticyclone

pressure tendency The change in atmospheric pressure that has taken place in a given time (generally three hours) at a certain place.

prevailing wind The direction from which the wind blows during the greatest proportion of the time.

prickly pear A flat-jointed cactus accidentally introduced into Australia, where it has become one of the principal agricultural problems in range management.

primary circulation The prevailing fundamental atmospheric circulation that occurs on a planetary scale because of the fact that in equatorial regions the earth's surface and the atmosphere gain heat from solar radiation, whereas in higher latitudes (beyond about 40 degrees) there is a net loss of heat by radiation from the earth and the atmosphere. This net equatorial heating and polar cooling lead to an interzonal circulation of air, in which heat energy is transferred from equator to pole by the actual transport of air. Due to the deflecting influence of the earth's rotation, this meridional circulation has a system of zonal circulations superimposed upon it; and the effect of land and sea distribution is to render the general circulation even more complex by setting up the semipermanent centers of action—such as the Aleutian Low and the Siberian High—with their resultant large-scale wind and pressure patterns.

primary economic activities Those activities of man concerned with the direct use of a resource: e.g., mining, lumbering, fishing, agriculture.

primary treatment The treatment that removes the material that floats or will settle in sewage. It is accomplished by using screens to catch the floating objects and tanks for the heavy matter to settle in.

primate A member of an order of mammals including man, apes, monkeys, marmosets, and lemurs.

primate city A concept developed by the geographer Mark Jefferson, according to whom each contiguous area (usually a nation) has one city that surpasses all others in size and is the focal point of most social and economic activities.

prime meridian The meridian through Greenwich, England, used as the line from which longitude is measured.

primitive area An area set aside to preserve the natural landscape. Most are managed by the U.S. Forest Service.

principal meridian 1. The central meridian of a map projection. 2. In the United States General Land Office Survey, the meridian upon which all surveys for a particular area are based.

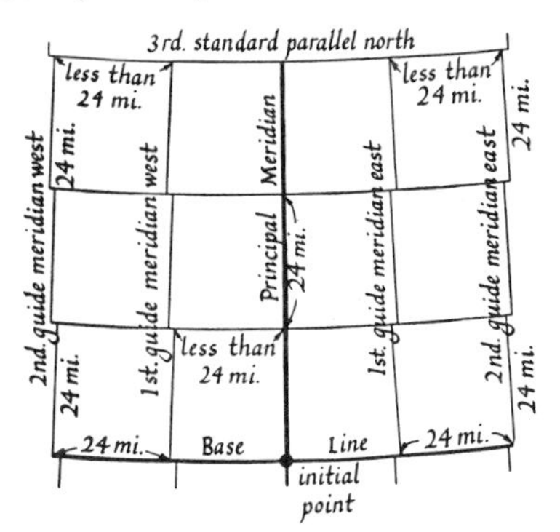

principal meridian

prismatic soil structure Prismlike structural aggregates with the vertical axes of the aggregates longer than the horizontal axes.

productivity With regard to soil, the present capability of a kind of soil for producing a specified plant or sequence of plants under a defined set of management practices. It is measured in terms of the outputs or harvests in relation to the inputs of production factors for a specific kind of soil under a physically defined system of management.

profile In earth study, a chart or diagram showing the characteristics that would be observed in a vertical slice through a part of the earth. See: soil profile; river profile

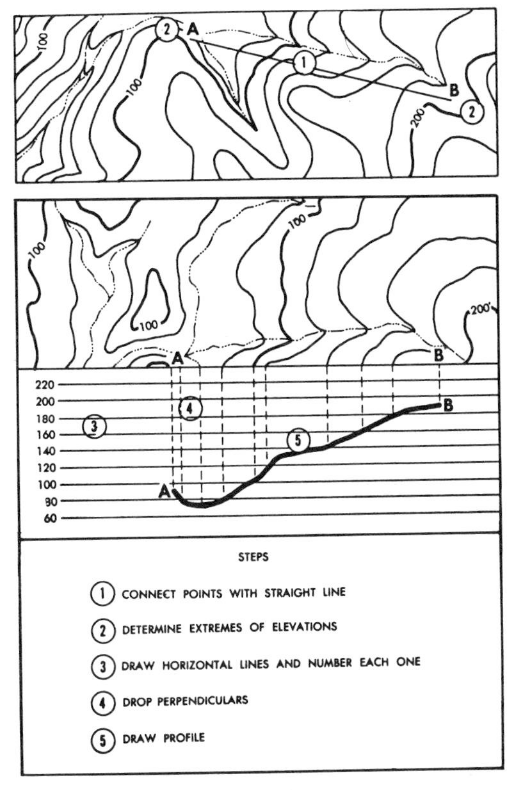

profile

progestin Any substance that prepares the lining of the uterus for implantation of the fertilized ovum.

projection See: map projection

projector (ceiling) See: ceiling light

prolamine One of a class of proteins. The distinguishing feature of prolamines is their solubility in concentrated alcoholic solutions. They are insoluble in water and in salt solutions. Only a few proteins belong to this class; namely, zein from corn, gliadin from wheat, and hordein from barley.

promontory Land projecting into the sea.

propagule Any reproductive structure that is capable of developing into an adult organism, such as some buds, shoots, and seeds.

propane A simple hydrocarbon found in petroleum and composed of three atoms of carbon and eight of hydrogen.

prop root A root extending horizontally away from a tree which tends to support it in marshy habitat.

protein Any of a group of nitrogen-containing compounds that yield amino acids on hydrolysis and have high molecular weights. They are essential parts of living matter and are one of the essential food substances of animals.

proton A positively charged elementary particle, identical to the nucleus of a hydrogen atom.

protoplasm The complex matter of which living things are composed.

protoplast The protoplasm of a single living cell.

protozoa Unicellular organisms that live in water.

pseudoadiabatic chart An adiabatic chart, to which two more sets of lines are added—curves of constant saturation mixing ratio and of pseudo-adiabats. The abscissa is temperature in degrees centigrade and the ordinate is the 0.288 power of the pressure in millibars. Dry adiabats are shown by the sloping, straight lines. On this chart, both dry and pseudo (or saturated) adiabatic motions of air may be determined, and many useful data derived, such as the dew point, condensation level, and a rough estimate of the possible precipitation to be derived from the air.

pseudoadiabatic process A process wherein all of the condensation that occurs in a rising element of moist air immediately falls out. This process differs from an adiabatic process in that it is nonreversible, and in the fact that the hail stage does not occur, since all the condensed moisture falls out; therefore, a rising element of air undergoing the pseudoadiabatic process and returning to its original position becomes somewhat warmer and drier than before. The rise in temperature is due to the heat of condensation, and the increased dryness is due to the loss of some of its original moisture by precipitation.

pseudoequivalent temperature The temperature that would be assumed by a parcel of moist air that had been moved upward dry-adiabatically to its condensation level, then moved upward pseudoadiabatically until all its water vapor content had been condensed, and finally returned dry-adiabatically downward to its original pressure.

psychrometer An instrument to determine the amount of atmospheric moisture. It consists essentially of two thermometers, one of which has its bulb covered with a closely fitting jacket of clean muslin that is saturated with water when an observation is about to be taken. Because of evaporation, the reading of the moistened thermometer is lower than that given by the dry bulb, and the difference between them in degrees is called the depression of the wet bulb. This, used with the atmosphere pressure, gives a measure of the humidity. By reference to appropriate tables the dew point, relative humidity, and vapor pressure may be obtained.

psychrometric chart A nomograph for graphically obtaining the relative humidity, absolute humidity, and dew point from wet- and dry-bulb thermometer readings.

psychrometric tables Tables used for obtaining the vapor pressure, relative humidity, and dew point from the readings of the wet- and dry-bulb thermometers.

public domain Property rights belonging to the community as a whole.

pulp 1. Any soft plant matter. 2. The mixture of rag or wood fibers of which paper is made.

pulsating or oscillating universe The subject of a theory that our universe goes through periods of expansion followed by contraction until all the galaxies come together, explode, and start expanding again.

pumice Volcanic rock made extremely light and porous by the sudden release of steam and gases as it solidifies.

pumped storage power plant A technological device designed to supply extra electric power during peak-load periods. At those hours (late evening, early morning) when extra power is available, water is pumped from a lower to an upper reservoir. During the periods of peak-load demand (beginning and end of the workday), the stored water is dropped through turbines to produce the needed power.

puna A bleak part of the high Andean plateau, between 10,000 and 13,000 feet above sea level.

purga (or poorga) A storm, similar to the blizzard of North America and the buran of south Russia and central Siberia, which rages in the tundra regions of northern Siberia in the winter. See: buran

purine One of a group of closely related compounds containing carbon, hydrogen, and nitrogen. Uric acid, an example, is formed from proteins as an end product of animal metabolism. Uric acid is the chief nitrogenous compound in the excrement of birds.

pusztas Hungarian steppes.

puy A French term denoting a hill that is the cone of an extinct volcano.

p wave An earthquake wave that progresses like a sound wave by alternately compressing and rarefying the substance through which it travels. Also called pressure wave, push wave, compressional wave or longitudinal wave.

pygmy One of a group of small people, less than five feet in height, living in the forests of equatorial Africa.

pyrheliometer An instrument for measuring the intensity of incoming radiation from the sun, the sky, or both.

pyrite Fool's gold. It is an iron and a source of sulfur, but it is not used as an iron source because of the difficulty of extracting the mineral from the ore.

pyroclastic rock Fragmental deposits of volcanic material that has been ejected explosively from a volcanic vent.

pyrotechnic generator A silver-iodide generator in which the silver iodide forms a part of a rocket's fuel mixture. The rockets are sent into clouds to accomplish seeding.

pyroxene A family of closely related minerals containing silica, calcium, magnesium, and iron. They have little commercial value.

quadrant A device to measure angles and altitudes.

quadrature In astronomy, the position of a heavenly body as seen from a point on earth when it is at a right angle to the sun. In particular, it is used to describe the condition when the rays of the sun and the moon are at right angles to one another in relation to earth.

quadroon The offspring of a mulatto and a white person.

quality of snow The fraction of the total weight of a snow sample that is in the form of ice, the remaining portion consisting of entrained liquid water.

quantum Unit quantity of energy according to the quantum theory. It is equal to the product of the frequency of radiation of the energy and 6.6256 x 10^{-27} erg-sec. The photon carries a quantum of electromagnetic energy.

quantum theory The statement according to Max Planck, German physicist, that energy is not emitted or absorbed continuously but in units or quanta. A corollary of this theory is that the energy of radiation is directly proportional to its frequency. See: quantum

quartz A form of silica. Quartz is one of the most common minerals in the rocks that form the earth's crust, and has many industrial uses.

quartzite One of the hardest metamorphic rocks, formed by the modification of sandstone under heat and pressure.

Quaternary See: Appendix I

quebracho A tropical tree with very hard wood that grows in Argentina and Paraguay. A source of tannin. See: tannin.

quicksand A mass of sand containing enough water so as to be semiliquid. It tends to retain objects mired in it.

Ra

Ra The sun god of the ancient Egyptians.

race A group of organisms somewhat alike yet not sufficiently different to be identified as a separate species.

rad A unit of absorbed radiation equal to a dose of 100 ergs of energy per gram of the absorbing material.

radial drainage A drainage pattern in which the streams radiate from a central point like the spokes of a wheel; characteristic of volcanoes.

radiant energy See: radiation

radiant heat Heat in transit in the form of electromagnetic radiation. Radiant heat may be reflected, refracted, and polarized, just as light.

radiation Electromagnetic waves traveling at 186,000 miles per second, many of which may be visible as light. Cosmic rays, gamma rays, X-rays, ultraviolet rays, visible light rays, infrared rays, and radio waves are some common types of radiation that vary in wave length from 0.0000000001 centimeter to 10,000,000,000 centimeters. Visible rays range from about 3.8 to 7.6 ten-millionths of a meter in wave length. The wave length emitted by an object decreases as the temperature of the object increases and a decrease in wave length signifies an increase in radiation frequency. Thus, a hot surface emits high-frequency radiation. The rate at which an object emits radiation is controlled by the temperature contrast between the object and its environment. Radiation travels best through a vacuum, but the quantity absorbed by a substance is controlled by the wave frequency and the density of the absorbing medium. High-frequency waves can penetrate dense media, whereas low-frequency waves may be absorbed by low-density media, especially by water vapor.

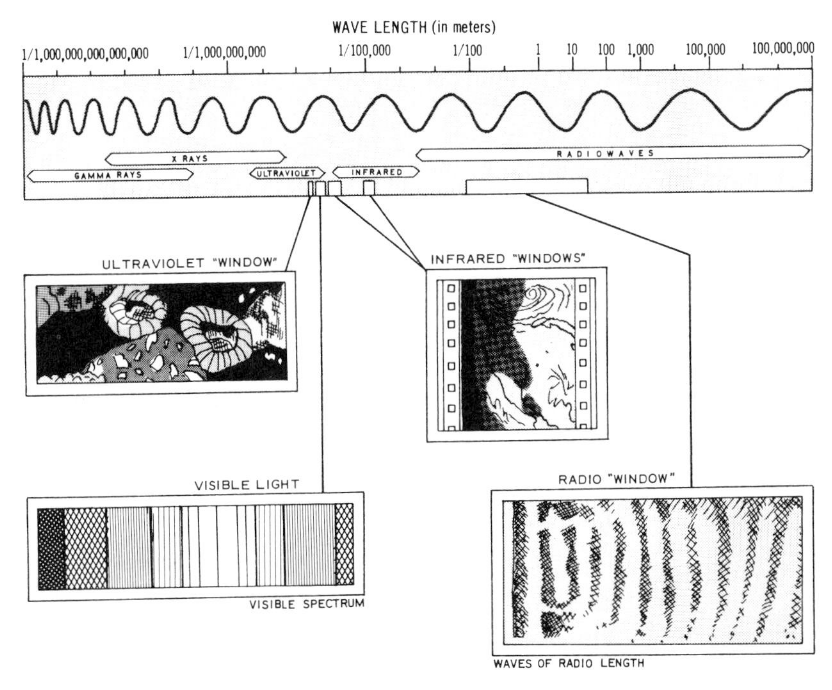

radiation

radiational cooling See: nocturnal radiation

radiation fog A shallow fog formed as a consequence of the cooling of the lower levels of an air mass. See: ground fog

radiation laws See: Kirchoff's law; Planck's law; Stefan's law; Wien's law

radioactive age determination Determination of the length of time that has passed since a mineral crystallized, through determination of the ratios between stable elements and the radioactive elements from which they may have formed. See: carbon 14; radioactive dating

radioactive contamination Deposition of radioactive material in any place where it may harm persons, spoil experiments, or make products or equipment unsuitable or unsafe for some specific use.

radioactive dating A technique for measuring the age of an object or sample of material by determining the ratios of various radioisotopes or products of radioactive decay it contains. See: carbon 14; radioactive age determination

radioactive decay The process by which an isotope of an element (the parent) loses particles from its nucleus to form an isotope of a new element (the daughter). The rate of decay is expressed in terms of an isotope's half-life, or the time that it takes for one-half the nuclei in a sample to decay.

radioactive isotope An element that has been exposed to radiation. Such isotopes may be intermingled with other substances and used as tracers to follow the flow of a fluid through a vessel. See: isotope; tracer techniques; radioactive decay

radioactive waste Highly radioactive material left over from fuels used in nuclear reactors. For disposal, some has been pumped deep into the interior of the earth, while other wastes have been put into concrete-lined barrels and dumped into the ocean.

radioactivity The property possessed by some isotopes of giving off alpha particles, beta rays, or gamma rays as their nuclei disintegrate and change into other isotopes.

radiology The science which deals with the use of all forms of ionizing radiation in the diagnosis and the treatment of disease.

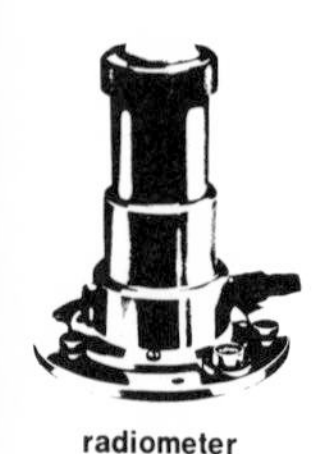
radiometer

radiometer An instrument for measuring radiant energy.

radioactive series A succession of nuclides, each of which transforms by radioactive disintegration into the next until a stable nuclide results. The first member is called the parent, the intermediate members are called daughters, and the final stable member is called the end product.

radioactive tracer A small quantity of radioactive isotope (either with carrier or carrier-free) used to follow biological, chemical or other processes, by detection, determination or localization of the radioactivity.

radionuclide Radioactive atoms that have the same atomic and mass numbers.

radiosonde An instrument equipped with elements for determining the pressure, temperature, and relative humidity of the upper air, and with radio units for automatically transmitting the measurements to ground stations. Radiosondes are carried aloft by balloons, and during such flights, signals are continually transmitted to indicate the conditions of pressure, temperature, and relative humidity prevailing in the air through which the instrument passes.

radium A radioactive metallic element with atomic number 88. As found in nature, the most common isotope has an atomic weight of 226. It occurs in minute quantities associated with uranium in pitchblende, carnotite and other minerals; the uranium decays to radium in a series of alpha and beta emissions. By virtue of being an alpha- and gamma-emitter, radium is used as a source of luminescence and as a radiation source in medicine and radiography.

rain Precipitation that reaches the earth's surface as water droplets. Rain may be classified as: (a) light, the rate of fall being from a trace to 0.10 inch per hour; (b) moderate, from 0.11 to 0.30 inch per hour; (c) heavy, over 0.30 inch per hour; (d) shower, characterized by the suddenness with which the rain starts or stops and by rapid changes in its intensity.

rainbow An arc of prismatic colors appearing in the sky opposite the sun caused by the refraction and reflection of the sun's rays in water droplets.

rain day A period of 24 hours, commencing at 9 a.m., in which 0.01 inch (or 0.2 mm) or more of rain is recorded. See: rainy day

raindrop A drop of liquid water formed in the atmosphere from cloud particles. Formerly a raindrop was thought to be merely an oversized cloud particle, produced by continued condensation and coagulation, but now it is believed to originate in a much more complex and still imperfectly understood fashion. The salient fact is that a raindrop reaches the ground before it evaporates, while a cloud particle does not.

rain factor A measure of rainfall effectiveness, obtained by dividing the amount of rain (expressed in millimeters) by the mean temperature (expressed in degrees centigrade).

rainfall 1. The amount of water falling from the clouds during a rainstorm. 2. Precipitation.

rainfall effectiveness. The ability of rainfall to promote vegetation. This is governed by the amount of rain absorbed by the soil and thus made available to the plant roots. Controlling factors are (a) the total amount and character of the rainfall, (b) its rate of fall and seasonal distribution, (c) the losses due to runoff and evaporation, and (d) the character and condition of the soil and of the vegetation.

rainfall excess Rain which falls at an intensity exceeding the infiltration capacity.

rainfall intensity The average rainfall for each rainy day, obtained by dividing the total amount of rainfall for any period by the number of rainy days in the same period. Thus, if the rainfall for a place for July was 2.40 inches and there were 12 rainy days, the rainfall intensity would be 0.20 inch.

rainfall intensity curve A curve that expresses the relation of rates of rainfall and their duration.

rainfall rate The amount of precipitation occurring in a unit of time, generally expressed in inches per hour.

rain forest Any tropical or temperate forest region characterized by large quantities of precipitation (over 100 inches). Rain forests contain large tall trees with dense canopy.

rain (or snow) gauge An instrument designed to measure the vertical depth of rain or snow (or its water equivalent). Because of difficulty of obtaining exact measurements from a simple pan with vertical sides, exposed during the period of precipitation, and also because of the loss of water by evaporation, most gauges are designed (a) to magnify the depth of rainfall, so that its real depth may be measured to the nearest hundredth of an inch or millimeter, and (b) to minimize evaporation.

rain gauge

rain gush The sudden and heavy downpour of rain that occurs soon after a sharp clap of thunder.

raininess 1. Rainfall intensity. 2. The average rainfall for rainy days, rainy day being defined in turn as one with 0.01 inch or more of rain or melted snow.

rainless (adj.) Without precipitation, or sometimes simply without rain. As far as records indicate there seem to be no places where precipitation (rain or snow) never occurs, even though some years may be absolutely dry. At Africa, Chile, seventeen years of records with only three measureable showers show a yearly average of only 0.02 inch; and in Upper Egypt as many as ten years pass without measurable rain, though some very light sprinkles occur. In parts of Antarctica no rain falls, since all moisture occurs in the form of ice and snow.

rainmaking The term popularly used to apply to all weather modification aimed at increasing the amount of precipitation that will fall from any cloud.

rain shadow The region of diminished rainfall on the lee side of a mountain or mountain range, where the rainfall is noticeably less than on the windward side. A good example of rain shadow in the United States is the country east of the Sierra Nevadas, for the prevailing westerlies deposit most of their moisture on the western slopes of the range. See: orographic rainfall

rain stage The interval in adiabatic ascent of air that extends from the point where saturation has been attained, when the dew point is above 32°F., to the point where the air has cooled to the freezing point at the saturation-adiabatic rate. See: adiabatic process

rain trees 1. Trees abounding in insects that secrete moisture and afterwards exude it, causing a falling mist under the trees. 2. Trees and bushes that in foggy countries catch copious amounts of moisture from the air, which then sometimes falls so abundantly to the ground that it amounts to a moderate shower; that is simply fog drip.

rainy day An expression technically defined as a day with 0.01 inch or more of rain, but popularly considered as a day with more or less continuous rain.

rainy season In the tropics, one of two seasons, the other being the dry season.

ramie A low, branching, shrub-like annual of the nettle family native to eastern Asia. Its fibers are used to make attractive and long-lasting textiles used in curtains and furniture upholstery.

random (adj.) 1. Completely irregular or without predictable pattern. 2. Characterizing a selection process in which each item of a set has an equal opportunity of being chosen.

randomize To design an experiment so that any bias is removed.

range 1. The difference between the largest and smallest values of a quantity, and so a measure of its variability. 2. The territory occupied by any group of organisms. 3. Land that produces primarily native forage plants suitable for grazing by livestock. Such land may have some forest trees.

raob An abbreviated term for radiosonde observation.

rapids A part of a river in whch the current is moving swiftly.

rapids

rasputitsa The brief spring period of Siberia when the ground is still snow-covered and lakes and rivers are frozen but melting is widespread. Transportation is difficult in such weather.

rating curve A curve that expresses graphically the relation between mutually dependent quantities: for example, in hydrology the relation between the stage of a stream and its discharge.

rating table 1. A table showing the relation between two mutually dependent quantities or variables over a given range of magnitude. 2. A table showing the relation between the gauge height and the discharge of a stream or conduit at a given gauging station. This is also called a discharge table.

ravine A long, narrow cut in the earth's surface. It is smaller than a valley but larger than a gully.

rawin A winds-aloft observation made by balloon and radio methods, without optical aid.

raw materials Unprocessed or partially processed material utilized in manufacturing industries; e.g., iron ore, wood pulp.

reach 1. The length of a channel uniform with respect to discharge depth, area, and slope. 2. The length of a channel for which a single gauge affords a satisfactory measure of the stage and discharge. 3. The length of a river between two gauging stations. 4. More generally, any length of a river.

reaction In soil study, the degree of acidity or alkalinity of a soil mass, expressed in either pH value or in words as follows:

	pH
Extremely acid	Below 4.5
Very strongly acid	4-5-5.0
Strongly acid	5.1-5.5
Medium acid	5.6-6.0
Slightly acid	6.1-6.5
Neutral	6.6-7.3
Mildly alkaline	7.4-7.8
Moderately alkaline	7.9-8.4
Strongly alkaline	8.5-9.0
Very strongly alkaline	9.1 and higher

recessional moraine The pile of glacial drift left by a glacier during a period of relative stability when it is melting away.

recessive trait A hereditary characteristic that does not appear in descendants when combined with a dominant trait.

recorded flood A flood on which reliable data are available, permitting a reasonably accurate determination of flows or stages. See: flood; historic flood

recrystallization The formation (in the solid state) of new mineral grains in a rock. The new grains may have the same chemical composition as the old, or a new mineral may be formed as a result of new conditions of temperature or pressure. Most metamorphic rocks are formed by recrystallization.

recharge See: ground-water recharge

reclamation 1. Industrial derivation of useful materials from waste. 2. The reconditioning of deserts or marshy or submerged land for cultivation or other use.

rectangular drainage A drainage pattern marked by right-angled bends in streams and their tributaries. It differs from trellised drainage in that it is more irregular; the side streams are not necessarily parallel to each other. See: trellised drainage

rectangular survey system See: General Land Office Survey

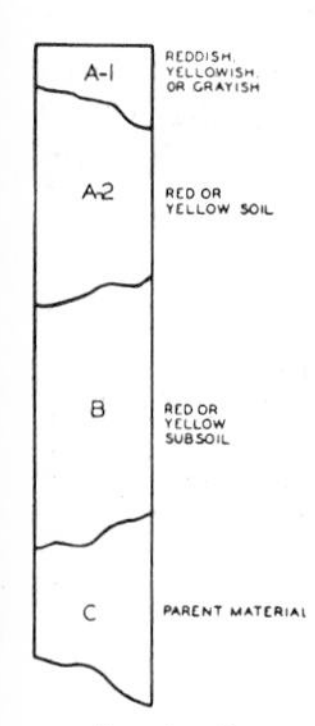

red and yellow podzolic soil

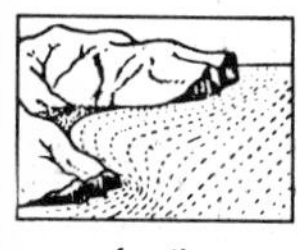
refraction

redwood

recurvature (or recurve) The action of a hurricane in curving away from its generally westward path and heading instead in a poleward and then an easterly direction.

recycle To reclaim waste materials for use.

red and yellow podzolic soil Soil formed under warm, humid, forested conditions. It is acid and reddish in color.

red pine A northern pine with long slender needles. It grows in Canada and the northern United States and is used as both an ornamental and timber tree.

red snow Snow colored red by the presence of either minute organisms or dust particles of that color.

red tides Sea water colored brownish-red by the presence of enormous numbers of microscopic flagellates.

reducing agent A chemical substance having the ability to deoxidize or change an element from a higher to a lower valence or combining power. For example, hydrogen reduces red-hot iron oxide to metallic iron.

reduction 1. A chemical change involving the removal of oxygen or its chemical equivalent. 2. A chemical change involving a decrease of positive valence or an increase of negative valence.

redwood The sequoia sempervirens, the world's tallest tree. It is an evergreen of the pine family and grows in coastal California and Oregon, where it is extensively used for lumber.

reef A ridge of rock lying at or near the surface of water; especially one formed by corals.

reflection The process by which part of the radiation impingent upon any interface separating two media of different densities is turned back into the first medium. If the surface is smooth, the reflection is regular or mirror (specular) reflection; if the surface is rough or discontinuous, it produces diffuse reflection or scattering.

reflectivity The ratio of the radiant energy reflected to the total amount incident. The many different surfaces of the earth reflect radiant energy in different amounts, varying from the high reflectivity of snow to the low reflectivity of woods and plowed fields. See: albedo

reforestation The planting of trees in an area where there was once a forest.

refraction The change in direction of propagation that occurs when sound or light waves pass obliquely from one medium to another of different density.

refrangible (adj.) In physics, capable of being deflected by refraction.

reg In the Sahara, extensive areas of flat, desert plain covered with small stones.

regelation A double phenomenon consisting of, first, the melting of contiguous surfaces of ice under pressure, and second, their refreezing when the pressure is reduced.

regime In climatological studies of precipitation, a term used in characterizing the seasonal distribution of rainfall at any place.

regimen of a stream The system or order characteristic of a stream: its habits with respect to velocity and volume, form and changes in channel, capacity to transport sediment, and amount of material supplied for transportation.

regional planning The process of inventorying the resource base of a region and attempting to guide its development.

regolith The unconsolidated mantle of weathered rock and soil material on the earth's surface; the loose earth materials above solid rock.

regolith

regosol An azonal group of soils that includes those without definite genetic horizons developing from deep unconsolidated or soft rocky deposits.

regression Reversion to an earlier or less advanced form or to a common or general type.

regression coefficient The coefficient of the independent variable in an equation designed to estimate the values of a corresponding dependent variable.

rejuvenated river An old river in which the gradient has been steepened as a result of uplift thus permitting it to begin cutting its channel again.

rejuvenation The process of restoring earlier powers.

relative humidity 1. The ratio of the amount of moisture in a given volume of space to the amount which that volume would contain were it in a state of saturation. 2. The ratio of the actual vapor pressure to the saturation vapor pressure.

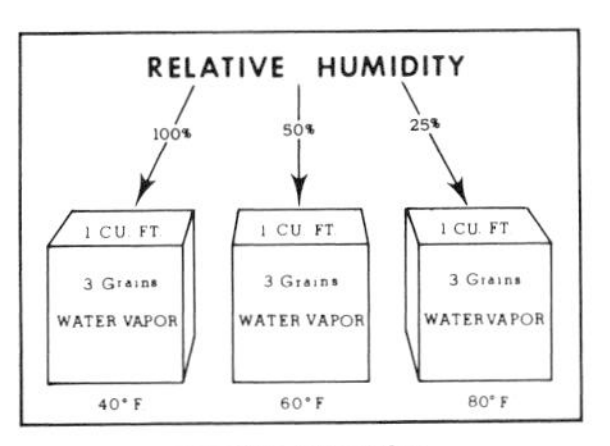

relative humidity

relief The differences in elevation of a land.

relief map A map showing the relief of an area by means of contour lines, special symbols, coloring, or shadows, or as a three-dimensional model.

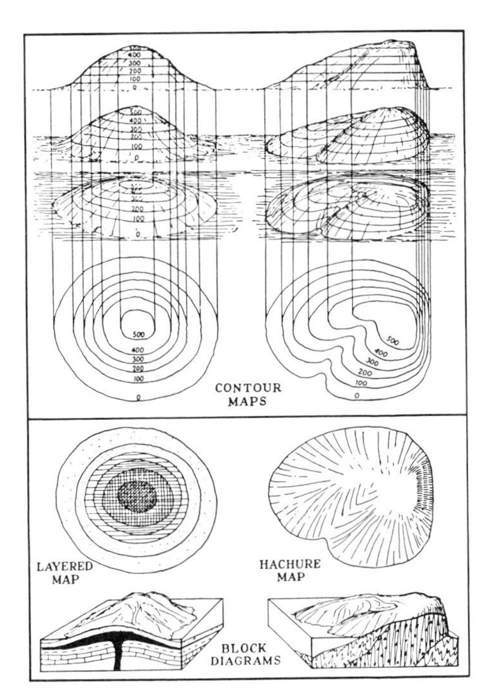

relief map

rem Acronym for roentgen equivalent man. The unit of dose of any ionizing radiation whch produces the same biological effect as a unit of absorbed dose of ordinary X-rays.

rendzina An intrazonal group of soils, usually with brown or black friable surface horizons, underlain by light gray or pale yellow soft calcareous material, developed under grass vegetation or mixed grass and forest vegetation in humid and semiarid regions.

renewable resource Any resource that can be replaced; more trees, for example, may be grown in a forested region if some are cut.

representative property In air masses, any property that characterizes an extensive region of the atmosphere adjacent to the point of observation.

reproduction rate See: crude birth rate, fertility ratio

reptile One of a class of cold-blooded animals with short legs or no legs at all, dry skin covered with scales or plates, and lungs.

reserve 1. A resource not normally called upon but available if needed. 2. A tract of public land set aside as a reservation.

reservoir A pond, lake, or basin, either natural or artificial, for the storage, regulation, and control of water.

residual The amount by which the value of a meteorological element at a given instant or for a given interval differs from the smoothed value or consecutive mean for the same instant or interval. Hence a residual represents the short-period fluctuations that are eliminated in the process of smoothing.

residual mass diagram A diagram or graph plotted with rectangular coordinates, each ordinate being equal to the summation of all preceding quantities in the series up to a given point, minus the arithmetical mean of the series, times the number of quantities in the series up to the given point, with the corresponding abscissa representing time, number of the item in the series, etc.

residual soil Soil that has remained in the place where it was formed.

resin Any of a variety of solid or semisolid substances, chiefly of vegetable origin, having the appearance of rosin. They are transparent or translucent and soluble in ether, alcohol, and other organic solvents, but not in water. Many are produced as exudates from plants, either alone or in admixture with essential oils. They are found in fossils after plants have decayed. Resins are used in making varnishes and resin soaps and in medicines. Also any of a large number of artificial products that possess most of the physical properties of natural resins.

resource See: natural resource

resource inventory The process of finding, listing, and evaluating all of those things useful to man within a given area.

resultant wind The vectorial average of all wind directions and speeds at a given place for a certain period, such as a month.

retention The part of the gross storm rainfall that is stored or delayed and thus fails to reach the concentration point during the time period under consideration.

return flow That part of a diverted flow that is not consumptively used and returns to its source or another body of water.

return seepage Water that percolates from canals and irrigated areas to underlying strata, raising the groundwater level and eventually returning to natural channels.

reverse fault An earth fault in which one block has been pushed up and over the other.

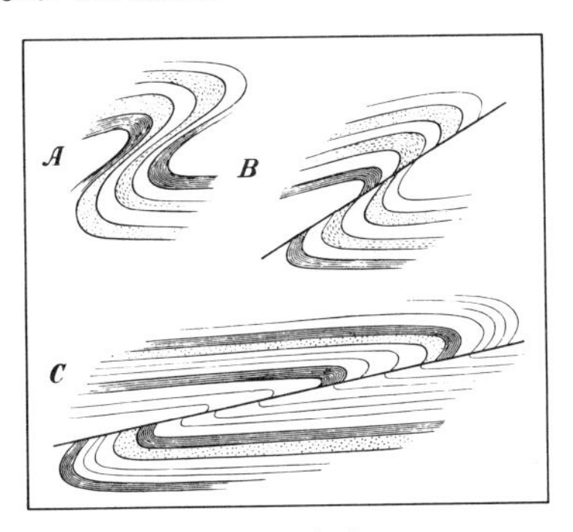

reverse fault

reversible cycle A sequence of changes in a thermodynamic system such that if the order in time of the changes is reversed, the only alteration in the corresponding changes in energy is reversal of sign.

revolution In geology, a time when folding and faulting produced major changes in part of the earth's crust. Also a time of formation of large bodies of igneous rocks.

Reynolds number A dimensionless number used as an index of fluid flow characteristics, equal to V (velocity) times D (diameter or depth) times ρ (mass density), divided by μ (dynamic viscosity).

rhizobia The bacteria that can live in symbiotic relations with leguminous plants within nodules on their roots. The normal result of the association is the fixation of nitrogen from the air into forms that can be used by living plants.

rhizome A horizontal stem wth nodes, buds, and branches that looks like a root and grows below ground.

rhizosphere The bounding surface of plant roots. The soil space in the immediate vicinity of the plant roots in which the abundance and composition of the microbial population are influenced by the presence of roots.

rhombohedron A six-sided prism whose faces are parallelograms.

rhombohedron

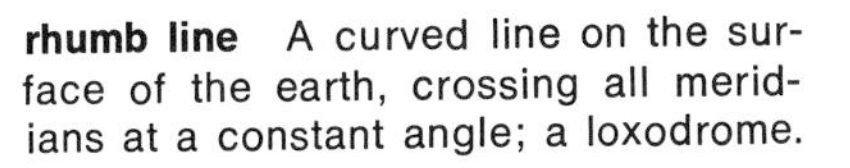

rhumb line A curved line on the surface of the earth, crossing all meridians at a constant angle; a loxodrome.

rhyolite A light-colored volcanic rock composed chiefly of quartz and feldspar with some hornblende, mica, or angite.

ria A long, narrow bay or inlet into the seacoast.

ria coast An embayed coastline formed by the submergence of the mouths of a number of parallel streams.

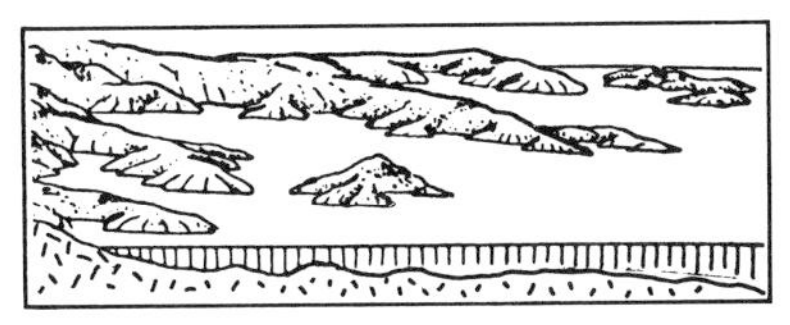

ria coast

ribbon lightning See: band lightning

riboflavin The essential vitamin B_2 occuring in many foods, such as meat, fruit, and vegetables.

ribose nucleic acid See: RNA

Richter scale A scale developed by Charles R. Richter of the California Institute of Technology to measure the magnitude of earthquakes. The ratings are generally expressed in Arabic numerals and are based on a logarithmic scale in which each whole number is approximately 31.5 times as great as the next lower number.

rickets A disease of early childhood characterized by alterations in the bones due to defective deposits of calcium salts at their growing ends. It is treated with sunlight and vitamin D.

ridge (of high pressure) An anticyclonic center that is greatly elongated Also the elongated extension of an anticyclone.

riffle A shallow rapids in an open stream, where the water surface is broken into waves by obstructions wholly or partly submerged.

rift valley A graben or valley formed by the downdropping of a part of the earth's surface between two nearly parallel faults. The best-known rift valley is located in East Africa.

rime A white or milky granular deposit of ice which forms on airplanes, fences, trees, telegraph poles, and other exposed objects at temperatures below the freezing point.

riparian Pertaining to the banks of a stream, lake, or other body of water.

riparian water right The legal right of the owner of land that borders or is contiguous to a natural stream or lake to take water from that source for use upon that land.

rip current The strong seaward undertow produced by waves breaking on a shore.

riprap A foundation or wall of broken rock thrown together irregularly, as to protect embankments from erosion; or the rock used for such purposes.

river A stream of water larger than a creek or brook that flows in a natural channel.

river basin A term used to designate the area drained by a river and its tributaries.

river capture or river piracy The procedure by which a river in enlarging its drainage area, "captures" some of the tributaries or the headwaters of another stream.

river profile In illustration, a section showing the slope of a river from its source to its mouth.

river stage The elevation of the water surface at a specified station above some arbitrary zero datum.

river terrace A level piece of land beside a river, representing a part of the former floodplain of the river.

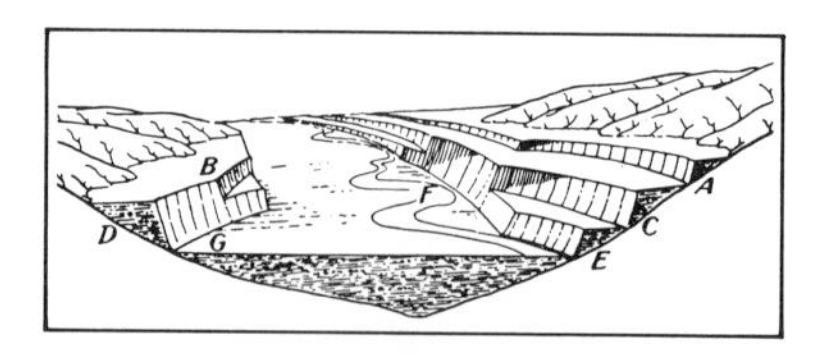

river terrace

RNA (ribose nucleic acid) Any nucleic acid found chiefly in the cytoplasm of cells. These acids aid in the synthesis of proteins.

roadstead An open anchorage for ships; not a harbor.

roaring forties The region of the southern oceans between 40° and 50° S. latitude, characterized by prevailing strong westerly winds.

rôches moutonnées Rock-cored hills associated with the scouring action of glaciers. From above they look like sheep asleep in a field.

rôches moutonnées

rock One of the solid materials of which the earth's crust is made, consisting of one or more minerals. See: igneous, sedimentary, metamorphic

rock cycle The sequence of events and processes involved in the creation of rock, its solution or disintegration, and its redeposition and lithification. Subcycles occur, as when igneous or sedimentary rocks become metamorphic rocks, which then become weathered and break up into their constituent parts.

rocker A crude box used in placer mining. In this procedure sand and gravel are placed in the box, which is then rocked or moved up and down while water is run over the contents.

rock flour Finely ground rock material produced by the abrasive action that takes place on the bed of a glacier.

rock stream A rock flow resembling a glacier in some respects. Occurring most frequently in subpolar regions, rock flows move slowly downslope, alternately freezing and thawing.

roentgen A unit of exposure to ionizing radiation. It is that amount of gamma or X-rays required to produce ions carrying 1 electrostatic unit of electrical charge (either positive or negative) in 1 cubic centimeter of dry air under standard conditions. Named after Wilhelm Roentgen, German scienist who discovered X-rays in 1895.

root That part of a plant that grows downward into the soil holding the plant in place and absorbing water and nutrients.

root crop A vegetable, such as a potato, carrot, or rutabaga, which produces most of its valuable nutrients in its root system.

root-mean-square error The square root of the average value of the square of the errors. See: standard deviation, error

rosin The resin remaining after distilling turpentine from the exudation of

rocker

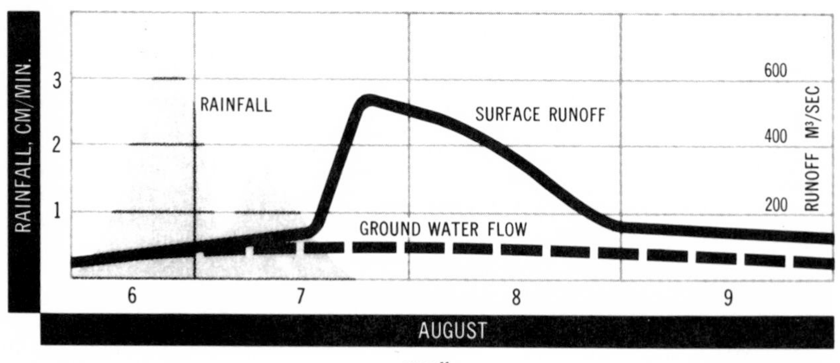

runoff

various species of pine, e. g., pinus palustris (longleaf).

Rossby diagram A diagram used in identifying air masses and named after its inventor, C. G. Rossby. The ordinate is the partial potential temperature on a logarithmic scale and the abscissa is the mixing ratio on a linear scale. Lines of constant equivalent-potential temperature appear as curves running diagonally across the diagram.

rotenone A white crystalline insecticidal compound occurring in many species of Derris, Lonchocarpus, Tephrosia, Mundulea, and Millettia, especially in the roots. The chief sources of rotenone insecticides are Derris elliptica from the East Indies and Lonchocarpus species from South America.

ruminant An animal, subsisting on plants, whose stomach has several compartments. Its food is swallowed whole and then returned to the mouth for chewing and digestion while the animal is resting. Cattle, bison, camels, and giraffes are ruminants.

run A natural channel of water.

runoff Precipitation falling upon a drainage area that is discharged to the sea as surface water in stream channels.

runoff cycle The portion of the hydrologic cycle between precipitation over land areas and discharge of this water through stream channels or evapotranspiration.

runoff rate The volume of water running off in a unit of time from a surface, often expressed in inches in depth of rainfall per hour or cubic feet per second.

rutile One of the principal ores of titanium, found in the United States, Norway, France, Switzerland, and Australia.

rye An important cereal plant of the grass family. It has been cultivated for over 2,000 years and is the most important grain crop on the southern margins of the taiga in North America and Europe. It is used as a green forage crop and for making bread flour and whiskey.

saddle 1. A col, the region of lower pressure between two highs. 2. A low point on a ridge or crest line.

safe yield With reference to either a surface-water or ground-water supply, the rate of diversion or extraction for consumptive use that can be maintained indefinitely within the limits of economic feasibility and under specified conditions of water-supply development. See: perennial yield

sage (sagebrush) A species of hardy perennial plant belonging to the mint family. It is a dominant species in coastal areas of California and in the Great Basin but is also grown as a cultivated plant for its leaves and oil.

sago palm A palm tree of the Asiatic rain forests, valued for its pith.

salina A salt-encrusted hollow in an arid or semiarid region. Also called a salt pan.

saline soil A soil containing enough soluble salts to impair its productivity for plants but not containing an excess of exchangeable sodium.

saline-alkali soil A soil having a combination of a harmful quantity of salts and either a high degree of alkalinity or a high amount of exchangeable sodium, or both, so distributed in the soil profile that the growth of most crop plants is less than normal.

salinity The relative concentration of salts, usually sodium chloride, in a given water. It is usually expressed in terms of the number of parts per million.

salpausselka A series of sand and gravel ridges extending in an east-west direction across Finland and thought to be a series of terminal moraines.

salt See: salts

salt balance A condition in which specific or total dissolved solids removed from a specified field, stratigraphic zone, political area, or drainage basin equal the comparable dissolved solids added to that location from all outside sources during a specified period of time.

salt dome or salt plug A mass of salt forced upward by subterranean pressures that bend the overlying strata upward and solidify them.

salt lake A lake without an outlet in an arid region. As years go by it becomes more salty as evaporation of the lake water leaves the minerals behind.

salt lick A place where evaporation of water from a salt spring has deposited salt on the rocks and ground.

sandbar

salt marsh Any marsh in an area of saline water. It may be inland or along the seacoast.

salt nucleus See: sea salt nuclei

salt pan See: salina

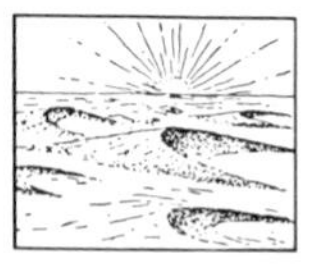
sand dune

salt plug See: salt dome

salts The products, other than water, of the reaction of an acid with a base. Salts commonly found in soils break up into cations (sodium, calcium, etc.) and anions (chloride, sulfate, etc.) when dissolved in water.

salt-water barrier A physical facility or system designed to prevent the intrusion of salt water into a body of fresh water.

salt-water intrusion The phenomenon occurring when a body of salt water, because of its greater density, invades a body of fresh water. It can occur in either surface-water or ground-water bodies.

salvaged water The part of a particular stream or other water supply that is saved from loss, in respect to quantity or quality, and is retained and made available for use.

samoon A hot, dry wind that blows off the mountains of Kurdestan in Persia. See: simoom

sampan A light cargo boat of the Far East, especially China, propelled either by long, heavy poles and oars or by a single sail.

sand Individual rock or mineral fragments in soils, having diameters ranging from 0.5 to 2.0 millimeters. Usually sand grains consist chiefly of quartz, but they may be of any mineral composition. The textural class name of any soil that contains 85 percent or more of sand and not more than 10 percent of clay.

sandbar A ridge of sand extending to or near the surface of a river or sea and formed by the action of currents or tides.

sand dune A mound of loose sand heaped up by the wind.

sandstone A porous sedimentary rock consisting of sand held together by such materials as silica or limestone.

sand storm A windstorm on the desert that raises and carries along quantities of sand often so dense as to render visibility practically nil.

sandy clay Soil of a textural class which contains 35 percent or more of clay and 45 percent or more of sand.

sandy clay loam Soil of a textural class which generally contains 20 to 35 percent clay, less than 28 percent silt, and 45 percent or more of sand.

sandy loam Soil of a textural class which generally contains 50 percent sand and less than 20 percent clay.

sandy soils A broad term for soils of the sand and loamy sand classes; soil material with more than 70 percent sand and less than 15 percent clay.

sansan A crest cloud in the Canadian Rockies.

Santa Ana In Southern California, a local name for a foehn wind which often occurs in winter. It is associated with the Santa Ana pass and river valley.

Santa Gertrudis Largest of the beef-cattle breeds. A cross between short horn and brahman, this breed is widely raised in subtropical regions.

Santa Gertrudis

sapling A young tree, usually between two and four inches thick.

saprophyte Any plant living on decaying organic matter.

sapwood The outer cylinder of light, soft wood that surrounds the heartwood of a stem, branch, or trunk of a tree.

sastruga (pl., sastrugi) A snow ridge found on the windswept polar plains, where the wind tends to blow constantly in one direction.

satellite Any object revolving about a larger one; thus, the moon is a satellite of the earth.

saturated-adiabatic lapse rate The rate of decrease of temperature of a parcel of saturated air as it rises. This rate is less than that for dry air because of the liberation of latent heat as condensation occurs. This rate is not constant, but varies inversely with the temperature and somewhat with change of pressure. See: adiabatic process; adiabatic gradient

saturated air Air containing all the water vapor that it is capable of holding at a given temperature and pressure.

saturated vapor pressure The pressure exerted by the vapor in a saturated space.

saturation The condition in which the pressure exerted by water vapor is equal to the maximum vapor pressure possible at the prevailing temperature.

saturation adiabat See: wet adiabat

saturation-adiabatic process See: adiabatic process

saturation curve A curve on an adiabatic chart giving for various temperatures the saturation moisture content of the atmosphere in grams of water vapor per kilogram of dry air, or the saturation vapor pressure corresponding to a given temperature.

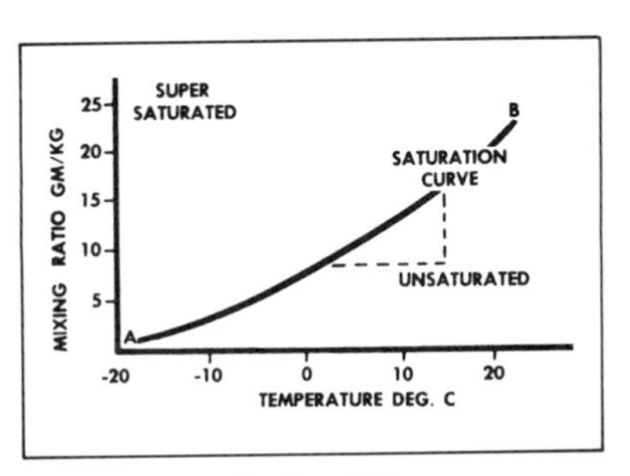

saturation curve

saturation deficit The difference between the actual vapor pressure and the saturation vapor pressure at the existing current temperature. Also known as the vapor pressure deficit.

Saturn 1. The Roman god of agriculture. 2. The planet second in size to Jupiter.

Saturn

savanna A tropical, open, level grassland of tall grasses, interspersed with trees and shrubs. The vegetation is able to survive a hot dry season of considerable length.

savanna climate A rainy climate similar to the tropical rain forest climate but differing from the latter chiefly in that it has a slightly higher annual range of temperature, a smaller annual amount of rainfall, and distinct wet and dry seasons.

saw timber Trees of a size and quality that will make logs suitable for sawing into lumber.

scalar A quantity having magnitude only. Typical scalar quantities are mass, volume, density, temperature, electric potential, and charge.

scarf cloud A thin, cirrus-like cloud, sometimes observed above a developing cumulus, caused by the elevation and consequent expansion and cooling of the air above the rising mass of the cumulus.

scarp An escarpment.

scaling The measurement of the volume of individual logs after trees have been felled.

scattering A diffuse reflection of light: it occurs when a beam of light falls upon an irregular surface, as a piece of paper, in which case the reflected rays are scattered in all directions. 2. A process that changes a particle's trajectory. Scattering is caused by particle collisions with atoms, nuclei, and other particles or by interactions with fields of magnetic force.

scenic highway Any highway so designated by the federal government. On such a highway certain regulations pertaining to billboards, etc., may be enforced.

schattenseite The shady side of an Alpine mountain. Also called a ubac.

schist A metamorphic rock that may be split into many thin plates.

scintillation The twinkling of stars and the analogous irregular fluctuation in brightness and color of distant terrestrial lights caused by the effect of winds and convection currents in bringing across the line of sight elements of air of different densities, and therefore of different refractive indices, in rapid succession.

scirocco British spelling of sirocco.

scoria A porous, cinder-like lava formed when gas bubbles are trapped as the rock cools. It is generally associated with basaltic flows.

Scotch mist A combination of thick mist and heavy drizzle occurring frequently in Scotland and in parts of England, where it is generally known by the same term.

scotch pine An Old-World species of tree with a crooked trunk and scraggly branches, planted as an ornamental in North America.

scour The erosive action of stream water in excavating and carrying away material from the bed and banks. Scour may occur in both earth and solid rock material.

scram The sudden shutdown of a nuclear reactor, usually by rapid insertion of the safety rods. Emergencies or deviations from normal reactor operation cause the reactor operator or automatic control equipment to scram the reactor.

scree Loose rock material lying on mountain slopes and at their foot.

screen An English term for a meteorological instrument shelter.

scrub Any growth of stunted trees or bushes.

scud Low, ragged, detached fragments of cloud, with elevation from 100 to 300 meters, usually associated with nimbostratus and a stormy sky.

scurvy Dietary disease due to a deficiency of vitamin C and characterized by hemorrhage, general debility, spongy gums, etc.

scutch grass See: Bermuda grass

sea In marine meteorology, a term equivalent to "state of the sea."

sea breeze The breeze that blows from the sea to the land on many coasts from about 10 or 11 a.m. to sunset on sunny days in summer, because of the diurnal heating of the shore and its surface layer of air. See: land breeze

sea cliff A wave-cut cliff lying along the ocean shore.

sea cliff

sea fog Fog formed at sea and caused by the transport of air from a warm-water surface to a cold-water surface, with subsequent cooling.

seal A carnivorous marine mammal hunted for its fur, hide, and oil.

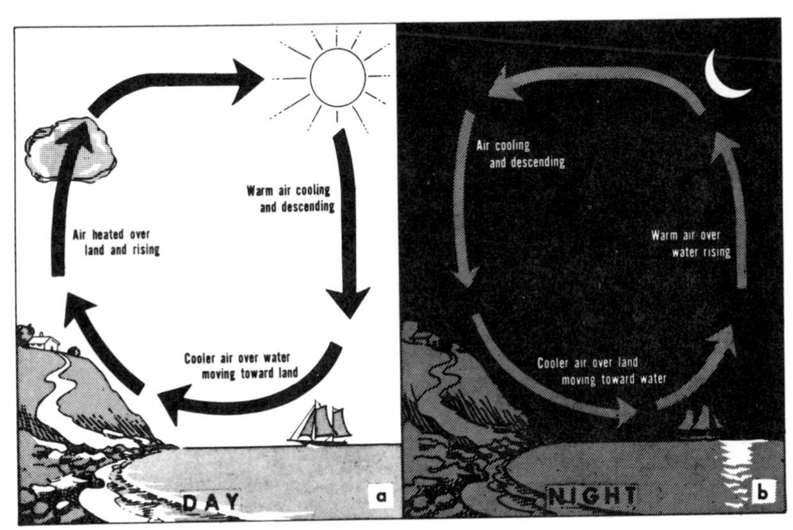

sea breeze

sea level A term commonly used informally in meteorology when mean sea level is actually meant. The latter is defined as the average of the actual heights of the sea surface over a long period.

sea lion One of a group of large, eared seals of various genera found along the shores of the Pacific. Their fur is of little value commercially.

seamount An isolated submerged mountain of volcanic origin. Flat-topped seamounts are called guyots.

sea salt nuclei Condensation nuclei of a highly hygroscopic nature formed by the desiccation of ocean spray.

sea smoke See: frost smoke

season A division of the year generally determined by some annually recurrent natural phenomenon, such as the state of vegetation or the meteorological conditions.

sea urchin A marine invertebrate, called the hedgehog of the sea because of its appearance. It is widely distributed in shallow water and protected locations.

sea wall A special form of breakwater consisting of solid material held together by concrete.

seca A Brazilian term for drought.

second 1. A unit of time equal to one-sixtieth of a minute. 2. A unit of latitude or longitude equal to one-sixtieth of a minute of degree.

secondary A small low-pressure center accompanying a cyclone. It usually originates as a wave formation on the cold front of the primary cyclone, and the first sign of formation is a distortion of the isobars, which may develop into a separate closed circulation.

secondary circulation A collective name for such wind systems as monsoons, tropical and extratropical cyclones, and anticyclones.

secondary cyclone See: secondary

secondary economic activities The processing or manufacturing of goods.

secondary front A term applied to the one or more fronts that not infrequently form behind and follow an active advancing front.

secondary low See: secondary

secondary recovery The injection of gas or water or the use of fire to obtain additional supplies of petroleum from an old oil field.

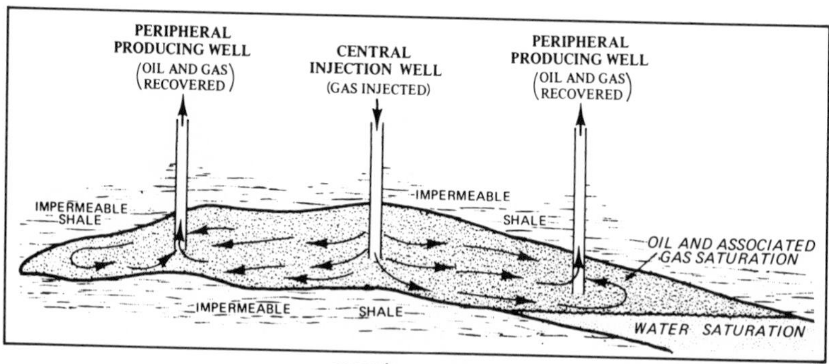

secondary recovery

second-foot An abbreviated expression for cubic foot per second (CFS).

second-growth forest Forest growth that comes up after removal of the old stand by cutting, fire, or other cause. In lumberman's parlance, either the smaller trees left after lumbering or the trees available for a second logging.

second law of thermodynamics The law that states that it is impossible to transfer heat from a colder to a warmer system without the occurrence of simultaneous changes in the two systems or in the environment.

section 1. A distinct part or subdivision of anything. 2. In the central and western parts of the United States, one of the 36 numbered subdivisions of a township, ordinarily one square mile or 640 acres in size. 3. A drawing showing the structures revealed if one could make a vertical cut through the object drawn.

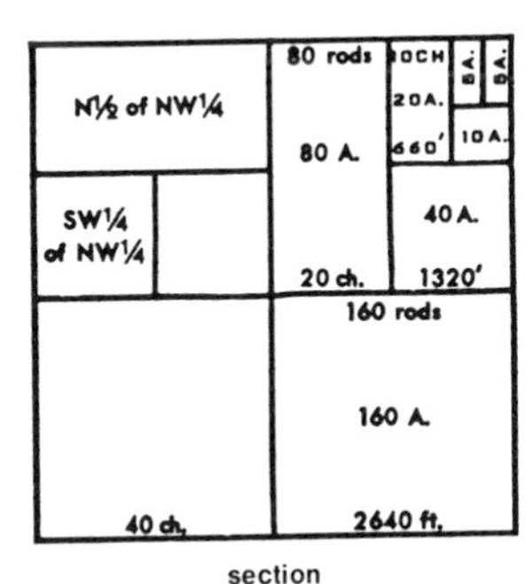

section

sector See: warm sector

secular Taking place slowly over long periods of time, i.e., hundreds of years.

sediment Fragmentary material that originates from weathering of rocks and is transported by, suspended in, or deposited by water or air or is accumulated in beds by other natural agencies.

sedimentary rock A rock composed of particles deposited from suspension in water. Chief groups of sedimentary rocks are conglomerates, from gravels; sandstones, from sand; shales, from clay; and limestones, from soft masses of calcium carbonate. There are many intermediate types. Some wind-deposited sands have been consolidated into sandstones.

sedimentation The process of subsidence and deposition of suspended matter carried by water, sewage, or other liquids.

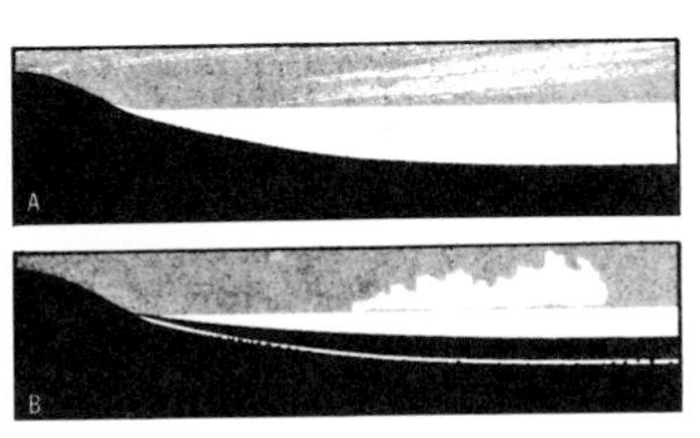

sedimentation

seed An ovule of a plant capable of developing by germination.

seed (and blanket) core A reactor core which includes a relatively small volume of highly enriched uranium (the seed) surrounded by a much larger volume of natural uranium or thorium (the blanket). As a result of fissions in the seed, neutrons are supplied to the blanket where more fission takes place. In this way, the blanket is made to furnish a substantial fraction of the total power of the reactor. Also called a spiked core.

seed crop Any crop grown chiefly to produce seeds as a cash crop.

seedling 1. Any tree that originates from a seed, in contrast to those originating as a sprout or root sucker, or from a cutting. 2. In applied forestry, such a tree under six feet in height.

seepage The percolation of water through the soil; infiltration.

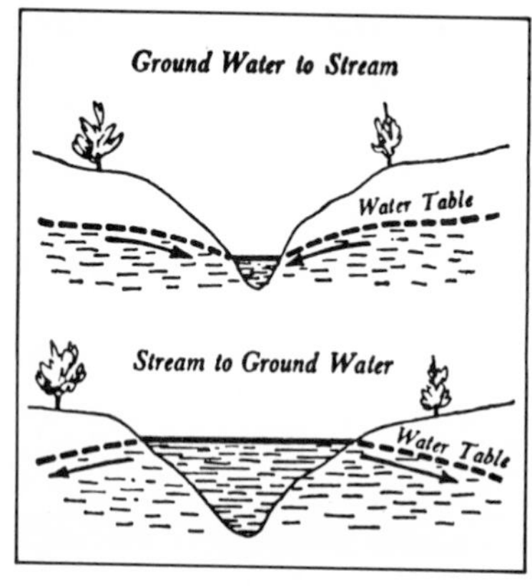

seepage

seiche An oscillation of the surface of a lake or other land-locked body of water, varying in period from a few minutes to several hours, and caused by minor earthquakes, winds, or variations in atmospheric pressure.

seif dune A longitudinal dune of great height and width oriented in the direction of the prevailing winds.

seismic (adj.) Pertaining to or produced by earthquakes or earth vibrations.

seismic wave 1. One of a group of elastic waves generated within the earth whenever a sudden displacement of rock material occurs. 2. A long-period wave in the ocean generated by a seismic disturbance. See: tsunami

seismograph A device that measures and records earthquake waves.

seismology The study of earthquakes.

Seistan A strong northerly wind that blows in summer in the Seistan Basin in eastern Persia, sometimes reaching gale force. It continues for about four months and is known as "the wind of 120 days."

selection Picking out, or culling; the choosing of the best of a group. Any process, natural or artificial, that results or tends to result in preventing certain individuals or groups of organisms from surviving and propagating and in allowing others to do so, with the result that the particular traits of the latter are given pronounced expression. See: natural selection

selective absorption The process in which a substance absorbs, either completely or partially, only certain wave lengths of incident radiation, and freely transmits the other regions of the radiation spectrum; e.g., water vapor absorbs low-frequency radiation but is transparent to high-frequency solar radiation.

selective breeding The process of careful selection and mating over many generations. In this way the traits considered to be most desirable may be developed: e.g., speed in race-horses.

selective harvesting The practice of taking only certain individuals in harvesting: e.g., in a forest, trees of a certain species, size, or condition.

selva A tropical rain forest.

semiarid climate A climate characteristic of the regions intermediate between the true deserts and subhumid areas. In a semiarid climate the precipitation-effectiveness (P-E) index ranges between 16 and 32. The upper limit of the average annual precipitation in cool semiarid regions is as low as 15 inches and in warm regions as much as 45 inches. The vegetation is close-growing or scattered short grass, bunchgrass, or shrubs.

semitropical (adj.) Pertaining to the climate prevailing at the tropical margins of the temperate zones. Subtropical.

sensible temperature The apparent temperature indicated by the sensations of the human body, as distinguished from the actual physical air temperature given by a thermometer. It is subjective, and varies with different people depending on their bodily condition, dress, and climatic environment, as well as upon four meteorological elements: dry-bulb air temperature, relative humidity, air movement, and radiation.

sequoia One of the largest and oldest living things, a member of the pine family growing on the slopes of the Sierra Nevada. The giant sequoia or big tree may be as large as 35 feet in circumference at the base and as old as 4000 years. The redwood is also a sequoia.

serein The phenomenon of fine rain falling from an apparently clear sky, the clouds, if any, being too thin to be visible.

series In soil classification, group of soils that have soil horizons similar in their differentiating characteristics and arrangement in the soil profile except for the texture of the surface soil. They are formed from a particular type of parent material.

serpentine A greenish mineral used as building stone and to make asbestos and firebrick.

severity A term often used in connection with weather and climate but having no precise technical meaning.

sewage The waste matter from industrial and domestic sources disposed of through sewers.

sex ratio The number of males per 100 females in a population.

sextant A device used to measure the angular distance between two objects.

shale A laminated, fine-grained sedimentary rock derived from clay muds.

shallow well A well of pumping head about 20 feet or less, permitting use of a suction pump.

shamal A northwesterly wind blowing in summer over the Mesopotamian plain. It is often strong during the daytime but decreases at night.

sequoia

sharecropper A tenant farmer who pays his rent by giving the owner of his farm a share of the crop.

shear The lateral deformation produced in a body by an external force.

shear of wind The rate of change of wind velocity (speed and direction) with distance.

shearing stress In the atmosphere, the tangential force that exists, by reason of the viscosity or internal friction of the air, between two layers of air that are in relative motion, i.e., in any layer of wind shear.

sheet erosion The type of erosion that occurs when water flows in a sheet down a sloping surface and removes material from the surface in a sheet of relatively uniform thickness.

sheet lightning A form of lightning, most conspicuous when seen in a distant thundercloud at night, which appears as beautiful illuminations, like great sheets of flame, that usually wander, flicker, and glow in exactly the same manner as does streak lightning, often for nearly an entire second, and occasionally even longer.

shelf ice A thick formation of freshwater ice, extending from the land and fed in part by one or more glaciers. It may be afloat or aground and is found in large bays or along a continental glacier at a place determined by prevailing storm winds.

shell In nuclear physics, one of a series of concentric spheres, or orbits, at various distances from the nucleus, in which, according to atomic theory, electrons move around the nucleus of an atom. The shells are designated, in the order of increasing distance from the nucleus, as the k, l, m, n, o, p, and q shells. The number of electrons which each shell can contain is limited. Electrons in each shell have the same energy level and are further grouped into subshells.

shelterbelt A row of trees arranged as protection against strong winds.

shield A large block of the earth's crust that has remained stable through much of geologic time, undergoing at most gentle warping but no severe folding or faulting. Shields are composed of Precambrian rocks and form the hearts of the continents.

shield volcano A volcano formed by successive flows of basic lava without violent eruptions. Its slope is gentle, usually between 4 and 10 degrees.

shift (v.) To change in direction, as the wind. (n.) 1. A change in direction. 2. The relative displacement of rocks on opposite sides of a fault.

shimmer Dry haze.

shingle Loose pebbles on the seashore.

shoal 1. A bank of sand or of rocks just below the surface of a water body, which may endanger navigation. 2. A large number of fish swimming near the surface of the sea.

shock wave A pressure pulse in air, water or earth, propagated from an explosion, which has two phases: in the first, or positive phase, the pressure rises sharply to a peak, then subsides to the normal pressure of the surrounding medium; in the second, or negative phase, the pressure falls

shield volcano

below that of the medium, then returns. A shock wave in air usually is called a blast wave.

shooting star See: meteor

shoreline See: coastline

Shorthorn Beef cattle bred originally in the British Isles but now raised in temperate climates throughout the world.

short-leaf pine An important southern timber tree.

short-wave radiation A term used to denote solar radiation, which has much shorter wave lengths than terrestrial or earth radiation.

shott Playa lakes and saline marshes of the Algerian Plateau and the valleys south of the Atlas Mountains.

shower Precipitation of a convective origin, and hence distinct from ordinary frontal or orographic precipitation. Showers are characterized by the suddenness with which the precipitation (rain, snow, snow pellets, etc.) starts or stops and by rapid changes of intensity.

shrub A woody perennial plant smaller than a tree and without a main trunk or bole.

sial The silicon-aluminum group of rocks, generally light in color and weight, whch lie beneath the continental masses; e.g., granite.

sidereal day The period of time during which a star completes a circle in its apparent journey around the pole star. This represents the period of rotation of the earth on its axis, and is equal to 23 hours, 56 minutes, and 4 seconds.

sidereal month The period of time it takes the moon to complete a revolution about the earth, as observed through its conjunction with a star (29.53 days).

sierozem Zonal soils having brownish-gray horizons that grade through lighter-colored material into accumulated calcium carbonate, developed under mixed shrubs in a cool temperate, arid climate.

Shorthorn

sierra A mountain range, particularly one with a sawtoothed profile.

silage Fodder that has been stored and allowed to ferment in a silo.

silica An important soil constituent composed of silicon and oxygen. The essential material of the mineral quartz.

silicate Any compound that contains silicon, oxygen, and one or more metallic elements.

siliceous Containing or made of silica. In igneous rocks, a rock containing a large amount of quartz.

sill 1. An intrusion of igneous rock approximately uniform in thickness, much wider than it is thick, and approximately parallel to the bedding of the sedimentary rock in which it has been emplaced. 2. A low point on the rim of a submarine basin, or the ridge separating two submarine basins.

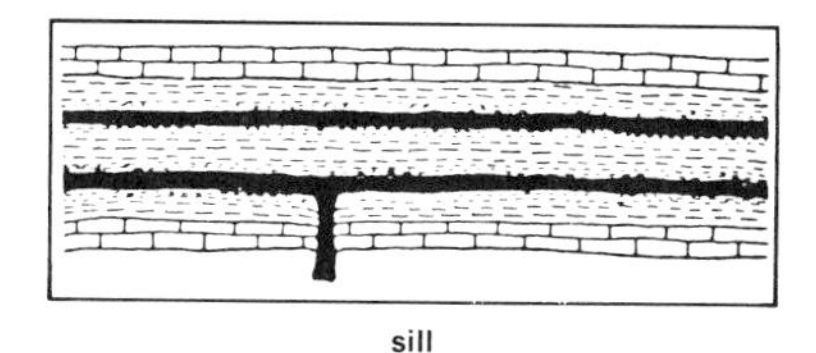

sill

silo A tall, round, airtight building for storing fodder.

silt 1. Individual mineral particles of soil that range in diameter between the upper size of clay, 0.002 millimeter, and the lower size of very fine sand, 0.05 millimeter. Soil of the textural class of silt contains 80 percent or more of silt and less than 12 percent of clay. 2. Sediments deposited from water in which the individual grains are approximately of the size of silt, although the term is sometimes applied loosely to sediments containing considerable sand and clay.

silt loam Soil material having (a) 50 percent or more of silt and 12 to 27 percent of clay or (b) 50 to 80 percent of silt and less than 12 percent of clay.

silty clay Soil of this textural class has 40 percent or more of clay and 40 percent or more of silt.

silty clay loam Soil of this textural class has 27 to 40 percent of clay and less than 20 percent of sand.

Silurian See: Appendix I

silver iodide The compound of silver and iodine, AgI, whose crystalline structure approximates that of ice crystals. It is useful in cloudseeding.

silver iodide generator A device used to generate smoke containing silver iodide. Most of them burn an acetone solution of AgI.

sink hole

silver thaw A deposit of ice on trees, shrubs, and other exposed objects. The term is synonymous with glaze and silver frost, but it is applied more particularly to a very thin coating of ice.

silviculture The science and practice of growing trees as a crop.

sima The dark, heavy silicon-magnesium rocks of the lithosphere underlying the sial; e.g., basalt.

simoom An intensely hot and dry wind of Asian and African deserts—the Sahara, Palestine, Syria, and the desert of Arabia—so laden with dust and sand as to be almost stifling, though some authorities state that it may be dust-free. Some alternative spellings are: samoom, samson, samoun, samum, samun, semoom, semoun, simoon, and simoun.

simulation analysis The study of the characteristics or behavior of a complex system (physical or nonphysical) by constructing and operating a physical or mathematical model of the system.

single-station analysis A technique of forecasting weather from the data available at just one station, based on surface observations and pilot balloon and radiosonde reports.

singularity Any phenomenon occurring with some regularity that represents a departure from the anticipated or average conditions.

sink A sink hole.

sink hole A funnel or saucer-shaped depression. Usually it is the result of the solution of a limestone bed, but it may be created by the collapse of a cavern roof.

sinking An optical phenomenon in which an object on or slightly above the geometrical horizon apparently sinks below it. It is the opposite of looming.

sinter A chemical deposit around a mineral spring. Travertine and geyserite are two rocks formed from such deposition.

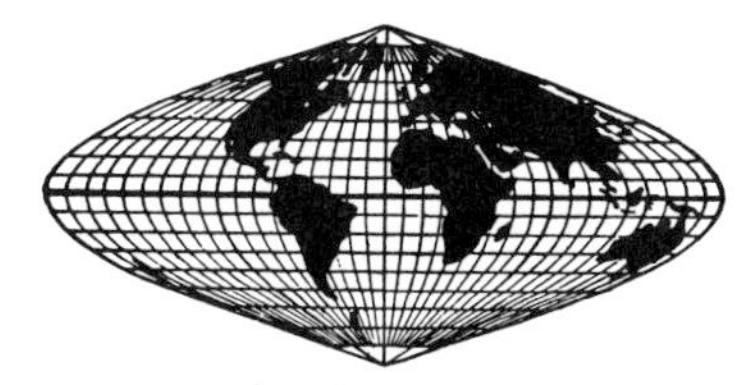
sinusoidal projection

sinusoidal projection An equal-area projection in which the equator, the parallels, and the central meridian are straight lines and all other meridians are sine curves.

sirocco 1. A name locally used in the Latin countries of southern Europe to designate a foehn wind. 2. The hot southerly wind in advance of a low center moving eastward across the Mediterranean Basin. At first it is dry and dust-laden (since it originates in the Sahara or the desert of Arabia), but it is rendered excessively humid by its passage across the sea.

skerry A rocky islet.

skewness 1. The lack of symmetry of an assymetrical frequency distribution. 2. A numerical measure or index of this lack of symmetry.

sky The meteorological term for sky condition, or state of the sky in respect to amount, kind, height, and direction of movement of clouds present. Skies are classified as clear, scattered, broken, and overcast, with many subdivisions and variants.

sky radiation That portion of the sun's radiant energy that comes indirectly to the earth from the air, or from the clouds and dust particles that float in the air.

slash Branches, bark, top, chunks, cull logs, uprooted stumps, and broken or uprooted trees left in the ground after logging of timber is completed; also, a large accumulation of debris after wind or fire.

slash pine A southern pine noted for its strong hard timber and as a source of turpentine.

slate Dense, fine-grained metamorphic rock produced from shale that has been subjected to heat and pressure.

sleeping sickness (trypanosomiasis) An African disease, generally fatal and characterized by fever, wasting, and progressive lethargy. It is caused by a parasitic protozoan carried by the tsetse fly.

sleet 1. Transparent, globular, hard grains of ice ranging in size from 1/25 to 4/25 inch. They are formed by the freezing of raindrops, and rebound when falling on hard surfaces. 2. As used by engineers and the general public in the United States, glaze: i.e., an ice coating formed by the instant freezing of rain on all exposed objects. 3. In Britain and colloquially in some localities of the United States, a mixture of snow and rain falling together.

slickensides Polished rock surfaces on either side of a fault plain, produced by friction between the two blocks.

sling psychrometer The most common type of psychrometer, which consists essentially of two thermometers mounted in a frame that can be rotated rapidly about an axis at right angles to its length.

slip The downslope movement of a mass of soil under wet or saturated conditions; a microlandslide that produces microrelief in soils.

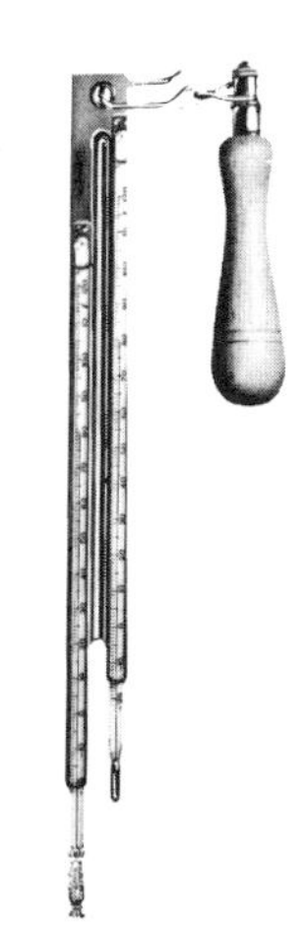
sling psychrometer

slip-off slope bank The bank of a meandering stream that is not eroded by stream action, and that may be built up gradually.

slope The incline of the surface of a soil. It is usually expressed in percentage of slope, which equals the number of feet of fall per 100 feet of horizontal distance.

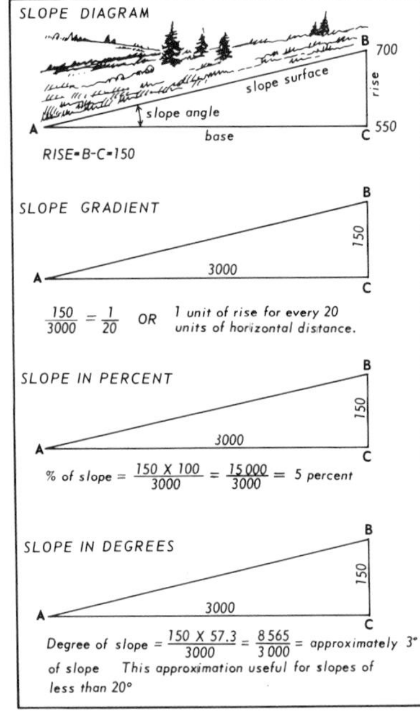

slope

slush Snow on the ground that has been reduced by a warm spell or by rain to a soft, watery mixture.

small circle

small circle Any circle on a globe representing the earth's surface that does not bisect the globe.

smallpox An acute contagious virus disease characterized by successive stages of skin eruptions.

smog A term coined in 1905 by Dr. Des Voeux to signify a mixture of smoke and fog, generally applied to all pollution.

smoke The presence of particles of foreign matter in the air resulting from combustion.

smoothing The process of eliminating, from tabular or graphical data, irregularities that are without significance or importance for the immediate object. It may be accomplished either graphically or by arithmetical formulas.

smudging A term used in connection with orchard heating in anticipation of frost. Properly, it means the production of heavy smoke, supposed to prevent cooling by radiation, but it is generally applied to the operations both of heating and of smoke production.

snap A brief period of cold weather, as in the phrase, " a cold snap."

snow A form of precipitation composed of ice crystals. Freshly fallen snow varies greatly in density; generally a depth of snow of from five to twenty inches equals one inch of water.

snow blindness Temporary blindness caused by the glaring light reflected from snow surfaces.

snow-board A sheet of thin white board, about 16 inches square, with a layer of cotton flannel tacked on to its surface, nap uppermost. It has been shown that more snow is caught on this board than with a can; hence it is useful for measuring snowfall.

snow course A line laid out and permanently marked on a drainage area along which the snow is sampled at definite distances or stations and at appropriate times, to determine its depth, water equivalent, and density during a snow survey.

snow cover Fallen snow that covers the earth's surface.

snow density The ratio between the volume of melt water derived from a sample of snow and the initial volume of the sample. This is numerically equal to the specific gravity of the snow.

snowfall The amount of snow, hail, sleet, or other precipitation occurring in solid form that reaches the earth's surface. It may be expressed in inches-depth as it falls, or in terms of inches-depth of the equivalent amount of water.

snow fence A board fence 5 to 10 feet high, placed about 50 feet away from and on the windward side of railway tracks and highways. Where snow conditions are severe, three or four lines of fences are used in parallel about 100 feet apart. By breaking the force of the wind, the fence causes the snow to be precipitated on the leeward side, i.e., between the fence and the track, leaving the latter comparatively clear.

snow field An area, usually at high elevation or in polar latitudes, where snow accumulates and remains on the ground throughout the entire year.

snowflake A single ice crystal, or, much more usually, an aggregation of them, formed by the condensation of atmospheric water vapor at some temperature below the freezing point.

snow gauge See: rain (or snow) gauge

snow garland A rare and beautiful phenomenon in which snow is festooned from trees, fences, etc., in the form of a rope of snow.

snow grains White, opaque particles like snow in structure and resembling snow pellets but more or less flat or oblong and generally less than 0.04 inch across in at least one direction.

snow line The altitude to which the continuous snow cover of high mountains retreats in summer—It is chiefly controlled by the depth of the winter snowfall and by the temperature of the summer.

snow melt The water that results from the melting either of snow fields or of snow on trees and shrubs. This water may evaporate, seep into the ground, or become a part of the runoff.

snow pack A term used locally in the Rocky Mountain region of the United States to mean a field of naturally packed snow, which gives a steady supply of water for irrigation throughout the growing season and is also useful in furnishing water power.

snow pellets White, opaque, round or seldom conical grains of snow-like structure, about 0.02 to 0.20 inches in diameter in all directions.

snowflake

snow roller A mass of snow shaped somewhat like a lady's muff, common in mountainous or hilly regions and occurring when the snow is moist enough to make it cohesive. It forms when the wind blows down a slope, picking up some snow and rolling it onward and downward until the roller either becomes too large or the ground levels off too much for the wind to propel it further. The area cleared of snow is V-shaped, accounting for the snow roller's peculiar shape, which is cylindrical with concave ends.

snow sampler Any device for collecting a vertical section of snow as it lies on the ground, which section may then be measured and the density of the snow sample thus determined.

snow shed A structure, formerly constructed of wood and now often of concrete, erected over railroad tracks as a protection not only against snow slides but also against snowstorms.

snow stake A wooden stick either driven into the ground or held upright by guy wires, used in regions of deep snowfall to indicate the depth of snow, which is read directly from graduations in inches on the stake.

snow survey A determination of the total amount of snow lying over a watershed or a particular area, by measurements of both the depth and the water content of the snow, for the purpose of predicting the amount of water that will be available for irrigation or other purposes.

soda ash A commercial anhydrous sodium carbonate. It is a relatively mild alkali. Its water solution will not attack fats but will react with the free fatty acids in crude fats. Therefore soda ash is used in refining fats and oils. It is also used in the manufacture of soap, paper, chemicals, paints, drugs, leather, textiles, and many other articles.

soft hail White, opaque, round pellets of snow.

softwood trees As commonly used, a conifer.

soil Earth material that has been modified by weathering and by plants and animals so that it will support the growth of rooted plants.

soil association A group of defined and named kinds of soil associated together in a characteristic geographic pattern. Except on detailed soil maps, it is not possible to delineate the various kinds of soil so that on all small-scale soil maps the areas shown consist of soil associations or two or more kinds of soil that are geographically associated.

soil characteristc A feature of a soil that can be seen and/or measured in the field or in the laboratory on soil samples. Examples include soil slope and stoniness as well as the texture, structure, color, and chemical composition of soil horizons.

soil climate The moisture and temperature conditions existing within the soil.

soil conservation district In the United States a local-federal agency formed by vote of the farmers in an area and organized to secure maximum utilization of the soil resources.

soil creep Slow movement of rock fragments down slopes, occurring most commonly when lower soil is nearly saturated with water.

soil erosion The wearing away of topsoil by natural forces.

soil-forming processes All of the processes that help to convert raw rock material into soil.

soil group A number of soils having similar internal characteristics.

soil horizon See: horizon (soil)

soil management The preparation, manipulation, and treatment of soils for the production of plants, including crops, grasses, and trees.

soil moisture or soil water Water diffused in the soil; the upper part of the zone of aeration from which water is discharged by the transpiration of plants or by soil evaporation.

soil nitrogen cycle The processes by which atmospheric nitrogen is incorporated into the soil.

soil profile The succession of zones seen in a vertical cut through a soil.

soil quality An attribute of a soil that cannot be seen or measured directly from the soil alone but that is inferred from soil characteristics and soil behavior under defined conditions. Fertility, productivity, and erodibility are examples of soil qualities (in contrast to soil characteristics).

soil series A group of soils having similar soil profiles.

soil structure The way that the individual particles in a soil hang together; in a soil with good structure the soil particles gather into groups or floccules.

soil survey A general term for the systematic examination of soils in the field and in the laboratories, their description and classification, the mapping of kinds of soil and the interpretation of soils according to their adaptability for various crops, grasses, and trees.

soil texture The size of the particles of which a soil is composed; clay, silt, and sand refer to particles of different size.

soil type A term used in the classification of soils with approximately the same meaning as soil series, except that the texture of the surface soil must vary only within narrow limits.

solano An east wind on the southeast coast of Spain, dreaded by the Spaniards because of its heat, dust, and moisture.

solar atmosphere The gaseous envelopes that compose the sun.

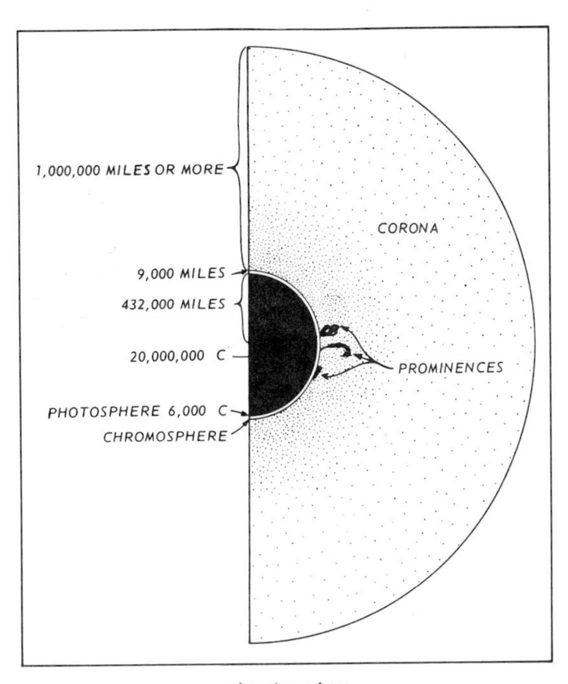

solar atmosphere

solar climate The hypothetical climate that would prevail if determined solely by the amount of solar radiation received according to latitude and season.

solar constant The rate at which solar radiant energy is received outside the atmosphere on a surface normal to the incident radiation, at the earth's mean distance from the sun. The value of the mean solar constant is two langleys per minute per square centimeter.

solar day A division of time equal to 24 hours and representing the period of time it takes the earth to complete one rotation on its axis.

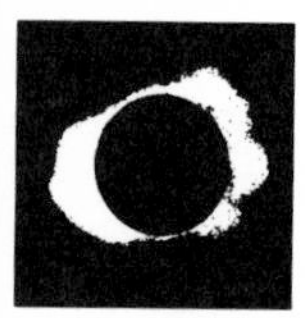

solar eclipse

solar eclipse The total or partial obscuring of the sun as the moon passes between it and the earth.

solar flare A large and brilliant eruption from the sun associated with bursts of energy emanating from within the sun. They stream millions of miles above the surface of the sun, sending free atoms out into space.

solar radiation Radiation received directly from the sun. As emitted from the sun, its spectral distribution is characterized by a maximum intensity in the blue-green part of the spectrum at 0.5μ; hence it is often called shortwave radiation to distinguish it from the predominantly longer wave terrestrial radiation.

solonetz soils

solar system A group of heavenly bodies consisting of a sun, its planets, and all other forms of matter in orbit around the sun.

solar time The system of time based on earth-sun relationships. See: mean solar time; apparent solar time; equation of time

solar-topographic theory The theory that climatic change may be attributed to variations in solar radiation, changes in orography, and the distribution of land and sea.

solar wind A cloud of protons moving out from the sun and affecting the magnetic fields of the earth.

solenoid In meteorology, the column of atmosphere bounded by two isobaric and two isoteric (constant specific volume) surfaces. The unit solenoid is the column bounded by two adjacent unit isobaric and unit isoteric surfaces.

solfatara The vent of a volcano that no longer ejects molten lava but gives off steam and various sulfurous gases.

solid waste The waste material that society cannot dispose of through sewers and which must be put into rubbish disposal sites.

solifluction The movement of soil down slopes.

solonchak soils An intrazonal group of soils with high concentrations of soluble salts in relation to those in other soils, usually light colored, without characteristic structural form, developed under salt-loving plants, and occurring mostly in a subhumid or semiarid climate. In soil classification, the term applies to a broad group of soils and is only approximately equivalent to the common term saline soil.

solonetz soils Soils developed under grass or shrub vegetation in subhumid or semiarid regions in which the surface horizon is underlain by hard, dark-colored, alkaline soil material.

solstices Points on the ecliptic midway between the equinoxes, or points where the sun attains its greatest north and its greatest south declinations.

The summer solstice, or the sun's most northern point of the ecliptic, occurs about June 22; and the winter solstice, or the sun's most southern point of the ecliptic, occurs about December 22.

soluble (adj.) Dissolvable. When a substance (solute) such as salt or sugar is placed in water (solvent), it apparently disappears because it is dissolved in the water. The solute is said to be soluble in the solven. The extent to which the solute dissolves is known as its solubility.

solum The upper part of a soil profile, above the parent material, in which the processes of soil formation are active. The solum in mature soils includes the A and B horizons.

sonic (adj.) Pertaining to sound.

sonic boom A loud noise generated by an aircraft moving at supersonic speeds.

sonnenseite The sunny side of an Alpine slope. Also called an adret.

Sonora storm A summer thunderstorm in the mountains and deserts of Lower and Southern California.

Sonora weather Summer weather characterized by showers frequently very intense and generally extremely localized in their effects.

sorghum A tall cereal grass, similar to corn.

soroche Spanish term for mountain sickness.

sound A narrow water passage.

sounding 1. In meteorology, the process of determining the conditions in the free atmosphere at various heights, commonly now by means of a radiosonde. Also, the graph that results when the values of temperature, pressure, and specific humidity thus determined are plotted on an adiabatic chart. 2. In nautical usage, the measurement of the depth of water under a ship.

soupy (adj.) Densely foggy or cloudy.

source region Any extensive area of the earth's surface characterized by essentially uniform surface conditions and so placed with respect to the general atmospheric circulation that an air mass may remain in contact with it long enough to acquire its characteristic properties.

southerly burster In Australia, a cold wind from the south. After a day or more of hot sultry weather with a northerly wind, there is a short lull; and then very suddenly, when the line of lowest pressure has passed, a strong, often violent, wind sets in from the south. A striking roll of cumulus cloud may accompany the burster, and there is usually heavy rain. Temperature drops suddenly, generally as much as 20°. The phenomenon is most frequent in spring and summer.

south frigid zone See: frigid zone

southern cross A cross-like constellation of four stars visible in the southern hemisphere. If the longer axis is extended almost five times its length, a point over the south pole is reached.

spall 1. To break off in chips or layers parallel to the surface. 2. A relatively thin, curved piece of rock produced by exfoliation.

spawn (v.) To lay eggs. The term is used especially in reference to fish and shellfish.

special nuclear material In atomic energy law, this term refers to plutonium-239, uranium-233, uranium containing more than the natural abundance of uranium-235, or any material artificially enriched in any of these substances.

specialization 1. The adaptation of a part of an organism to a particular function. 2. The adaptation of an organism to its environment.

species A group of plants or animals with similar structures and traits, so that their various forms can successfully interbreed for successive generations.

specific gravity A measure of the relative density of a substance. For solids the density of water is taken as a standard and for gases the density of hydrogen.

specific heat The amount of heat required to raise the temperature of unit mass of a substance by unit amount; it varies slightly with temperature and may depend greatly on the conditions under which the heat is added. In the case of gases, the only specific heats usually considered are the specific heat at constant volume and the specific heat at constant pressure. The specific heat of water expressed as the amount of heat required to raise one gram of water from 15° to 16°C is the standard calorie.

specific humidity The mass of water vapor in a unit mass of moist air, usually expressed as so many grams per gram or per kilogram of moist air. Specific humidity must be distinguished from mixing ratio, which is the mass of water vapor per unit mass of absolutely dry air.

specific power The power generated in a nuclear reactor per unit mass of fuel. It is expressed in kilowatts of heat per kilogram of fuel. See: power density

specific yield The quantity of water that a unit volume of permeable rock or soil, after being saturated, will yield when drained by gravity. It may be expressed as a ratio or as a percentage of volume.

spectroradiometer The instrument for determining the radiant energy distribution in a spectrum.

spectrum A dispersion or separation into a linear sequence, according to wave length, of the individual simple

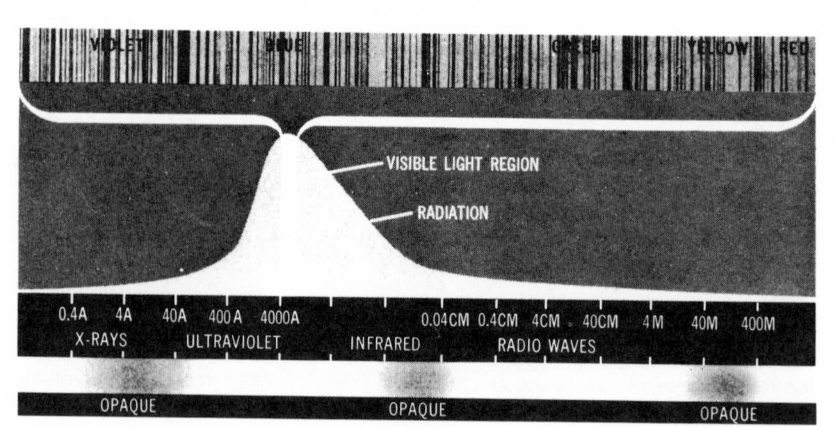

spectrum

waves into which any complex wave disturbance may be resolved.

speed The numerical magnitude of a velocity. Speed is a scalar quantity, whereas velocity involves direction as well as magnitude, and is a vector quantity.

speleology The study of caves.

spell A period during which fairly constant weather conditions prevail in a region where variable weather is ordinarily expected. Thus, the long protracted periods of uniform weather in the tropics are not considered spells, but, in the temperate zones, there may be spells of warm, or rainy, or cold weather, and so on.

spelunker A person who explores or studies caves.

sperm The male reproductive cell.

sphagnum A group of mosses that grow in moist places. By annual increments of growth, deep layers of fibrous and highly absorbent peat may be built up. Sphagnum grows best in cool, humid regions.

spherical coordinates A system of curvilinear coordinates in which the position of a point in space may be identified by reference to the celestial sphere and the celestial pole.

spiked core A seed core.

spillway A channel for carrying off excess water from a lake or reservoir.

spit A narrow strip of sand and gravel projecting into the sea.

spodosol Soil of the cool, humid forested regions of the earth; a podzol with siliceous parent material.

spontaneous fission Fission that occurs without an external stimulus. Several heavy isotopes decay mainly in this manner; examples: californium-252 and californium-254. The process occurs occasionally in all fissionable materials, including uranium-235.

spontaneous generation A theory, now discarded, that living organisms found in decayed or dead organic matter were generated from such matter.

spore A single reproductive cell that develops into a separate organism without fertilization.

spring 1. A season of the year, which in the north temperate zone is commonly regarded as including the months of March, April, and May. 2. In astronomy, the period extending from the time of the vernal equinox to the summer solstice. 3. A natural flow of water from the ground.

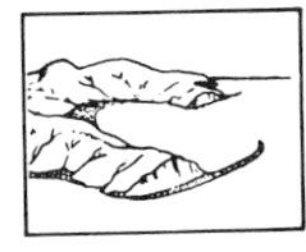

spit

spring equinox See: equinox

spring tides Tides of great amplitude developed when the earth, sun, and moon are nearly in a line.

sprouting The generation of new plants as a consequence of new vegetative growth appearing from roots or stumps of plants.

sprouting

squall A wind of considerable intensity caused by atmospheric instability. It comes up and dies down quickly.

A squall is often named after the special weather phenomenon that accompanies it: thus there are, for example, rain, snow, and hail squalls.

squall cloud A small cloud sometimes formed below the front edge of a thunderstorm cloud.

stalactite

squall line 1. The line of discontinuity at the forward edge of an advancing cold air mass that is displacing warmer air in its path. The most notable characteristic of the squall line is the sharp change of wind, from the southerly to the westerly or northwesterly quadrant. In certain cases, severe squalls and thunderstorms may extend along it for hundreds of miles. 2. A line of squalls other than that occurring at a severe cold front, often found in the warm sector some 50 to 200 miles ahead of and roughly parallel to the front. Hence it may also be called a prefrontal or precold front squall line.

stability A state of vertical equilibrium in which the vertical distribution of temperature is such that an element of air will resist displacement from the level at which it is in equilibrium with its environment.

stack A rocky islet near a coastline.

staff gauge A scale set so that a portion of it is immersed in the water at all times to measure the height of the water.

stage The height of a water surface above an established datum plane. See: gauge height

stage-capacity curve A graph showing the relation between the surface elevation of the water in a reservoir, usually plotted as ordinate, against the volume below that elevation, plotted as abscissa.

stage-discharge curve (rating curve) A graph showing the relation between the gauge height, usually plotted as ordinate, and the amount of water flowing in a channel, expressed as volume per unit of time, plotted as abscissa.

stalactite A conical or cylindrical mineral deposit hanging from the roof of a cave, and usually composed of calcite or aragonite.

stalagmite A column or ridge of mineral material rising from the floor of a cave. It is usually formed of calcite or aragonite deposited from water dripping from stalactites above. A stalactite and a stalagmite may meet, forming a solid column from floor to roof of the cave.

stamen A flower organ consisting of a stalk or filament and an anther in which pollen develops.

standard atmosphere An idealized atmospheric structure in the middle latitudes up to the 700 kilometer level, portrayed in the U.S. Standard Atmosphere (1962). It is defined in terms of temperatures at certain fixed heights; between these levels temperatures are considered to vary linearly and other properties such as density, pressure, and speed of sound are derived from the relevant formulas.

standing cloud See: cap cloud

standard deviation A measure of the extent to which individual values of a series are scattered about their average value, computed by taking the square root of the arithmetic mean of the squares of the departures of the various items from the arithmetic mean of the group. The square of the standard deviation is known as the variance or mean square.

standard gravity Standard gravity is a conventional value of the acceleration of gravity widely adopted for use as a reference basis in barometry and other physical fields. The value of gravity adopted by the Service International of the International Bureau of Weights and Measures, called the standard value, is 980.665 cm/sec^2.

standard pressures Adopted values of pressure used for specific purposes; the value of a standard pressure of one atmosphere, used in the determination of gas densities, is defined as the pressure produced by a column of pure mercury 76.0 centimeters in height at a temperature of 0°C. under standard gravity.

standard time Time used at a place, in accordance with a plan agreed upon by an international conference in 1884.

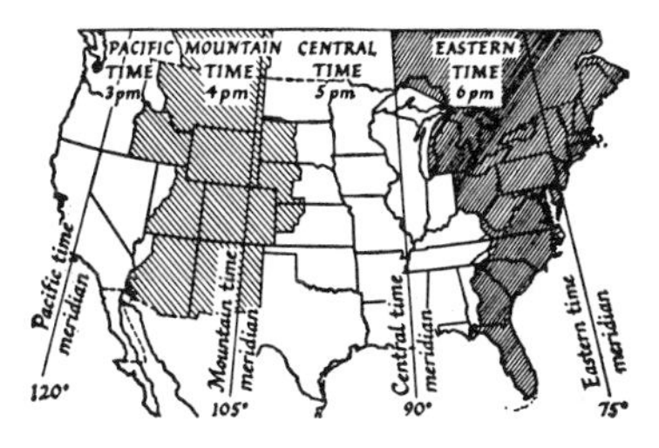

standard time

This plan states that standard meridians are to be used at intervals of 15 degrees (or one hour) east and west of Greenwich, England, and the ideal standard time at any place is the local civil time of the standard meridian nearest that place; i.e., when it is noon at any standard meridian, it is conventionally noon at all places 7½ degrees east of and 7½ degrees west of that standard meridian. In actual practice, however, the standard time of a place is the time of the meridian chosen for convenience.

star Any of the great light-giving bodies of the universe of which the nearest is the sun. Their energy is derived mainly from the conversion of hydrogen to helium. Their brightness and their color is related to the rate at which nuclear reactions take place.

starch Polysaccharide carbohydrates present in seeds, tubers, and other parts of plants.

starling A bird of European origin introduced into this country in 1890 to curb insects; they have multiplied so rapidly that they have become major urban pests.

state of the sky The cloud cover aspect of the sky. The state of the sky is fully described when the amounts, kinds, directions, and heights of all clouds are given.

static Electromagnetic waves that interfere with radio reception.

stationary cyclone A large, more or less stationary and semipermanent low-pressure area, such as the aleutian low.

stationary front A front along which one of the two air masses separated by it is not displacing the other.

statute mile See: mile (statute)

steady-state theory The theory that the universe is in a state of dynamic equilibrium and does not go through a cycle of explosion and contraction.

steam In meteorology, visible condensed vapor rising from ground or water, best exemplified by the steaming of rivers in times of intense frost, and of asphalt roads when the sun shines after a sudden cold shower, and the rising of clouds of condensed vapor over the open sea in the arctic regions.

steam fog Fog formed when cold air having a low vapor pressure passes over warm water. Steam fogs are observed in the arctic where they are also called arctic sea smoke and over inland lakes and rivers where they interfere with aviation.

steam mist The result of steam forming over open water, such as lakes and rivers. It often occurs in the United States on clear, autumn mornings. In the arctic it is called arctic sea smoke.

steering The directing influence exercised on the trajectories of surface disturbances by the flow patterns at 10,000 feet or some other upper level.

Stefan's law Sometimes called Stefan Boltzmann's law, which states that the amount of energy radiated per unit time from a unit surface of a black body is proportional to the fourth power of the absolute temperature of the black body.

St. Elmo's fire A luminous brush discharge of electricity from pointed objects on the earth to the air, a phenomenon often seen under stormy conditions.

stem flow The dripping of water down the trunk of a tree or down branches of a shrub during foggy or rainy periods.

steppe A physiographic and climatic term, first applied by the Russians to their grasslands. It is a semiarid region characterized by short clump grasses.

stereographic projection A conformal map projection in which lines emanate from a point directly opposite a point where a plane touches the globe.

stereophotogrammetry The process of preparing a relief map from aerial photographs of the area.

sterol Any of a class of solid higher alcohols, such as cholesterol and phytosterol, widely distributed in animals and plants. Sterols in general are colorless crystalline compounds of very complex nature and play important physiological roles not yet fully understood. The sterols are neutral and comparatively stable substances that occur partly in the free condition and partly esterified with higher fatty acids.

Stevenson screen An English term for the standard housing for meteorological thermometers.

stikine wind A local term for the severe, gusty (williwaw) east-northeast wind named after the Stikine River near Wrangell, Alaska.

stock A small-sized batholith.

stock resource Resources that are capable of being used up or consumed.

Stokes' law A law in physics that gives the steady rate of fall or terminal velocity of small spheres falling through a viscous fluid. According to this law, a raindrop of 0.001 centimeter diameter has a limiting velocity of fall in the atmosphere of approximately 0.0091 centimeters per second, and hence would take about three hours to settle one meter.

stoma (pl., stomata) Minute openings, chiefly on the surface of the leaves of plants, through which water is evaporated and through which gaseous exchange takes place. Stomata are physiologically regulated by the plant.

Stone Age The period when men used stone weapons and tools.

stone-river A stream of boulders and rock fragments that moves down a valley slowly.

stooping An optical phenomenon in which rays from the base of an object are curved down by atmospheric refraction much more rapidly than those from the top, with the obvious result of apparent vertical contraction, and the production of effects quite as odd and grotesque as those due to towering.

storage 1. Water artificially impounded in surface or underground reservoirs for future use. 2. Water naturally detained in a drainage basin, such as ground water, channel storage, and depression storage.

storage capacity The amount of water that can be stored in the soil for future use by plants and evaporation.

storage ratio The net available storage divided by the mean flow for one year.

storm In general, a disturbance of the ordinary average conditions of the atmosphere, which, unless specifically qualified, may include any or all meteorological disturbances, such as wind, rain, snow, hail, thunder, etc.

storminess A term with no recognized precise meaning; generally it refers to that quality of climate that is evidenced by the frequency of cyclones or their violence and may also include factors of cloudiness and precipitation.

storm loss Infiltration plus depression storage. The term also indicates interception loss in some cases.

storm paths Lines drawn on maps indicating the paths of the centers of all the lows for a certain period, or the average tracks or paths for each of the various types of lows. They are also called storm tracks, tracks of lows, etc.

storm seepage That part of precipitation that infiltrates the surface soil and moves toward the streams as ephemeral, shallow, perched ground water above the main ground-water level. Storm seepage is usually part of the direct runoff.

storm tide See: tidal wave

storm tracks See: storm paths

storm wave A rise of the sea over low coasts not ordinarily subject to overflow. It is sometimes called a tidal wave, though it is caused primarily by wind and has no relation to the tide brought about by gravitational forces except that the two may combine.

straight blow A wind of considerable force that blows for a long distance (perhaps more than 100 miles) in a straight direction. It is known by various names, such as line blow, straight-line gale, long-path, or geostrophic wind.

strait A narrow stretch of water connecting two ocean areas.

stratification 1. A structure produced by deposition of sediments in beds or layers. 2. The separation of the atmosphere into more or less nearly horizontal layers that mark separate steps in the change of some particular element with height. Thermal stratification is an example.

stratified (adj.) Composed of, or arranged in, strata or layers. See: stratum

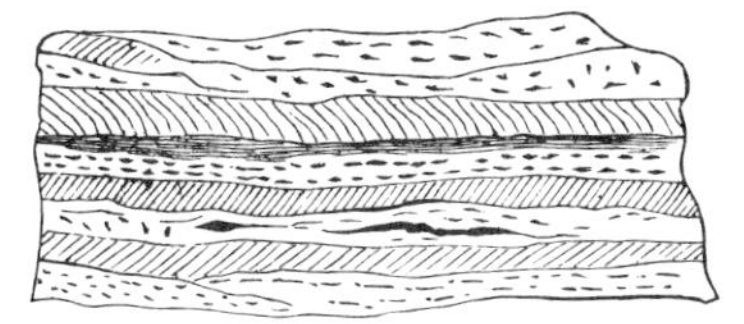

stratified

stratiform (adj.) Pertaining to stratus clouds or clouds of similar form.

stratigraphy The study of the formation, composition, sequence, and correlation of the stratified rocks of the earth's crust.

stratocumulus A cloud layer or cloud patches composed of laminae, globular masses, or rolls. The smallest of the regularly arranged elements are fairly large, soft, and gray with darker parts. These elements are arranged in groups, in lines, or in waves, aligned in one or in two directions. Very often the rolls are so close that their edges join. When they cover the whole sky, they have a wavy appearance.

stratopause The transition layer between the stratosphere and the mesosphere.

stratosphere The region of the upper atmosphere characterized by little or no temperature change with altitude; there may even be a slight increase of the temperature upward.

stratovolcano A relatively large volcanic cone built of alternating layers of lava and pyroclastic materials.

stratum (pl. strata) A layer of sedimentary rock, or a group of layers consisting throughout of approximately the same material. Stratum, bed, and layer are approximately equivalent terms.

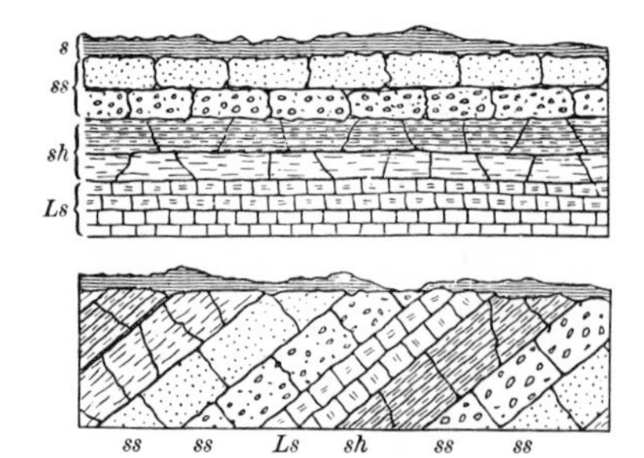

stratum

stratus A low uniform layer of cloud resembling fog but not resting on the ground. When this very low layer is broken up into irregular shreds it is designated fractostratus.

streak lightning The ordinary sinuous form of lightning.

stream A general term for a body of flowing water. In hydrology the term is generally applied to the water flowing in a natural channel as distinct from a canal. More generally, as in the term stream gauging, it is applied to the water flowing in any channel, natural or artificial. Streams may be classified as ephemeral, intermittent, or perennial, in relation to time. In relation to ground water, they may be:

Gaining. A stream or reach of a stream that receives water from the zone of saturation.

Insulated. A stream or reach of a stream that neither contributes water to the zone of saturation nor receives water from it. It is separated from the zones of saturation by an impermeable bed.

Losing. A stream or reach of a stream that contributes water to the zone of saturation.

Perched. A perched stream is either a losing stream or an insulated stream that is separated from the underlying ground water by a zone of aeration.

stream-channel degradation The removal of channel bed materials and the downcutting of natural stream channels.

streamflow The discharge that occurs in a natural channel. The term is more general than runoff, as streamflow may be applied to discharge whether or not it is affected by diversion or regulation.

stream gauging 1. Measuring the velocity of a stream of water in a channel or open conduit, and the average of cross section of the water, for the purpose of determining the discharge. 2. A discharge measurement expressed numerically and in appropriate units as the discharge determined.

stream gradient A general slope, or rate of change in vertical elevation per unit of horizontal distance, of the water surface of a flowing stream.

streamlines Lines that are everywhere parallel to the instantaneous direction of motion in a fluid. The streamline field is customarily constructed in such a way that the spacing of the streamlines at a given point is inversely proportional to the speed of the motion at this point.

stream order A method of numbering streams as part of a drainage basin network. The smallest unbranched mapped tributary is called first order, the stream receiving the tributary is called second order, and so on. It is usually necessary to specify the scale of the map used. A first-order stream on a 1:62,500 map may be a third-order stream on a 1:12,000 map. Tributaries that have no branches are designated as of the first order, streams that receive only first-order tributaries are of the second order, larger branches that receive only first-order and second-order tributaries are designated third order, and so on, the main stream being always of the highest order.

stream regimen The condition of a stream and its channel as it relates to their erosive characteristics. A stream or conduit is in regimen or in physiographic balance if its channel has reached a stable form as the result of its flow characteristics.

stream system A river and its tributaries.

stream table A laboratory tool used to examine the way in which streams act. It consists of a table, a bed of sand, and a water source.

stream terrace The gently sloping elevated surfaces found in stream valleys which represent old floodplain levels.

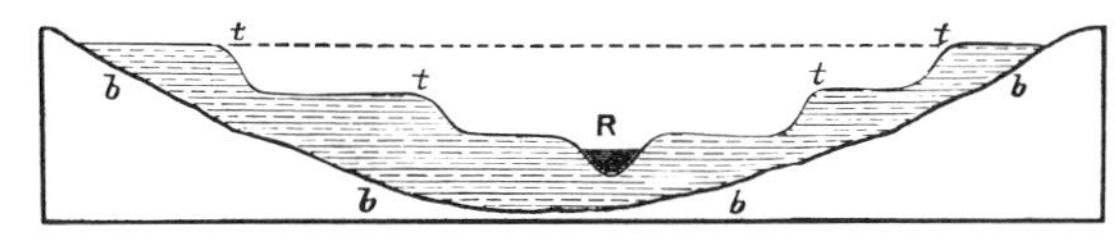

stream terrace

stream underflow Percolating water flowing parallel to the surface stream, in the permeable bed of a stream.

stress 1. The total energy with which water is held in the soil, including tension of soil moisture and additional effects of salts in the soil water. It can be expressed in any convenient pressure unit. 2. The force exerted on one thing by another.

striae Grooves worn on the surfaces of rocks by rock fragments dragged by a glacier.

strike The bearing of the outcrop of an inclined layer of rock perpendicular to the dip.

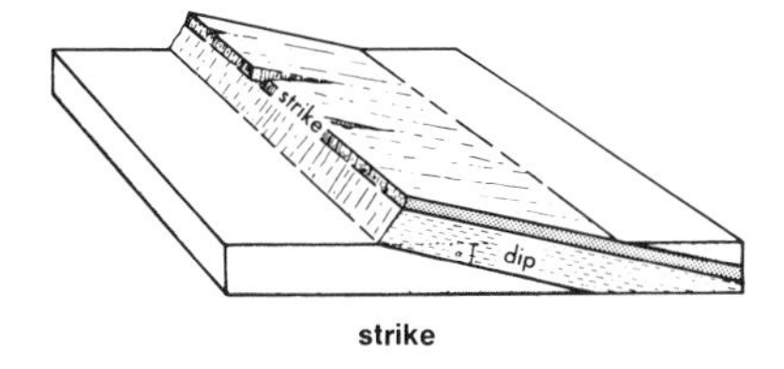

strike

strip cropping (or farming) The practice of growing crops in a systematic arrangements of strips or bands. Commonly cultivated crops and sod crops are alternated in strips to protect the soil and vegetation against running water or wind.

strip mining The mining of ores near the surface by stripping off the overburden to get at the minerals.

strontium 90 The radioactive isotope of strontium. It is the principal form of fallout from nuclear bombs and causes tumors if taken in food or drink.

structure 1. In geology, features produced in rock by movements that take place after deposition. 2. The arrangement and interrelation of subdivisions of the atmosphere in respect to their physical characteristics such as temperature, pressure, density, humidity, and movement. The structure of the atmosphere is, therefore, the instantaneous picture of the static and dynamic conditions at any given time. 3. In soil study, the arrangement of primary soil particles into compound particles or clusters that are separated from adjoining aggregates and have properties unlike those of an equal mass of unaggregated primary soil particles. The principal forms of soil structure are platy, prismatic, columnar (prisms with rounded tops), blocky (angular or subangular), and granular. Structureless soils are (a) single grain—each grain by itself, as in dune sand—or (b) massive—the particles adhering together without any regular cleavage as in many claypans and hardpans.

St. Swithin's Day A meteorological key day or control day, July 15. An old proverb says that if it rains on St. Swithin's Day, rain will continue for forty days.

stubble mulch A mulch consisting of the stubble and other crop residues left in and on the surface of the soil as a protective cover during the preparation of a seedbed and during at least part of the growing of the succeeding crop.

stumpage The value of timber as it stands uncut in the woods; in a general sense, the standing timber itself.

sturgeon A valuable food fish. Their roe is also used for caviar. There are both marine and fresh-water species.

Stüve diagram The form of the adiabatic or pseudoadiabatic diagram that has the absolute temperature on a linear scale as abscissas, and the pressure plotted on an exponential scale decreasing upward in terms of $p^{0.228}$ as ordinates. This particular form of the adiabatic diagram has the advantage of making dry adiabatic (isentropic) processes appear as straight lines.

subhumid climate A climate intermediate between semiarid and humid with sufficient precipitation to support a moderate to heavy growth of short and tall grasses, or shrubs, or of these and widely spaced trees or clumps of trees. The precipitation-effectiveness (P-E) index ranges from about 32 to 64. The upper limit of rainfall in subhumid climates may be as low as 20 inches in cold regions and as high as 60 inches in hot regions.

subirrigation Irrigation through controlling the water table in order to raise it into the root zone. Water is applied in open ditches or through tile until the water table is raised enough to wet the soil. Some soils along streams are said to be naturally subirrigated.

sublimation The transition of a substance directly from the solid state to the vapor state, or vice versa, without passing through the intermediate liquid stage.

sublimation nucleus One of a group of particles in the atmosphere about which an ice crystal will form by sublimation.

subpolar region The region lying equatorward of the polar zone. This region includes the taiga.

subsequent stream A tributary to consequent stream, flowing along the strike of inclined sedimentary beds.

subsidence The slow settling or sinking of a stagnant mass of air, generally accompanied by divergence in the lower layers. In its slow movement downwards, the air is compressed and warmed at the dry adiabatic rate, so that its thermal structure is changed and its stability enhanced.

subsidence inversion An increase in temperature vertically through a layer of the atmosphere, caused by subsidence.

subsistence farming Agriculture in which most of the crops produced are utilized on the farm.

subsoil The B-horizons of soils with distinct profiles. In soils with weak profile development, the subsoil can be defined as the soil below the plowed soil (or its equivalent of surface soil), in which roots normally grow. Although a common term, it cannot be defined accurately. It has been carried over from early days when "soil" was conceived only as the plowed soil and that under it as the "subsoil."

subsoiling The tillage of the soil below the normal plow depth, usually to shatter a hardpan or claypan.

subspecies A division of a species.

subsurface flow Water that infiltrates the soil surface and moves laterally through the upper soil layers until it enters a channel.

subsurface water All water below the land surface, including soil moisture, intermediate zone water, capillary fringe water, and ground water.

subterranean (adj.) Underground.

subterranean waters In legal parlance those subsurface waters whose courses are well defined and reasonably ascertainable and whose existence is not temporary.

subtropical (or subtropic) (adj.) 1. Pertaining to a type of climate found at the tropical margins of the temperate zones. 2. Pertaining to the belts of high pressure and of calms or variable winds in the general vicinity of 30°N. and 30°S. latitude.

subtropics Subtropical regions, which lie at the equator side of the temperate zones.

succession The process of replacement of one plant community by another until the climax is reached. Each community in turn changes the temperature, moisture, and other factors of the environment; these new

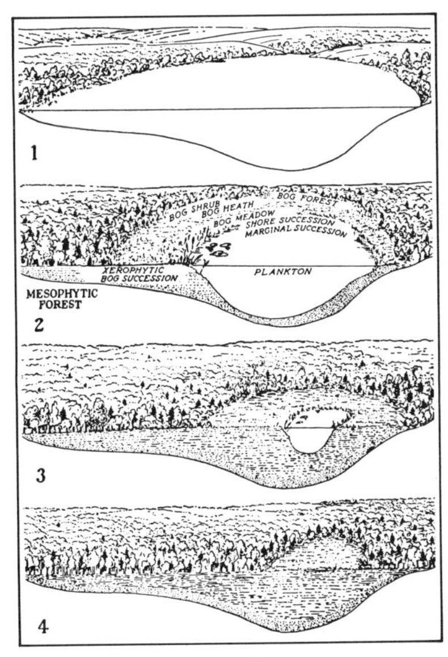

succession

conditions hinder the community that brought them about and favor a new one, which becomes the next step in the succession.

sucrose Sugar obtained from sugar cane, the sugar beet, and sorghum. Sucrose is designated $C_{12}H_{22}O_{11}$.

sudd Floating vegetable matter in the White Nile. It forms an impediment to navigation.

Suestado A storm with southeast gales, caused by intense cyclonic activity off the coasts of Argentina and Uruguay, which affects the southern part of the coast of Brazil in the winter.

sugar Any of a class of sweet, soluble compounds comprising the simpler carbohydrates. The simple sugars are those that are not reduced to simpler or smaller molecular units by hydrolysis. Complex sugars are formed by the union of molecules of two or more of the simple sugars.

sugar beet A tuberous vegetable from which sugar is extracted.

sugar cane A tall annual tropical plant belonging to the grass family. It is a principal source of sugar.

sugar pine The largest American pine, grown in the mountains of Oregon and California. It is an important lumber tree.

sulfur A common mineral, pale yellow in color. It is used to make matches, gunpowder, insecticides, and rubber goods.

sulfur dioxide A colorless, nonflammable suffocating gas, formed when sulfur burns. It is designated SO_2.

sultry (adj.) Term used to describe weather that is hot and humid.

sumatra In the Malacca Straits, a squall with violent thunder, lightning, and rain, which blows at night usually during the southwest monsoon and is intensified by strong mountain breezes.

summer A season of the year—the warmest in temperate and polar regions, and the dry season in the tropics. Meteorologically, the summer is most commonly regarded as including the months of June, July, and August; in astronomical practice, however, it extends from the summer solstice, about June 21, to the autumnal equinox, about September 22.

sun The star that is the central body of our solar system and about which the planets revolve and from which they receive light and heat.

sun drawing water The popular phrase for the phenomenon of crepuscular rays.

sunrise The phenomenon of the sun's appearance on the eastern horizon as a result of the earth's rotation.

sunset The phenomenon of the sun's disappearance below the western horizon as a result of the earth's rotation.

sunshine Light received directly from the sun.

sunshine recorder An instrument designed to record the duration of sunshine, without regard to its intensity. It may operate by reacting to either the heat energy or the photographic effect of the sun's rays.

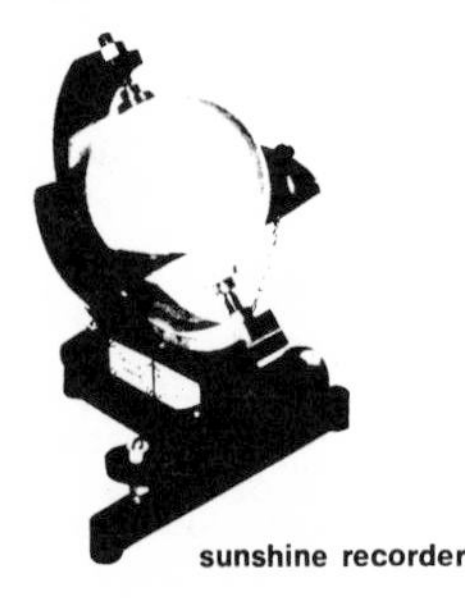

sunshine recorder

sunspot A dark spot on the surface of the sun. Sunspots are thought to be violent eruptions of gases cooler than the surrounding surface areas.

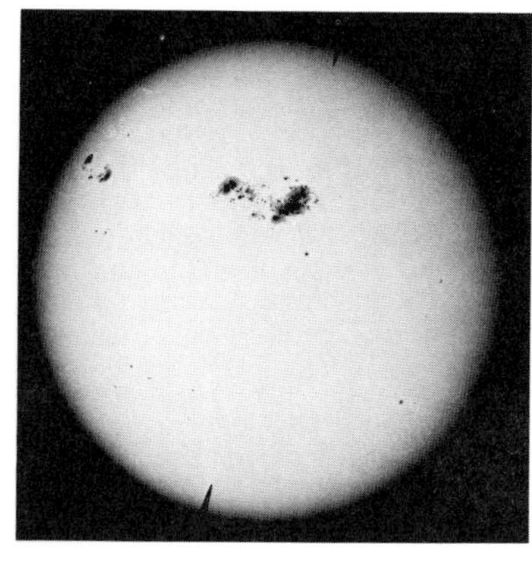

sunspot

sunspot numbers The numbers of sunspots apparent on the sun at different times.

sunstroke A summer malady affecting those whose body mechanism does not accommodate itself to the excessive heat. The first danger sign is the cessation of perspiration, resulting in fever and often coma.

superadiabatic lapse rate A temperature lapse rate in the free atmosphere such that the potential temperature decreases with height, or such that any air particle that is displaced adiabatically upward or downward from its initial position finds itself increasingly warmer or colder, respectively, than the surrounding atmosphere.

supercooled cloud A cloud composed of water droplets with temperatures below 0°C.

supercooled water droplet A liquid water droplet whose temperature is below 0°C. These have been observed in fogs having temperatures as low as –40°C and in clouds nearly as cold. Supercooled water seems to be dependent upon the undisturbed condition of the water molecules, for when they are disturbed, as by contact with a solid surface, solidification takes place at once.

superheating The heating of a vapor, particularly saturated (wet) steam, to a temperature much higher than the boiling point at the existing pressure. This is done in power plants to improve efficiency and to reduce condensation in the turbines.

supernova An explosion of a star with such extraordinary intensity that it is observable on earth.

superposed stream A river whose present course has been established on young rocks burying an old surface. With uplift, the stream cuts into the older surface.

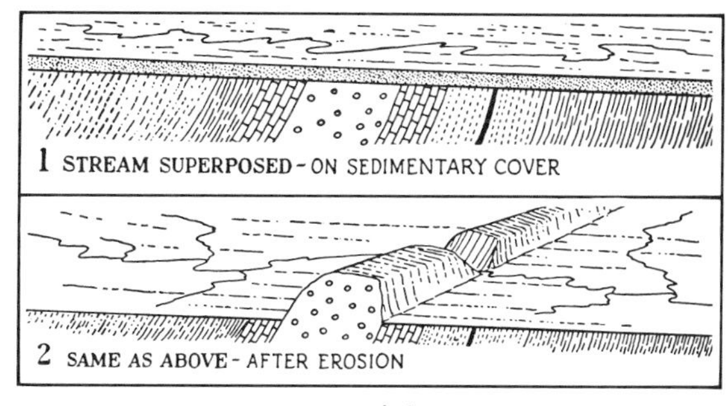

superposed stream

supersaturation The condition existing in a given space when it contains more water vapor than is needed to cause saturation.

supersonic See: ultrasonic

surf Waves breaking into foamy water on the shore or around rocks.

surface detention That part of the rain that remains on the ground surface during rain and either runs off or infiltrates after the rain ends.

surface runoff That part of the runoff that travels over the soil surface to the nearest stream channel. It is also defined as that part of the runoff of a drainage basin that has not passed beneath the surface since precipitation.

surface tension A phenomenon peculiar to the surface of liquids, in which the surface molecules seem to have a greater cohesion for one another than do the molecules in the body of the liquid, so that the surface acts like a stretched elastic film.

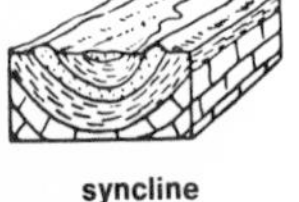
syncline

surge 1. Any change of pressure over a wide area, the precise cause of which is unknown, but which is not due to the passage of a low or a high pressure area or to diurnal barometric variation. 2. In reference to wind, a variation of its mean velocity occurring about once in a minute or some longer period, the mean velocity being calculated for one minute or the longer period.

survival of the fittest A tenet of Darwin's theory of evolution that states that only the best adapted individuals survive to breed succeeding generations.

suspended sediment The very fine soil particles that remain in suspension in water for a very considerable period of time without contact with the bottom.

suspensoids Colloidal particles that remain in suspension under all conditions and will combine or react with the liquid in which they exist to only a limited extent.

sustained yield A relatively constant level of production from a renewable resource.

swale A shallow, marshy depression in generally level ground or a shallow depression in a glacial moraine.

swamp A water-saturated tract of land covered with dense vegetation.

swell An ocean wave caused by wind.

symbiosis The living together of two different organisms with a resulting mutual benefit. A common example includes the association of rhizobia with legumes; the resulting nitrogen fixation is sometimes called symbiotic nitrogen fixation.

synchronous Changing in the same way at the same regular times.

syncline Layers of rock folded so that they slope upward and outward from the center line of the fold. See: anticline

synclinorium A broad regional trough within which are a number of minor folds.

syndrome A characteristic group of symptoms that indicate a certain physical condition or disease.

synergism Cooperative action of discrete entities such that the total effect is greater than the sum of two effects taken independently.

synodic period The time between one conjunction of two bodies and the following conjunction.

synoptic (adj.) Atmospheric conditions existing at a given time over an extended region: e.g., a synoptic weather map, which is drawn from observations taken simultaneously at a network of stations over a large area, thus giving a general view of weather conditions.

synoptic analysis The process of analyzing the atmospheric conditions by studying synoptic charts arranged in chronological sequence.

synoptic chart 1. A map of a limited region of the earth, which contains data of weather conditions at many observation points taken simultaneously, or nearly so. 2. A map showing the change of one or more of the meteorological elements at a number of points from one hour of observation to another, or showing the distribution of accumulated precipitation during the period between observations.

synoptic meteorology A branch of meteorology concerned with the problem of interpreting collective meteorological observations made simultaneously at the surface or aloft at a number of places over a large area of the earth.

synthesis The combination of simple molecules to form another substance: for example, the union of carbon dioxide and water under the action of light in photosynthesis.

system Objects or groups of objects, including both the materials and the energy in them (e.g., an automobile).

system analysis The process of evaluating the inputs and costs required by a program as well as evaluating the outputs, service, benefits, and profits. This is also known as system approach and system management.

syzygy The time at which two heavenly bodies are in conjunction or opposition.

t

Tablecloth A local name for a form of crest cloud that occurs over Table Mountain in South Africa.

table iceberg An iceberg that has broken off from shelf ice, which is formed along a polar coast in shallow bays and inlets. A table iceberg often reaches the bottom and becomes fastened to the shore, and may grow hundreds of miles out to sea.

tableland An elevated and generally level region of considerable extent bounded by steep cliffs on at least one side; a plateau.

tabular iceberg A mass of ice calved from shelf ice. It has a flat upper surface. The upper portion is formed of stratified snow or névé.

taconite A low-grade iron ore that contains from 20 to 30 percent iron in the form of magnetite.

taiga A Russian word applied to the cold forested region of the north (particularly Siberia) which begins where the tundra leaves off, and also to like regions in Europe and North America.

tail race A channel that conducts water from a water wheel.

Taku wind Local name given to a strong, gusty east-northeast wind that can occur in the vicinity of Juneau, Alaska, between October and March.

talc Hydrous magnesium silicate. It is the softest mineral known and has many commercial uses, such as filler in making paper, roofing, paint, soap, and cosmetics.

tall-grass prairie Grasslands on the margins of forested regions with grasses three to six feet tall. Most prairies are now cultivated land.

talus A mass of rock debris that collects at the bottom of a cliff.

talus creep The movement of rock fragments down a talus slope.

tank One of a group of lakes or ponds formed by mud dams built across the valleys of small streams in India. They fill during the rainy season and, in the Deccan plateau, provide the sole source of water.

tannin One of a group of astringent, aromatic, acidic compounds found in the bark, roots, wood, foliage, or fruit of various trees and other plants. Tannin is used to convert hide substance (collagen) into leather.

tapeworm A parasitic flatworm that can live in the digestive tract or liver of every vertebrate, including man.

target area In weather modification, the area within which the effects of seeding are expected to be found.

tarn A small mountain lake occupying a cirque.

taro A root crop which is a principal source of food for Polynesians and other inhabitants of the Pacific islands.

taxonomy The systematic classification of living things into kingdoms, phyla, classes, orders, families, genera, and species.

Taylor Grazing Act An act passed by Congress in 1934 to bring unreserved public domain under proper management. Since 1946 this land has been under the jurisdiction of the Bureau of Land Management.

tea A shrub native to southeast Asia. The leaves are used to make a beverage.

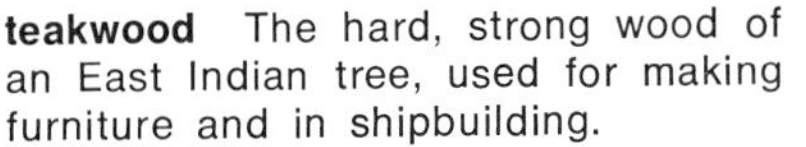

teakwood The hard, strong wood of an East Indian tree, used for making furniture and in shipbuilding.

tectonic (adj.) Pertaining to the forces and processes of earth movement: folding, faulting, etc.

Tehuantepecer A violent north wind, frequent in the winter in the region around the Gulf of Tehuantepec in Mexico. It is similar to the bora and mistral. It is caused by the outpouring of air that has travelled from the North American continent across the Gulf of Mexico and through the seventy-mile-wide gap in the high cordillera that exists in the Isthmus of Tehuantepec.

telethermoscope Usually, an electrical-resistance thermometer whereby temperatures at a distance may be determined but not recorded. It consists essentially of two parts, the thermic element and the indicating apparatus. At meteorological stations, the thermic element of this instrument is placed in the instrument shelter and the indicating apparatus is placed in the office of the observer.

temblor A Spanish word meaning trembling, used in the southwestern United States to mean an earthquake.

temperate rain forest See: rain forest

temperate zone The climatic regions in the mid-latitudes, characterized by cool or cold winters and warm or hot summers.

temperature 1. The thermal state of a substance with respect to its ability to communicate heat to its environment. 2. The measure of this thermal

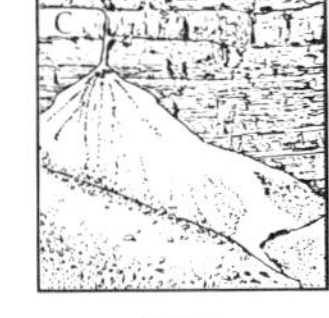

talus

state on some arbitrarily chosen numerical scale.

temperature anomaly The difference between the mean temperature of a place and that of the parallel of latitude on which it is situated.

temperature (or thermal) efficiency A climatic element in Thornthwaite's classification of climates, corresponding to precipitation effectiveness. It expresses the degree to which the temperature of a place favors plant growth, and ranges from zero on the polar limit of the tundra to a maximum in the tropics.

temperature gradient The rate of change of temperature with distance in any given direction at any point. The average rate of change of temperature between two points is the difference between their temperatures divided by the distance between them. The vertical temperature gradient, or lapse rate (the preferred term), is the rate of change of the temperature of the air with elevation.

temperature inversion An increase of temperature with altitude.

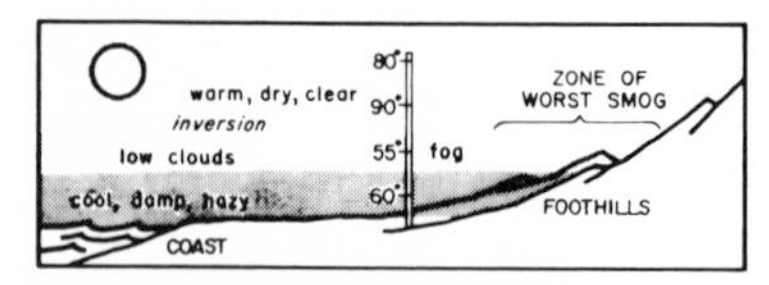

temperature inversion

temperature lag 1. The continued increase in temperature after the sun has passed its position directly overhead at noon. Incoming radiation continues to exceed re-radiation from the earth until after 2 p.m. 2. Similar increases in temperature in the northern hemisphere after the summer solstice (June 22), when the sun is at its highest position. The long hours of daylight and relatively direct radiation cause the summer temperatures to continue increasing into July and August.

temperature scale An arbitrary numerical scale of thermal states for measuring temperature. The scale of temperature in common use in meteorology are the Fahrenheit and Celsius; their fiducial points are the freezing and boiling points of water under standard conditions. The conversion from Celsius to Fahrenheit temperatures and vice versa may be accomplished by using the following formulae: $C^\circ = 5/9\,(F^\circ - 32^\circ)$, and $F^\circ = 9/5C^\circ + 32^\circ$.

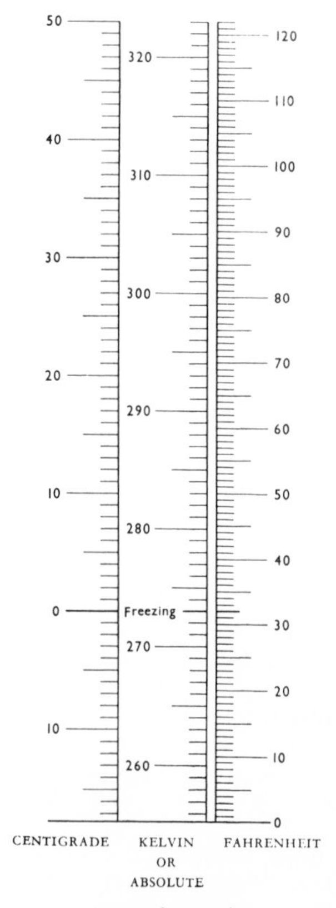

temperature scale

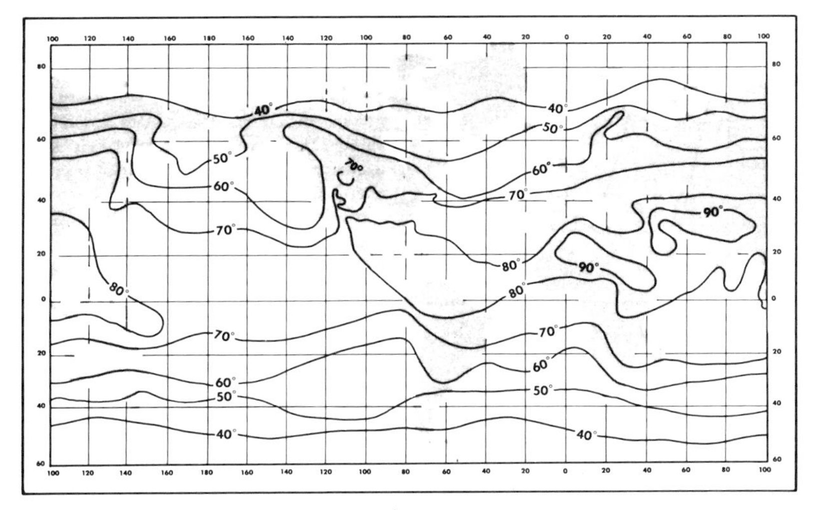

temperature zone

temperature zone An area or latitudinal belt on the earth delimited by given temperature conditions.

temporales Strong southwest winds blowing in summer on the Pacific coast of Central America.

tenant farmer A farmer who rents from a big landowner as much land as he and his family can cultivate.

tendency See: barometric tendency

Tennessee Valley Authority (TVA) The federal agency that has built dams and aided in the economic development of the Tennessee River Valley.

tensiometer A device for measuring the tension with which water is held in the soil. It is a combination of a porous cup and a vacuum gauge.

tephigram A thermodynamic diagram used in estimating the quantity of available convective energy in an overlying air column, for example, in forecasting the probability of thunderstorms. The ordinate is entropy or the logarithm of potential temperature, and the abscissa is absolute temperature.

terminal moraine The pile of glacial drift marking the point of farthest advance of a glacier.

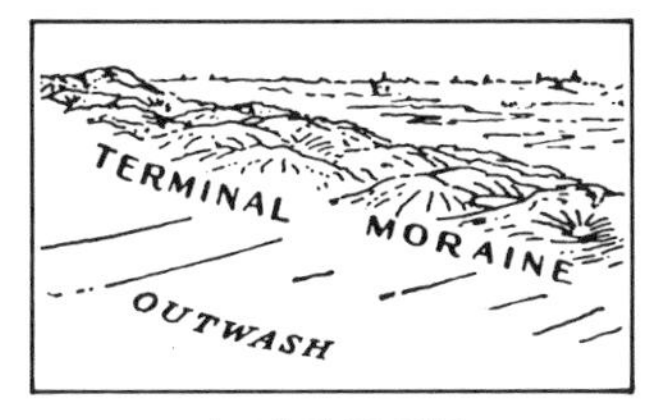

terminal moraine

terminator The line across the moon separating the illuminated from the unilluminated part.

tern Any of several birds resembling a seagull.

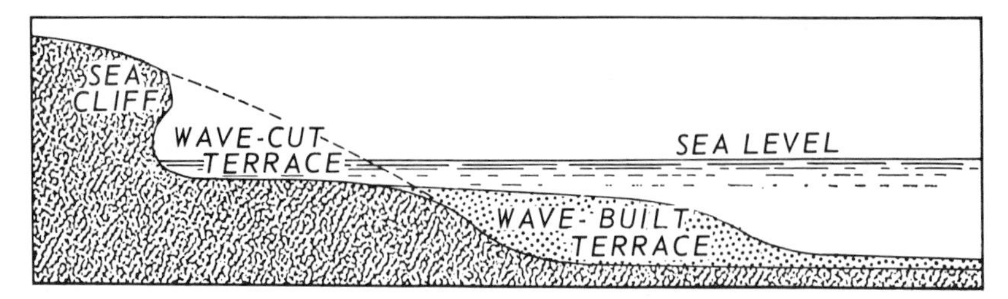

terrace

terpene A hydrocarbon having the general formula $C_{10}H_{16}$, found in essential oils, resins, and other vegetable aromatic compounds, of which turpentine is an example.

terrace A relative flat earth surface, usually long and narrow, bounded by an ascending slope on one side and a descending slope on the other.

terracette A small terrace often seen in groups on steep grassy slopes.

terral The land breeze of the west coasts of Peru and Chile.

terra rossa Reddish clay found in limestone regions where a Mediterranean climate prevails.

terra roxa A reddish Brazilian soil with a high humus content, used for coffee production.

terrestrial radiation The energy that is radiated by the earth and its atmosphere. Black-body radiation at terrestrial temperatures of roughly 300°A will be contained within the limits 3μ and 80μ, having its maximum intensity at 10μ. Since this range of wavelengths is widely separated from that of incoming solar radiation, it is customary to call it long-wave radiation, to distinguish it from direct solar radiation, which is called shortwave radiation.

terrestrial equator The great circle of the earth midway between the poles. It is the zero of all latitude measurements.

territoriality The behavior of animals in defending their home territory.

territorial waters The zone of waters along a coast that a country considers to be within its jurisdiction.

tertiary circulation A term used in some classifications of wind movements that subdivide atmospheric motions into the primary or planetary circulation, comprising the permanent winds; the secondary circulation, comprising cyclones, anticyclones, and monsoons; and the tertiary, comprising the numerous variable winds, land and sea breezes, thunderstorm winds, etc.

tertiary economic activities Trade and

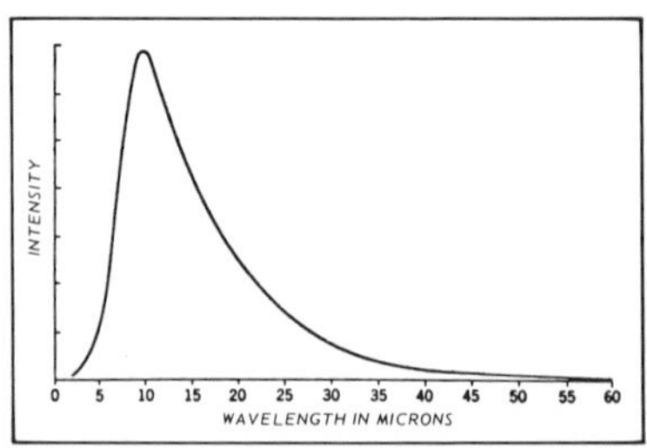

terrestrial radiation

commerce: the disposal of the products of primary and secondary economic activities.

tertiary treatment In sewage, the additional treatment of effluent beyond that of secondary treatment, in order to obtain a very high quality of effluent.

tetragonal prism A prism with four sides.

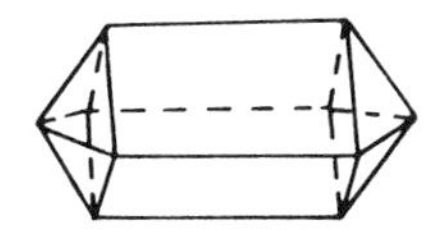

tetragonal prism

textural class The classification of soil material according to the proportions of sand, silt, and clay. The principal textural classes in soil, in increasing order of the amount of silt and clay, are as follows: sand, loamy sand, sandy loam, loam, silt loam, silt, sandy clay loam, clay loam, silty clay loam, sandy clay, silty clay, and clay. These class names are modified to indicate the size of the sand fraction or the presence of gravel, cobbles, and stones. For example, terms such as loamy fine sand, very fine sandy loam, gravelly loam, stony clay, and cobbly loam, are used on detailed soil maps. These terms apply only to individual soil horizons or to the surface layer of a soil type.

texture In soil study, the relative proportions of the various size groups of individual soil grains in a mass of soil. Specifically, the term refers to the proportions of sand, silt, and clay.

thalweg 1. The lowest trough along the axial part of a valley. 2. A subsurface ground-water stream percolating beneath and in the general direction of a surface stream course or valley. 3. The middle or chief navigable channel of a waterway.

thaw 1. (v.) To melt. When ice or snow melts, it is said to thaw. 2. (n.) A weather condition in which ice and snow melt, as in the "January thaw" known in parts of the New England and middle Atlantic states.

theodolite An optical instrument in which a telescope rotates around vertical and horizontal axes and is equipped with graduated circles to measure horizontal and vertical angles.

theodolite

theoretical meteorology See: dynamic meteorology

theory A proposed explanation whose status is still conjectural.

thermal 1. (adj.) Pertaining to heat. 2. (n.) An ascending air current of thermal origin. Used in connection with gliding.

thermal belt or zone The portion of a slope above a valley floor that is marked by relatively high night temperatures.

thermal capacity The number of calories required to raise the temperature of a body 1°C, or the quantity of heat that is given up when the temperature is lowered 1°C, provided in both cases that there is no change of state.

thermal climate The climatic type in a system of classification based on actual temperatures and consisting of a number of temperature zones, delimited by selected annual isotherms instead of parallels of latitude.

thermal cracking A process in which complex molecules are broken down into simpler ones by the action of heat, usually in the presence of catalysts: for example the production of gasoline from petroleum.

thermal efficiency The ratio of the electric power produced by a power plant to the amount of heat produced by the fuel; a measure of the efficiency with which the plant converts thermal to electrical energy.

thermal equator The region of the earth inclosed within the annual isotherms of 80°F.

thermal gradient The rate of variation of temperature from the surface of the earth downward.

thermal island See: heat island

thermal pollution The ejection of heated water into the environment, usually aquatic ecosystems, raising the temperature above normal limits.

thermal stratification See: stratification

thermal zone See: thermal belt

thermocline A layer of water in which there is an abrupt decrease of temperature with depth.

thermocouple A device used to measure temperature. It is made by joining two dissimilar metallic conductors, such as copper and iron.

thermodynamic diagram Any diagram that satisfies the condition that the area on it represents work or energy. Such a diagram is used in meteorology to compute the energy available for convective activity in the atmosphere if an aerographic sounding is available.

thermodynamics The branch of physics that considers the processes involved in the transformation of heat into mechanical or other forms of energy, and of other forms of energy into heat.

thermogram The trace made by a thermograph.

thermograph An instrument designed to make an automatic record of temperature.

thermograph

thermometer An instrument for measuring temperature. It is most commonly based on the change in volume of a given substance with changes in temperature.

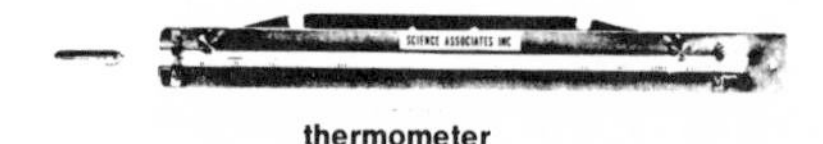

thermometer

thermometer screen Synonymous with thermometer shelter.

thermometer shelter A structure in which thermometers are exposed in order that they may attain, as closely as possible, the same temperature as that of the free air.

thermometry The science of measuring the temperature of any substance by employing any of the various types of thermometers.

thermonuclear reaction A reaction in which very high temperatures bring about the fusion of two light nuclei to form the nucleus of a heavier atom, releasing a large amount of energy. In a hydrogen bomb, the high temperature to initiate the thermonuclear reaction is produced by a preliminary fission reaction.

thermosphere That part of the upper atmosphere above the mesosphere in which the temperature increases continuously with elevation.

thiamine That part of the vitamin B complex that helps to keep the nerves, muscles, and digestion elements of higher animals healthy. Lack of thiamine produces beriberi.

thin (adj.) A term used in weather reports in describing the cloudiness whenever the solar or lunar disk or stars are faintly visible through them.

thinning A cutting made in an immature stand of trees for the purpose of increasing the rate of growth and improving the form (or quality) of the trees that remain and increasing the total production of the stand.

Thor The Scandinavian god of thunder.

Thor

thorn forest A deciduous forest of thorny scrub growth on the margins of the tropical forests.

threatening (adj.) A term popularly applied to weather which seems to betoken the appearance of a storm.

threshold temperature The temperature at which plant growth begins. It varies according to the species but for many temperate plants it is about 43°F.

throughfall In a vegetated area, the precipitation that falls directly to the ground or the rainwater or snowmelt that drops from twigs or leaves.

thrust fault (reversed fault) A fault in which the upper rock layers have been pushed forward over the lower ones.

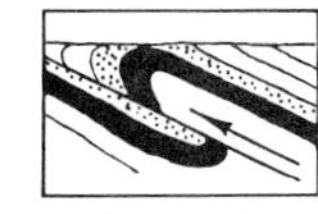

thrust fault

thunder The sound accompanying lightning. It is caused by the action of the lightning in heating and ionizing (and therefore rapidly expanding) the air along its path and sending out a compressional wave.

thunderbolt A lightning discharge accompanied by thunder. In ancient times it was thought that a lightning flash was accompanied by a bolt or dart which was responsible for the damage.

thundercloud A cumulonimbus or well-developed cumulus cloud; a thunderstorm cloud mass. See: cumulonimbus

thunderhead A rounded mass of cumulus cloud with shining white edges, often appearing before a thunderstorm.

thundersquall The wind that rushes out from the lower part of the thunderstorm squall cloud. It is experienced in all well-developed thunderstorms.

thunderstorm A local storm accompanied by lightning and thunder, often by strong gusts of wind and heavy rain, and sometimes by hail. Thunderstorms are of short duration: usually about two hours.

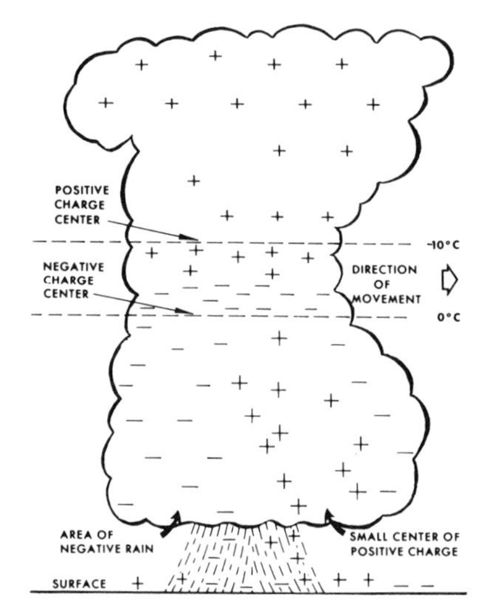

thunderstorm

tidal bore A tide-produced wave that flows up a river estuary or into a narrow, funnel-shaped bay.

tidal current The tidal movement of water in a bay. Water moves into it at floodtide and out at ebbtide.

tidal range The average difference in water level between high tide and low tide.

tidal wave 1. In meteorological and popular usage, a large, isolated, travelling ocean wave that suddenly inundates the land, most frequently caused by a seismic disturbance; or a rapid abnormal rise in sea level caused by the strong winds associated with a hurricane or severe gale, often reinforced by the astronomical tide, and also known as a storm wave. 2. In astronomical usage, the periodic variations of sea level produced by the gravitational attractions of the sun and the moon.

tides The regular rise and fall of sea level, caused chiefly by the gravitational pull of the moon.

tierra caliente The lowest of four zones of altitude in the mountain areas of Central and South America. This zone is below 3,000 feet and is hot and humid.

tierra fria In the mountains of Central and South America, the zone of altitude lying between 7,000 and 10,000 feet. It is the highest and coolest zone of vegetation.

tierra nevada In the mountains of Central and South America, the snow zone, above 10,000 feet in altitude.

tierra templada In the mountains of Central and South America, the zone lying between 3,000 and 7,000 feet. Temperatures are moderate and the annual range small.

till Unsorted, unstratified sediments carried and deposited by a glacier.

till

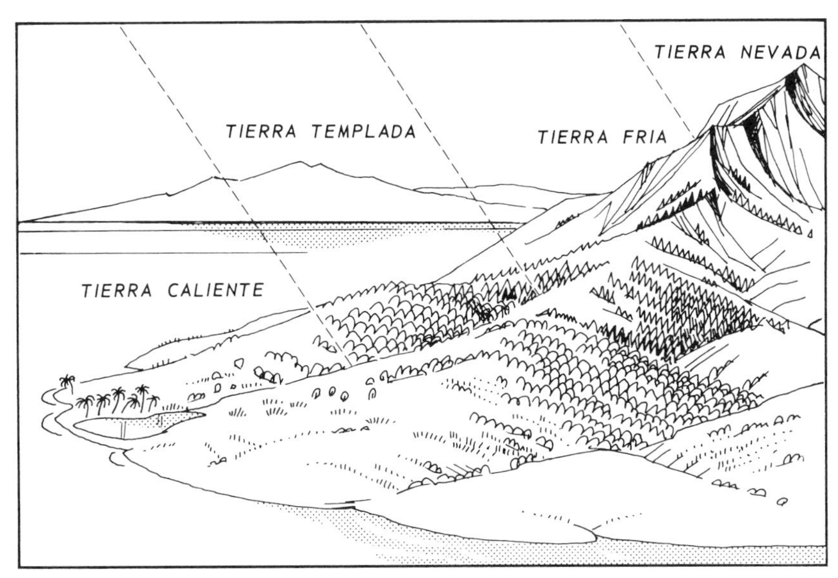

tierra (caliente, fria, nevada, templada)

tillage The operation of implements through the soil to prepare seedbeds and rootbeds.

tilth The physical condition of a soil in respect to its fitness for the growth of a specified plant or sequence of plants. Ideal soil tilth is not the same for each kind of crop nor is it uniform for the same kind of crop growing on contrasting kinds of soil.

timber line On mountains and in frigid regions, the line above which there are no trees.

tolerance The ability of an organism to withstand a given condition.

tombolo A sandbar that joins an island to the mainland or joins two islands.

topographic map A map showing the topography of an area, usually by means of contour lines.

topographic profile A drawing made to show the topographic shape of the land along a given line.

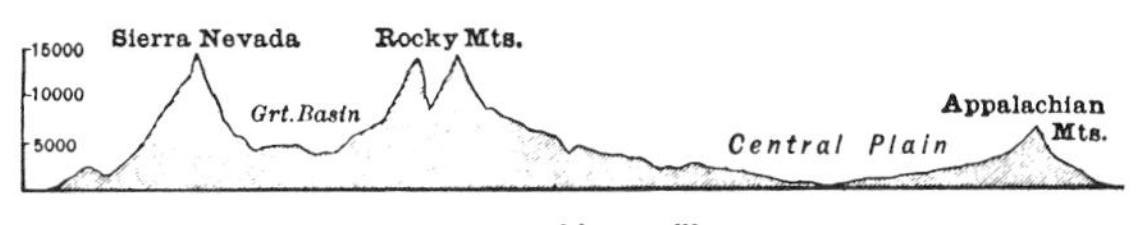

topographic profile

topography All the physical features of an area that can be represented on a map, especially the relief and contour of the land.

topsoil A general term used in at least four different senses: (a) a presumed fertile soil or soil material, usually rich in organic matter, used to topdress roadbanks, lawns, and gardens; (b) the surface plow layer of a soil, and thus a synonym for surface soil; (c) the original or present dark-colored upper soil, which ranges in depth from a mere fraction of an inch

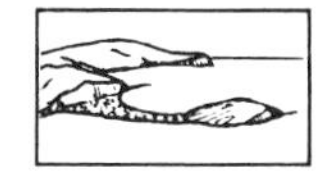
tombolo

to two or three feet on different kinds of soil; and (d) the original or present A-horizon, varying widely among different kinds of soil. Applied to soils in the field, the term has no precise meaning unless defined as to depth or productivity in relation to a specific kind of soil.

tornado A rotary storm of small diameter. It is one of the most violent types of storm known, and travels across the country leaving great devastation along a narrow path. In the central United States, where it most frequently occurs, it is also known popularly as a twister and as a cyclone.

tornado belt That portion of the central United States where tornadoes are most frequent.

Torricelli's experiment An experiment first performed in 1643, in which Torricelli filled with mercury a tube 33 inches long and closed at one end. He then inverted the tube in a dish of mercury and found that the mercury in the tube stood at a height of about 30 inches, leaving a vacuum above.

torrid zone The climatic zone lying between the tropics, and hence also called the tropic or tropical zone. It is the largest of the climatic zones, including nearly one-half the earth's area.

tourism The act of touring or the business of caring for the needs of tourists.

township A unit in the rectangular survey system ideally containing 36 sections, although many do not because of the necessity of rectifying survey lines.

trace 1. A term taken from chemistry, and meaning in meteorology the fall of less than 0.005 inch of rain or less than 0.05 inch of snow: i.e., an amount too small to be measured. 2. The record made by any self-registering instrument.

trace element A chemical element needed in minute quantities by plants and animals.

tracer, isotopic An isotope of an element, a small amount of which may be incorporated into a sample of material (the carrier) in order to follow (trace) the course of that element through a chemical, physical, or biological process.

tracer technique The use of small amounts of radioactive isotopes to follow normal elements. The tracer is readily detected and measured by its radioactivity.

trade inversion A sharp increase in temperature, or at least a rapid decrease in moisture, through a layer of air in a subtropical high-pressure cell, marking the boundary zone between the dry subsiding air aloft and the moist tropical air below.

trade-wind desert A desert, such as the Sahara, formed by the drying ac-

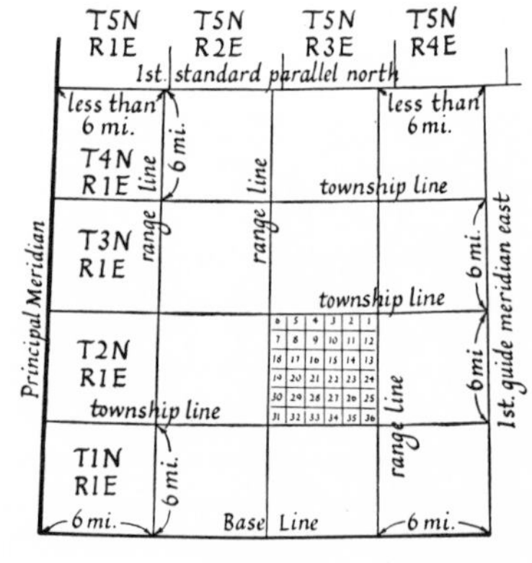

township

tion of the trade winds and characterized by almost negligible cloudiness and extreme range of temperature, both annual and diurnal.

trade winds The planetary winds that blow from the belts of high pressure centered at about 30°N and 30°S latitude toward the equator. Thus, there are two belts of trade winds: the northeast trades in the northern hemisphere and the southeast trades in the southern.

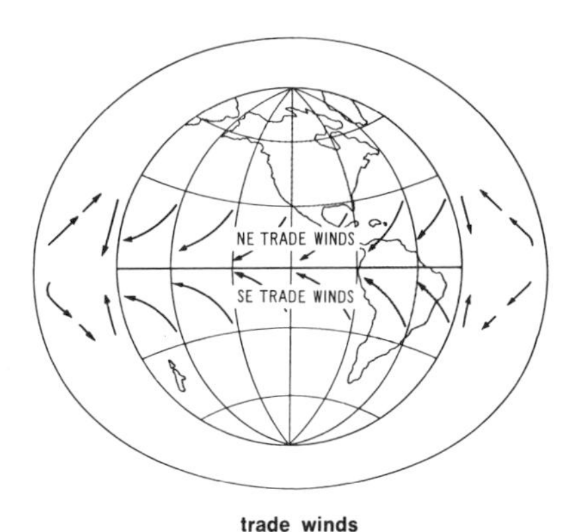

trade winds

traffic pan A subsurface layer in soil that has been so compacted by the application of weight (e.g., machines, tractors, etc.) that the penetration of water and roots is lessened. It is also known as a pressure pan.

trait A characteristic.

tramontana A local name for a northeasterly or northerly wind that in winter is prominent on the west coast of Italy and fairly prevalent off the north of Corsica. It is a fresh wind of the fine-weather mistral type and does not often reach gale force. It is associated with a depression over the Adriatic simultaneous with an anticyclone farther west.

transhumance The practice of moving herds of animals between two regions of different climate in mountainous areas.

translucent (adj.) Partially transmitting light rays but diffusing them so that objects cannot be clearly identified.

transmissibility The flow capacity of an aquifer in gallons per day per foot width, equal to the product of permeability times the saturated thickness of the aquifer.

transmissibility coefficient The number of gallons of water a day that percolates through each square mile of water-bearing bed for each foot thickness of bed.

transparent (adj.) Transmitting energy rays readily. Thus the atmosphere is said to be transparent to solar radiation but relatively opaque to terrestrial radiation.

transpiration The process by which water vapor escapes from a living plant (principally the leaves) and enters the atmosphere.

transpiration ratio The number of pounds of water required for transpiration per pound of dry plant tissue produced.

transport capacity The ability of a stream to transport a suspended load. It is expressed in terms of the total weight of the particles transported.

transverse profile A section across a stream valley, mountain range, etc. See: longitudinal profile

transverse valley A valley cutting across a mountain range.

transverse wave An earthquake wave in which the motion is perpendicular to the direction in which the wave travels. Also called an S wave.

trap Any geologic structure in which a pool of oil collects. See: anticline; fault

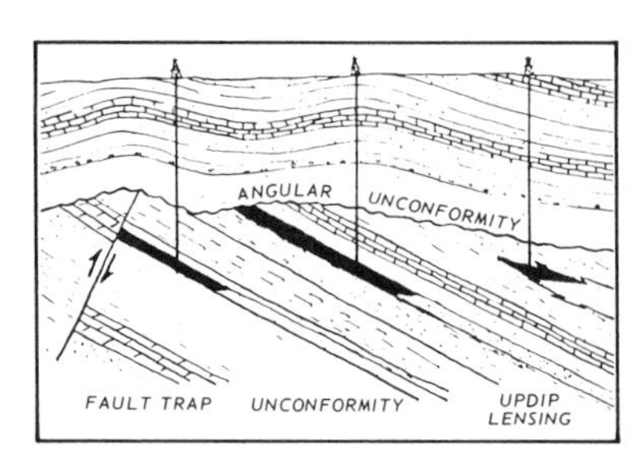

trap

trash fish Undesirable fish, such as carp, which crowd game fish out of lakes and ruin sport fishing.

travertine A calcite rock material deposited where calcium-bearing waters cool or evaporate around springs.

tree farm A plantation in which trees are the principal crop. See: silviculture

tree line See: timber line

trellis drainage A rectangular drainage pattern formed in belted sedimentary rocks.

trench A deep, V-shaped canyon on the floor of the ocean. Most are adjacent to mountain or island chains and are over 30,000 feet deep.

trend A statistical term referring to the direction or rate of increase or decrease in magnitude of the individual members of a time series of data when random fluctuations of individual members are disregarded.

triangular facets Truncated ends of mountain ridges generally associated with faulting.

triangulation Determination of the exact location of a point by measuring the angles to it from the two ends of a base line of known length.

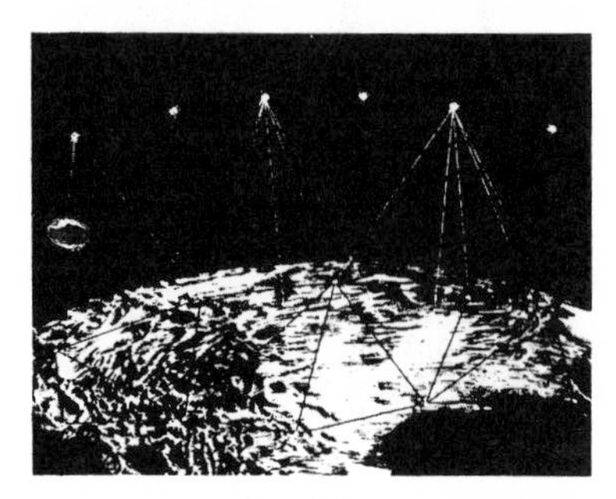

triangulation

Triassic Period See: Appendix I

tributary A stream or other body of water that contributes its water to a larger body of water.

triclinic prism Any prism in which there are three unequal and mutually oblique axes characterized by a total lack of symmetry.

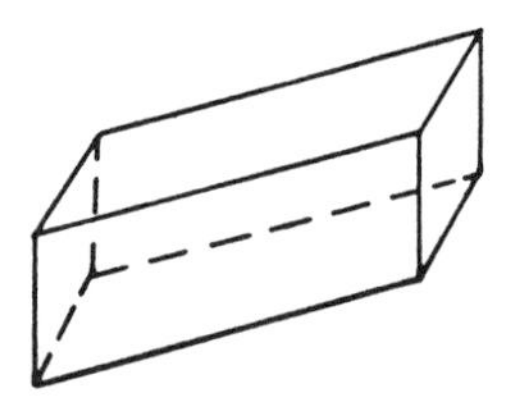

triclinic prism

trigger action The processes necessary to set off convection in an unstable condition of the atmosphere such as surface heating or sudden forced lifting at mountain ranges or frontal surfaces.

trilobite A marine invertebrate which was the dominant form of life on earth 500 million years ago.

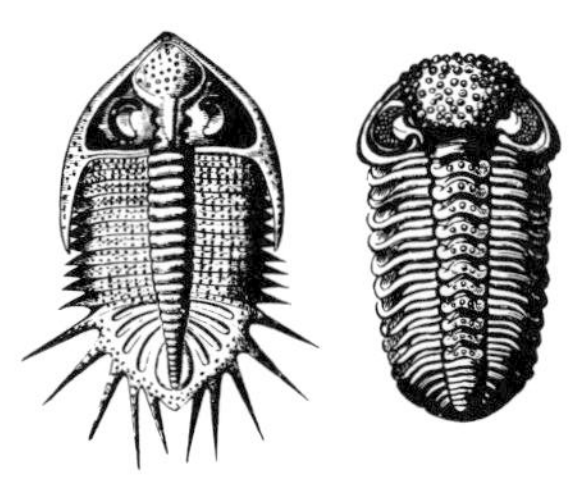

trilobite

trophic (adj.) Pertaining to the use of food by animals and man to produce growth and development.

tropical air Warm air formed in the subtropical anticyclones. There are two types of tropical air masses: tropical continental (cT) and tropical maritime (mT). See: air mass; source region

tropical continental air (cT) Any air mass originating over a land area in low latitudes, characterized by extreme dryness and warmth, and convective instability.

tropical cyclone A cyclonic storm of great intensity that originates in the tropics over the oceans. It first moves westward and then recurves to the northeast (toward the southeast in the southern hemisphere). The central pressure is usually below 28 inches (949 millibars), but in one instance a minimum pressure of 26.19 inches (886.8 millibars) was observed. These low pressures result in very high winds, which in the northern hemisphere blow counterclockwise and spirally toward the center. These storms vary from 25 to 600 miles in diameter. At the outer edge of the storm the wind is moderate, but it increases toward the center, where a velocity as high as 150 miles per hour (67.1 m.p.s.) has been recorded. At the center, there is an area averaging about fourteen miles in diameter called the eye of the storm, where the winds are very light, the seas are confused and mountainous, the sky is often clear, and drizzle may occur.

tropical grassland See: savanna

tropical maritime air (mT) An air mass that originates over an ocean area in the tropics, especially over the broad reaches of the Atlantic and the Pacific between 20° and 45°N. The summer type differs somewhat from that of winter, because when mT air invades a warm continent in summer, instability develops rapidly; the air changes from a warm mass to a cold mass. In winter, on the other hand, the stability is increased and deep layers of fog (advection fog) may cover large areas.

tropical rain forest The dense forest of the humid tropics. Also called a selva.

tropical rain forest climate A type of climate prevailing in the zone between 5° and 10° N. and S. latitudes, the two distinguishing characteristics of which are uniformly high temperatures and heavy precipitation distributed throughout the year, so that there is no marked dry season.

tropical scrub forest The deciduous forests on the margins of the selva.

Tropic of Cancer The parallel at 23½ °N. that marks the northernmost limit of the area in which the sun appears directly overhead at noon.

Tropic of Capricorn The parallel at 23½ °S. that marks the southernmost limit of the area in which the sun appears directly overhead at noon.

tsetse fly

Tropics 1. A name commonly given to both the Tropic of Cancer and the Tropic of Capricorn. 2. The area or belt of the surface of earth bounded by the Tropics of Cancer and Capricorn, also called the torrid zone.

tropopause The zone of transition between the troposphere and the stratosphere. Its height is variable: It is highest (about 17 to 18 kilometers) over the equator, and lowest (about 6 to 8 kilometers) over the poles. Its height also changes with the seasons and with the passage of cyclones and anticyclones.

tropophyte A plant acting as a hydrophyte in one season and a xerophyte in another: for example, a tree in the savanna.

troposphere The region of the atmosphere extending from the surface up to the tropopause. It is characterized by convective air movements and a pronounced vertical temperature gradient, in contrast to the convectionless and almost vertically isothermal stratosphere above the tropopause.

trough An elongated area of relatively low pressure, extending from the center of a cyclone.

trough line A line in the trough of a low-pressure area. It is the locus of points in the isobars where the curvature is at a maximum.

truck farming Intensive production of vegetables and fruit for urban markets located some distance away.

trypanosome A flagellate protozoan of the genus Trypanosoma, parasitic in the blood of vertebrates and in insects, which transmit them. Causes sleeping sickness and other diseases in man, horses, and cattle in Africa.

tsetse fly A blood-sucking African fly that transmits sleeping sickness (trypanosomiasis).

tsunami An enormous sea wave produced by an earthquake under the floor of the ocean or along a seacoast. It is sometimes wrongly called a tidal wave.

tufa A sedimentary rock composed of silica or calcium carbonate, precipitated from solution in the water of a spring or lake or from percolating groundwater.

tuff A rock formed by the compression of volcanic fragments generally less than one-sixth of an inch in diameter.

tuna A large saltwater fish belonging to the mackerel family, found in all tropical and temperate waters. In commercial value it is the most important commercial fish food in the United States.

tundra A treeless plain between the shores of the polar seas and the taiga. It has a growth of mosses, lichens, sedges, ceres, and small shrubs such as birches and willows.

tundra climate A cold climate peculiar to the northern parts of Eurasia and North America. The mean temperature of its warmest month is more than 32°F but less than 50°F.

tungsten A metal used as a ferroalloy and in the filaments for light bulbs.

turbidity A condition of the atmosphere caused by wind and vertical currents, and characterized as more or less hazy because of smoke, dust, haze, and clouds, or peculiar optical qualities of the atmosphere brought about by the presence of layers of air with different densities and hence different indices of refraction.

turbidity current A current produced by the higher density of muddy or turbid water, as compared to the surrounding water.

turbulence 1. Irregular motion of a moving fluid, caused by an impediment in the stream, by friction, or by vortex action. 2. In the atmosphere, the irregular local transitory variations in the general airflow, which, when vigorous (as in a thunderstorm), are manifested by bumpiness, updrafts, and downdrafts, and, when less intense, as gustiness. Such irregular air motion is made up of a number of small eddies that travel with the general air current, superimposed on it.

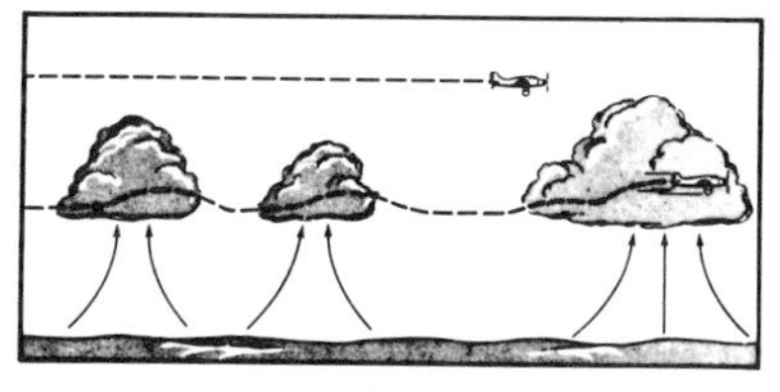

turbulence

turbulence inversion An inversion of temperature in the atmosphere between a turbulent layer and the layer immediately above which is unaffected by turbulence.

turbulent flow A fluid flow in which successive flow particles follow independent path lines.

turbulent mixing The vertical stirring by turbulence of a layer of the atmosphere, producing a turbulence inversion.

TVA See: Tennessee Valley Authority

twilight The intervals of incomplete darkness following sunset and preceding sunrise. The time at which evening twilight ends or morning twilight begins is determined by arbitrary convention. The duration of twilight depends upon latitude and time of year.

twister A tornado.

typhoid An infectious disease caused by bacteria taken into the system in food or water. It is frequently fatal.

typhoon See: tropical cyclone

typhoon squall A form of waterspout similar, in its excessive violence, to a real typhoon but seldom more than a few hundred yards in diameter and a few minutes in duration.

typhus An acute infectious disease transmitted by the bite of lice or fleas and characterized by severe nervous symptoms and a peculiar eruption of reddish spots on the body.

typology The systematic classification or study of types.

tyrannosaurus A genus of dinosaurs living 120 million years ago.

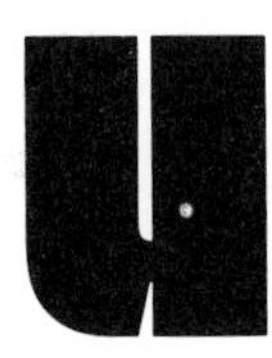

ubac The shady side of an Alpine mountain. Also called the schattenseite.

Ullea's ring A halo surrounding a point in the sky diametrically opposite the sun, and sometimes described as a white rainbow.

unconformity

ultisol A soil of the humid, forested subtropics: red-yellow podzols, red-brown laterites, etc.

ultraphonic See: ultrasonic

ultrasonic (adj.) Pertaining to vibrations and waves whose frequencies are greater than those that affect the human ear (20,000 per second). Also called ultraphonic or supersonic.

ultraviolet (adj.) Pertaining to the radiation wave band just beyond the violet end of the visible spectrum and extending to the X rays.

umbra The dark central portion of the earth's or moon's shadow in an eclipse. Also the dark central portion of a sunspot.

unconformity A surface separating younger strata from older strata and showing signs of erosion or weathering.

underdevelopment Any condition in which full industrialization and the full use of the resource base is not occurring.

underground stream A stream underground that has all the characteristics of a watercourse on the surface—a definite channel with bed and banks, a definite pathway of water, and a definite source or sources of supply.

understory That portion of the trees in a forest that is below the level of the main canopy. Also, the trees forming such a layer.

unearned increment The increase in

property value resulting from such factors as population growth rather than from labor or capital improvements by the owner.

ungulate Having hoofs.

uniformitarianism A concept suggesting that the processes now operative on earth to produce change were operative in the past, only at different rates.

unsaturated fatty acids Fatty acids that contain less than the maximum proportion of hydrogen relative to the amount of carbon present. These hydrogen-deficient fatty acids are very reactive. Their occurrence in fats has a pronounced depressing effect on the melting point and storage life of the fat.

unsettled (adj.) A term used in forecasts to describe weather that may be fair at the time but is liable to develop into a rainy, cloudy, or stormy condition.

updraft An upward-rushing air current caused by convection and usually found in the front portion of a heavy-to-violent thunderstorm.

upland The higher land of a region.

uplift The results of the warping or lifting of a portion of the earth's crust.

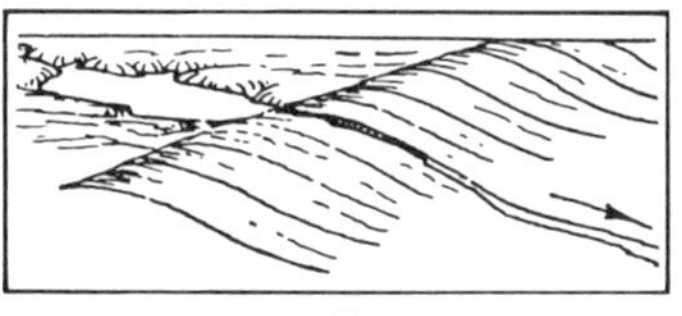

uplift

upper air chart A chart depicting the weather conditions at various levels such as 700, 500, and 300 millibars. Such charts are drawn much the same as charts of surface conditions and exhibit all data obtained from radiosondes. They are used in conjunction with the surface maps in analyzing atmospheric conditions.

upper front A front at some level in the free air instead of at the surface. As it passes aloft the upper front often produces at the surface some of the characteristic phenomena of a frontal passage at the ground, such as cloudiness, pressure changes, and precipitation.

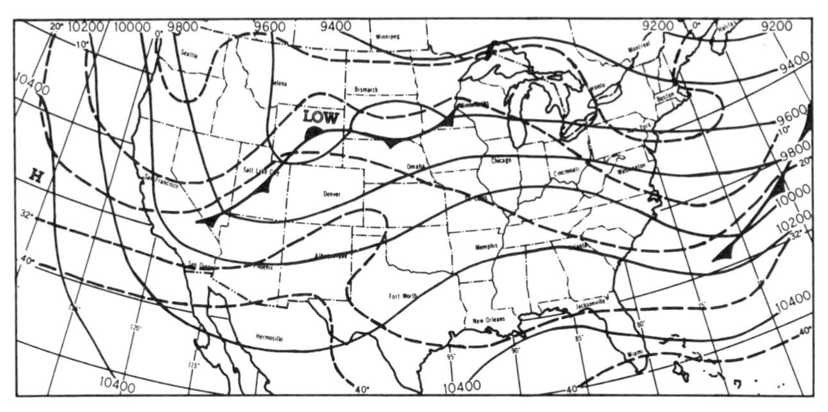

upper air chart

upslope fog Fog produced by the slow movement of air up a gradual slope. The air is cooled by expansion, resulting in fog.

upwelling The movement of cold bottom water to the surface as a consequence of the effects of winds blowing parallel to or away from the coastal zones. It results in abnormally low surface-water temperatures.

uranium A white, lustrous, radioactive element used in photography and in coloring glass. Various isotopes (uranium 235, 238, and 239) are utilized in nuclear reactors.

urban blight Deterioration of portions of cities.

urbanization Processes associated with the growth of urban areas.

urban sprawl The scattering of urban structures over a far greater area than is necessary to satisfy the needed urban functions.

U-shaped valley The valley left as a consequence of mountain glaciation.

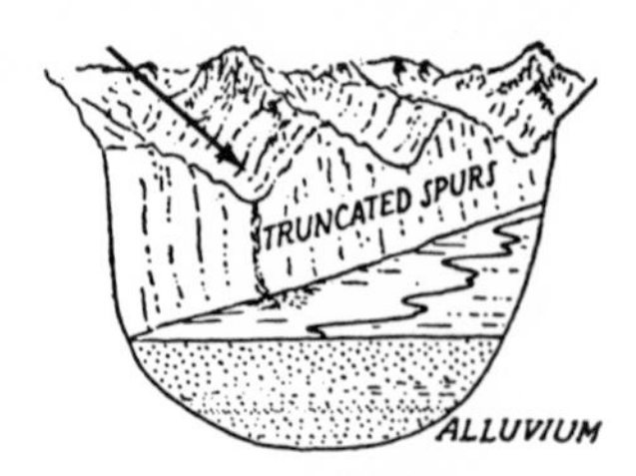

U-shaped valley

uvala A relatively large depression (although generally less than a mile in diameter) formed by the coalescence of several dolines or sinkholes in a karst region.

vadose water Water occupying the pore spaces in the soil above the water table.

valence The combining capacity of atoms or groups of atoms. Sodium (Na^{+}) and potassium (K^{+}) are monovalent, while calcium (Ca^{++}) is divalent.

valley A long narrow depression in the earth's crust. It may be erosional (due to the work of running water) or structural (due to folding or faulting).

valley

valley breeze A gentle wind blowing up a valley or mountain slope in the absence of cyclonic or anticyclonic winds. It is caused by the warming of the mountainside and valley floor by the sun.

valley train A rock deposit carried by a stream flowing from a melting glacier and left as a feature of the landscape when the stream has ceased to flow.

Van Allen Belts A pair of radiation belts that encircle the earth. They consist of a mixture of high-energy protons and electrons permanently trapped in the earth's magnetic field.

vane See: wind vane

vapor Any substance existing in a gaseous state at a temperature lower than that of its critical point. Thus a vapor can be liquified if sufficient pressure is applied to it.

vapor blanket The layer of air that overlays a body of water and has, due to its proximity to the water, a higher content of water vapor than the surrounding atmosphere.

vaporization The process by which water changes from the liquid to the gaseous state.

vapor pressure (or tension) 1. The pressure of the vapor of a liquid kept in confinement so that the vapor can accumulate above it. 2. The partial pressure of the water vapor in the atmosphere. Its symbol is e.

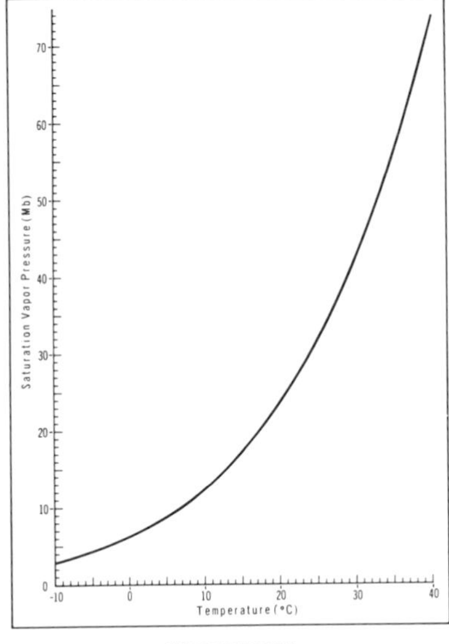

vapor pressure

vapor trail See: condensation trail

variability The scattering of the values of a frequency distribution, often measured by the standard deviation.

variance A measure of the amount of spread or dispersion of a set of values around their mean, obtained by calculating the mean value of the squares of the deviations from the mean, and hence equal to the square of the standard deviation.

variation The manner and degree of change in value of a meteorological or climatological element throughout any given period such as a day or a year.

varve The annual deposit of sediment on the floor of a lake; the spring and summer deposits consist of coarse sediments in contrast to the fall and winter deposits which are finer and darker in color.

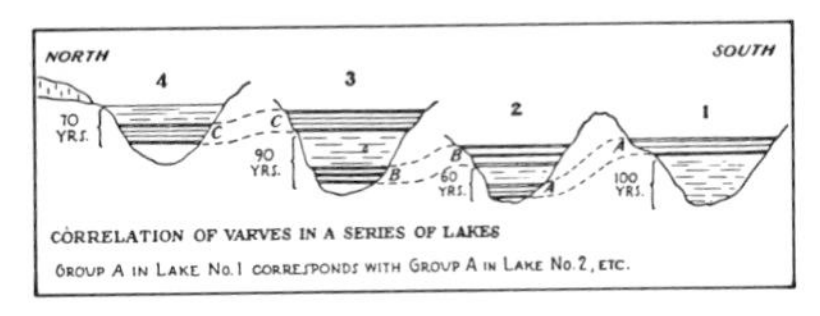

varve

vascular system Any set of vessels that aids the flow of liquids in an organism.

vector A quantity that has magnitude and direction represented by a line.

veer With respect to the wind, to shift in a clockwise direction in the northern hemisphere and in a counter clockwise direction in the southern. See: back

vega In Spain, irrigated land yielding only one crop per year.

vein A more or less regular body of ore.

veld or veldt Elevated open country bearing grasses and shrubs suitable for grazing; part of the great tableland of South Africa.

velocity 1. The vector-time rate of change of position, including both direction of motion and rapidity. 2. Often, though inexactly, speed, which is the magnitude of the time rate of motion without reference to direction.

velo cloud A type of high fog frequent in the United States along the seacoasts, especially in southern California.

vendavales Strong southwesterly winds in the Straits of Gibraltar and on the east coast of Spain.

veneer A thin sheet of wood produced by rotating a log or bolt against a knife in a lathe or by sawing or slicing.

ventifact A stone that has been shaped by windblown sand so that its surface exhibits flat facets with sharp edges.

Venturi tube A tube designed to measure the rate of flow of fluids. It is used in waterflow meters and also in measuring the speed of aircraft.

veranillo A short dry season that breaks the rainy season.

verano A long dry season.

verification The determination, generally by statistical methods, of the degree of accuracy of a forecast.

vernal equinox See: equinox

vernier An instrumental device applied to any graduated scale, linear or circular. It provides a means, when the index (a part of the vernier) is not exactly opposite one of the graduation marks of the scale, for estimating the portion of the division indicated by the index.

vertebrate An organism with a backbone composed of vertebrae. The vertebrates include fish, amphibians, reptiles, birds, and mammals.

vertical zonation The variation of vegetation with elevation. This variation is related to changes in temperature and moisture available for plant growth.

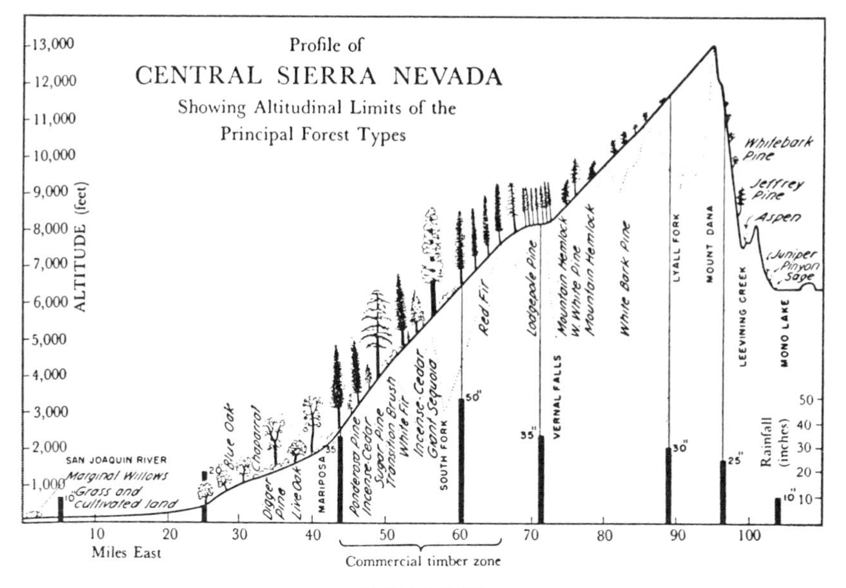

vertical zonation

vertisol Soils formed under the grasslands of the tropics and subtropics. These alluvial soils include a high proportion of swelling clays. They tend to expand, shrink, and crack under varying moisture conditions.

viability The ability to live, grow, and develop.

vicuña A wild animal of the camel family, native to South America and found in the Andes Mountains of Peru, Ecuador, and Bolivia. A very soft woolen cloth is made from its fine fleece.

virazon The sea breeze of the west coasts of Peru and Chile.

virga Wisps or falling trails of precipitation, frequently seen hanging from altocumulus and altostratus clouds.

virgin flow Streamflow unmodified by man.

volcanic bomb

virgin forest A mature or overmature forest growth essentially uninfluenced by human activity. Virgin forests are also referred to as old-growth forests.

virgin soil A soil that has not been significantly disturbed from its natural environment.

virtual temperature The temperature of a mass of dry air with the same density and pressure as an equivalent mass of moist air.

virus An infectious agent reproducing in living cells.

viscosity of fluid The property of stickiness of a liquid or gas, due to its cohesive and adhesive characteristics.

visibility The distance that a given standard object can be seen and identified by the naked eye. As reported by a weather station, it represents the average of conditions prevailing over at least half the horizon.

vitamin One of a group of chemical constituents of foods in their natural state, of which very small quantities are essential for the normal nutrition of animals and plants. They are accessory food factors that must be supplied for the nutrition of an organism and that exert an important action in the control and coordination of physiological functions by virtue of specific molecular structure.

viticulture The raising of grapes.

volatilization The evaporation or changing of a substance from liquid to vapor.

volcanic ash Fine lava particles ejected from a volcano during an eruption.

volcanic bomb A solid mass of rock material originally ejected from a volcano. It may have solidified during its flight through the air.

volcanic cinders See: lapilli

volcanic dust Fine solid particles ejected by a volcano.

volcanic neck The erosional remnant formed of the material that once filled the central vent of a volcano. It is left isolated as the weaker surrounding material erodes.

volcanic neck

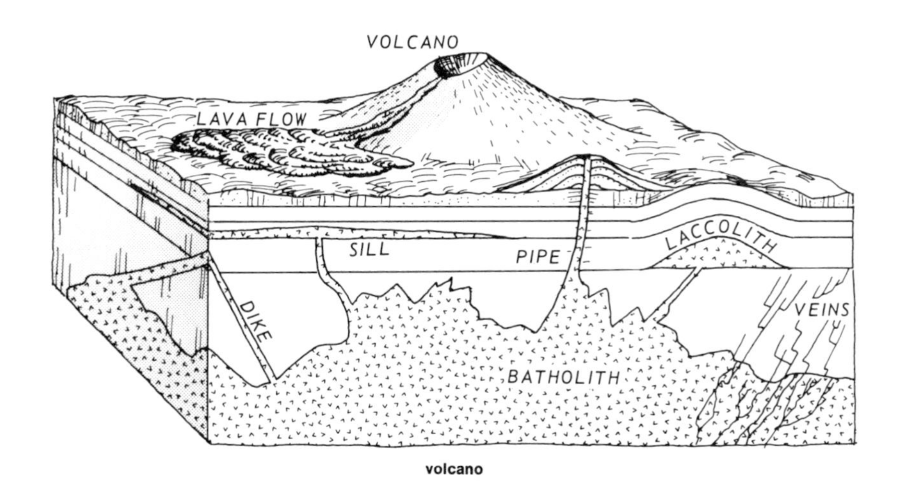

volcano

volcanism Phenomena associated with volcanic activity.

volcano 1. Any opening in the earth's crust from which lava, gases, or other materials are ejected. 2. A mountain built up from such materials.

vorticity In a general sense, rotational circulation of air about a center, the axis of rotation being in any direction whatsoever; usually applied, however, to the circulation in whirling storms, such as cyclones and tornadoes, and in anticyclones. In the northern hemisphere, when the movement is clockwise, the vorticity is considered negative, and when counterclockwise, positive.

V-shaped depression A cyclone in which the isobars are more or less V-shaped, the point of the V usually being towards the south or southwest, and the trough-line being curved with the convex side toward the east.

V-shaped valley A young river valley with steep sides.

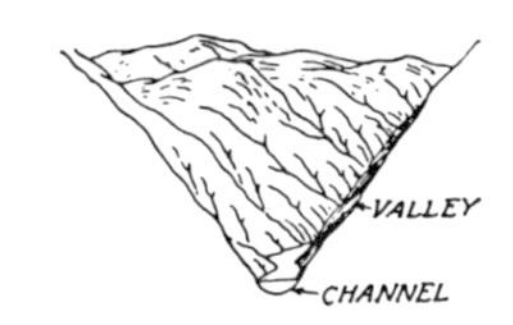

V-shaped valley

vulcanism See: volcanism

wadi In North Africa and parts of the Middle East, the channel of a watercourse that is dry except after a rainstorm. Similar to an arroyo.

wailer A fallen tree lying in the fork of another tree and causing a howl or wail, produced by friction, when the wind produces movement of the two trees.

Wallace's Line The line separating the Oriental and Australian plant realms, suggested by Alfred Wallace, an English naturalist.

walrus A large mammal of the Arctic Ocean valuable for its hides, tusks, and oil.

Wankel engine A European-developed power plant which uses a triangular rotor with fixed port valving. The engine is one-third to one-half the size of the conventional engine of the same horsepower. A small number of automobiles in Europe and the United States use the Wankel engine as a standard power plant.

warm-blooded (adj.) Pertaining to an animal whose body temperature is ordinarily above that of the environment in which it lives.

warm cloud A cloud in which temperatures are above 0°C.

warm fog A warm cloud on or near the ground. The term is often used interchangeably with advection fog.

warm front The line of discontinuity along the earth's surface, or a horizontal plane aloft, where the forward edge of an advancing current of relatively warm air is replacing a retreating colder air mass.

warm-front occlusion An occlusion formed when the air in the rear of a front is somewhat warmer than the air in advance of the front.

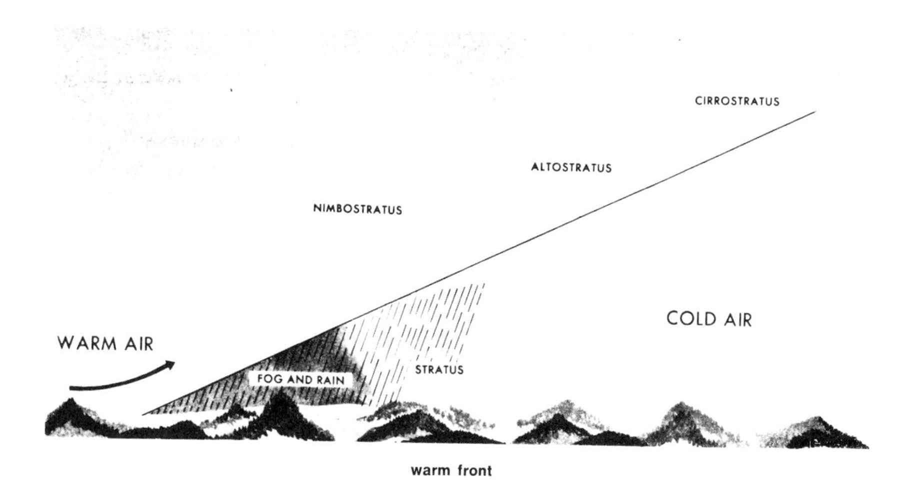

warm front

warm rain Rain from a cloud that is warmer than 0°C.

warm sector The area at the earth's surface bounded by the cold and warm fronts of a cyclone over which relatively warm air is present. The sector gradually narrows and ultimately disappears in the process of occlusion.

Wasatch Range A block-mountain range that marks the eastern margin of the Great Basin of the western United States.

wash load In a stream system, the relatively fine material in near-permanent suspension which is transported entirely through the system without deposition.

waste treatment plant A series of tanks, screens, filters, and other processes by which pollutants are removed from water.

water A chemical compound (H_2O) in which each molecule consists of two hydrogen atoms and one oxygen atom. It is found in nature in the liquid, solid (ice, snow, etc.), and gaseous (water vapor) states.

water analysis The determination of the physical, chemical, and biological characteristics of water. Such analyses usually involve four different kinds of examination: bacterial, chemical, microscopic, and physical.

water balance The complete accounting for the inflow, outflow, storage, and transformation of water within an arbitrarily defined system (e.g., a thunderstorm, a river basin, or a continent).

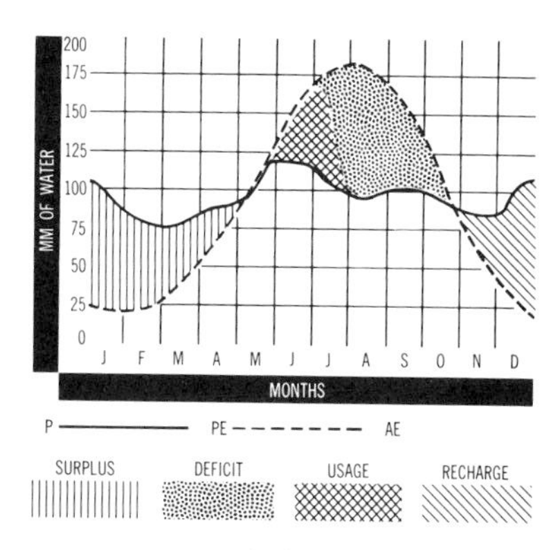

water balance

water budget See: water balance

water cloud A cloud consisting entirely of liquid water.

watercourse See: channel

water cycle See: hydrologic cycle

water equivalent of snow The amount of water that would be obtained from snow that was completely melted. On the other hand, the water content may be merely the amount of liquid water in the snow at the time of observation.

waterfall A steep flow or fall of water over a precipice of some height.

waterfall

water gap A narrow gorge cut by a stream through a ridge.

water hemisphere That half of the globe consisting mainly of water (generally south of the equator).

water hemisphere

waterlogged Pertaining to soil in which both large and small pore spaces are filled with water.

water power Energy obtained from falling water.

water quality The chemical, physical, and biological characteristics of water in respect to its suitability for a particular purpose. The same water may be of good quality for one purpose or use and bad for another.

water requirement of plants Generally the amount of water required by plants for satisfactory growth during the season. More strictly, the number of units of water required by a plant during the growing season in relation to the number of units of dry matter produced. The water requirement varies with climatic conditions, soil moisture, and soil characteristics. Factors unfavorable to plant growth, such as low fertility, disease, and drought, increase the water requirement.

water right A legally protected right to take possession of water in a water supply and to divert that water and put it to beneficial use.

watershed In the United States, the total area above a given point on a stream that contributes water to the flow at that point. Synonyms are drainage basin or catchment basin. In some other countries, the term is used for the topographic boundary separating one drainage basin from another.

waterspout A small whirling storm over an ocean or inland body of water. Its chief characteristic is a funnel-shaped cloud extending in a fully developed spout from the surface of the water to the base of a cumulus cloud.

water spreading Controlled application of water to land for the purpose of recharging ground-water aquifers.

water table The upper limit of the part of the soil or underlying rock material that is wholly saturated with water. In some places an upper, or perched, water table may be separated from a lower one by a dry zone.

water vapor The gaseous form of water; one of the most important constituents of the atmosphere.

water year The twelve-month period from October 1 through September 30. The water year is designated by the calendar year in which it ends. Thus the year ending September 30, 1959, is called the 1959 water year.

water yield The runoff from the drainage basin, including ground-water outflow that appears in the stream plus ground-water outflow that bypasses the gauging station and leaves the basin underground. Water yield is the precipitation minus the evapotranspiration. It is also known as the water crop or runout.

wave 1. Any propagated disturbance in a continuous medium: e.g., light waves, sound waves, ocean waves, and the localized deformation of a front. 2. A spell, as in the expressions cold wave and heat wave.

wave-built terrace A marine platform built seaward from the shoreline by deposition of loose rock material carried by waves.

wave cloud A billow cloud.

wave-cut terrace A marine platform cut into solid rock by wave action.

wave cyclone An entratropical cyclone. It is called a wave cyclone because it first develops as a wave along a front. See: cyclone wave

wave length The least distance between particles situated in the same phase of vibration in wave motion. The wave length is measured in the direction of propagation of the wave, usually from the midpoint of a crest or trough to the midpoint of the next crest or trough.

wave motion Any undulatory motion, whether periodic or aperiodic.

wax One of many natural products, of which beeswax is best known, having some properties similar to those of fats. They contain esters and often free fatty acids, free alcohols, and higher hydrocarbons. They occur as insect secretions and protective coatings on the cuticles of the leaves or fruits of plants, and rarely as constituents of cells.

weather 1. The state of the atmosphere, defined by measurement of the six meteorological elements: air temperature, barometric pressure, wind velocity, humidity, clouds, and precipitation. 2. The state of the sky: clear or cloudy, rainy or fair, etc.

weather control Purposeful and predictable weather modification. There is some question whether this is or ever will be feasible.

weather glass Formerly the popular name for a mercurial barometer.

weathering The breakdown of rock material by exposure to the atmosphere.

weathering

weather map A map showing the weather conditions prevailing over a considerable area, constructed on the basis of weather observations taken at the same time at a number of stations. It is thus also known as a synoptic chart.

weather modification The intentional or inadvertent change in weather conditions by human action. See: weather control

weather type A set of large-scale atmospheric conditions identifiable by certain prominent characteristics, recurring with sufficient frequency to be recognized and to allow future developments to be anticipated from it.

wedge An elongated high-pressure area, or an extension of a high between two lows. Its analog in a low-pressure area is called a trough.

weed Any wild plant with little or no commercial value that grows on cultivated ground and excludes or damages the desired plants.

weighting A statistical method of adjusting the results of observations by taking into consideration the fact that not all the data may be of equal reliability; e.g., in processing data, the results of observations made under the least favorable conditions may be counted only once, while those made in more favorable circumstances may be counted several times.

weir A low dam on a river generally used to raise the level of the water.

well An underground water source made accessible by drilling or digging.

well hydrograph A graphical representation of the fluctuations of the water surface in a well (plotted as ordinates) against time (plotted as abscissas).

western hemisphere The half of the earth's surface lying west of the prime meridian.

western hemisphere

wet adiabat A temperature-height or temperature-pressure curve along which a parcel of saturated air will travel.

wet adiabatic lapse rate The rate at which rising air cools while condensation is occurring, in mid-latitudes about 2½ to 3°F per 1000 feet. Adiabatic cooling by expansion is partially compensated for by the added heat of condensation.

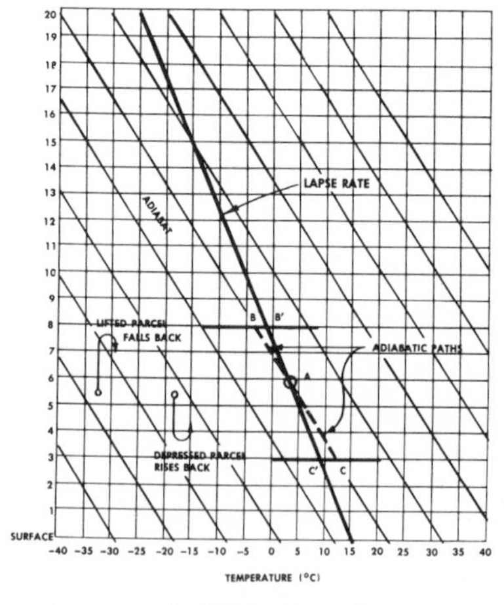

wet adiabatic lapse rate

wet-bulb temperature The lowest temperature to which air can be cooled by evaporating water into it at constant pressure, when the heat required for evaporation is supplied by the cooling of the air.

wet-bulb thermometer One of the two thermometers comprising the sling (or whirling) psychrometer. Its bulb is covered with muslin that is wetted just before an observation is taken; hence its name.

wetland Inland, any area that is more or less regularly wet or flooded, where the water table stands at or above the land surface for at least part of the year. Coastal, land types such as salt marshes, tidal marshes, and brackish marshes subject to saline and/or tidal influences.

wetted perimeter The length of the wetted contact between a stream of flowing water and its containing conduit or channel, measured in a plane at right angles to the direction of flow.

whale The largest existing animal. Whales are an order of mammals, sought chiefly for the fats and oils derived from their carcasses.

wheat A genus of the grass family, and one of the most widely used bread grains.

wheat belt Any agricultural area of considerable size in which wheat is the predominant crop.

wheat crescent A wheat belt in Argentina extending around the dry western margins of the pampas.

whey The serum or fluid part of milk that separates from the curd or solid part when milk is curdled in making cheese. It is thin, watery, and light yellow in color.

whirlies Small violent storms from a few yards to 100 yards or more in diameter. They are frequent in Antarctica near the time of the equinox.

whirlpool A circular eddy in any body of water.

whirlwind Any revolving mass of air, including at one extreme the hurricane and at the other the dust whirl of street corners, but usually applied to small wind eddies of local origin such as dust or sand whirls, often seen during a dry spell anywhere and especially on level deserts.

white oak One of America's most important timber trees, valued for its strong and durable wood and found throughout the eastern United States.

white pine A western pine valuable for its soft, straight-grained wood and used extensively for the interior finishing of houses.

white squall A type of squall encountered in tropical and subtropical regions. It is so-called either because it makes the ocean foam or because it occurs under a bright, cloudless sky.

Wien's law A radiation law that states that the wave length of maximum radiation intensity is inversely proportional to the absolute temperature of the radiating body.

wilderness An uninhabited region largely left in its natural state.

Wilderness Act A law passed in 1964 in the United States establishing a natural wilderness system.

wild lands See: wilderness

wild species Plants and animals that have not been domesticated.

williwaw A violent squall in the Strait of Magellan.

willy-willies Revolving storms that originate over the Timor Sea and move first southwest and then southeast, across the interior of Western Australia. They are similar to hurricanes.

wilting point The point at which the quantity of water in a given soil is insufficient to maintain plant growth. When the quantity of moisture reaches this point, the leaves begin to droop and shrivel up. In any given soil the quantity is nearly constant for all plants, but it increases with a decrease in the size of soil particles.

wind 1. In general, air in natural motion relative to the surface of the earth, in any direction whatever and with any velocity. 2. In meteorology, the component of air motion parallel to the earth's surface, the direction of which is indicated by the weather vane, wind cone, etc., and the speed of which is measured commonly by the anemometer.

windbreak A clump or row of trees planted on level farmlands as protection against winds and erosion.

windbreak

wind divide A range of hills or mountains separating air flowing in different directions.

wind gap A narrow gorge through which air flows.

wind erosion The removal, transportation, and deposition of rock material by wind.

windfall A tree knocked down by the wind, or an area of such trees.

wind pressure The pressure exerted on the exposed surface of an object by moving air. It is usually expressed, for engineering purposes, in pounds per square foot of surface normal to the wind.

wind rose A diagram that indicates, at a given station, the average percentage of winds coming from each of the principal compass points, together with the percentage occurrence of calm air.

wind shear The rate of change of wind velocity (speed and/or direction) with distance. Eddies and gusts form in areas of wind shear, thus producing turbulent flying conditions. Wind shear may occur in either the vertical or horizontal plane.

wind-shift line A line that may be drawn through a low-pressure system, particularly marked in the case of the type of low known as the V-shaped depression, east of which the winds are warm and southerly, and west of which they are northerly and cold. It is identical with the squall line or the cold front.

wind structure The nature, arrangement, and interrelation of the detailed variations in the physical characteristics of a wind stream, including all the mechanical properties of wind, such as its variations in speed and direction from the surface to higher levels and the origins and effects of eddies and turbulence.

wind vane An instrument used to indicate wind direction and known by various other names, weather vane and weather cock among them. The commonest form consists of an arrow mounted at its center of gravity on a vertical axis. The arrow points into the wind because the area (hence resistance) of the barbed end is smaller than that of the feathered end.

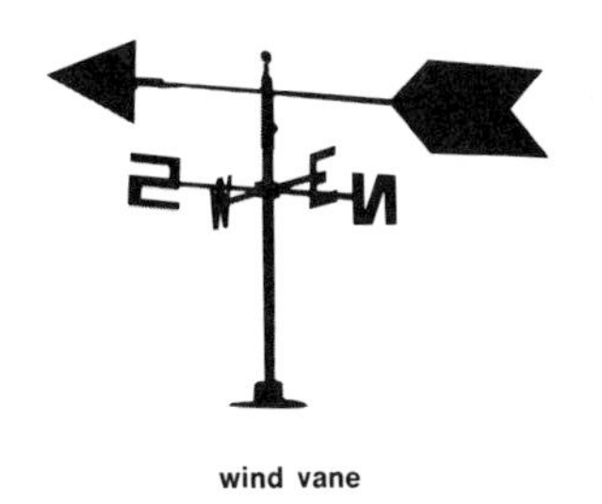

wind vane

windward The side or direction that faces the wind.

wind zones Geographical areas delimited by the prevalence of certain wind systems; for example, the zone of the trade winds.

winnow (v.) To sift or blow away the chaff, or husks, from grain.

winter One of the four seasons recognized in the temperate zones. It is popularly regarded as comprising the months of December, January, and February in the north temperate zone, though the astronomical winter extends from the winter solstice, about December 22, to the vernal equinox, about March 22.

wolframite A heavy, dark brown or black mineral. It is the chief ore of tungsten.

wollastonite A light, white, or gray mineral used to make rock wool, an insulating material.

woolpack A name applied to cirrocumulus and altocumulus clouds because they have a fleecy appearance and resemble flocks of sheep or lambs.

x-axis The horizontal axis in a two-dimensional cartesian coordinate system, or one of the axes in a three-dimensional system.

X chromosome The chromosome that determines the sex of an organism. Each female cell has two X chromosomes; each male has one X and one Y chromosome.

x-coordinate See: abscissa

xerochore A waterless desert.

xerophyte A plant that grows in an arid area. See: hydrophyte; phreatophyte; mesophyte

X rays Electromagnetic waves of extremely short wavelengths. They may be produced by high-speed electrons striking a heavy metal, as in an X-ray tube. X rays are of particular value because they pass through many materials that visible light cannot penetrate.

xylem Plant tissue that conducts water from roots to leaves.

y-axis The vertical axis in a two-dimensional cartesian coordinate system, or one of the axes in a three-dimensional system.

yazoo stream A tributary that is prevented from joining the main stream because the latter has developed high levees. The yazoo stream thus is forced to flow parallel to the main stream for a long distance before joining it.

Y chromosome The partner of an X chromosome in the cells of males. See: X chromosome

y-coordinate See: ordinate

year The period between two successive passages of the sun in its apparent motion around the celestial sphere, through an adopted co-ordinate reference point. Several kinds of years—tropical year, sidereal year, anomalistic year—are therefore distinguished according to the reference point chosen.

year-climate 1. The climate of a region during a particular year, as distinguished from its average or prevailing climate during a long series of years. 2. The name given to what should properly be termed a climatic year.

yeast Microscopic, one-celled fungi that reproduce vegetatively by budding or fission. Best known are those used in bread-making and for converting sugar to alcohol in the brewing and distilling industries.

yellow fever An infectious tropical disease caused by a virus transmitted by the mosquito and characterized by jaundice, hemorrhages, and vomitting.

yellow snow Snow given a golden or yellow appearance by the presence in it of pine or cypress pollen.

yerba maté An evergreen shrub that grows in South America. Its leaves are used to make a tea called maté.

Yosemite Valley That portion of the Merced River Valley that was occupied by glaciers during the Pleistocene.

young river valley A V-shaped river valley produced by a stream with a steep gradient.

yurt A dome-shaped felt or woolen tent used by the Mongols.

young river valley

Z

zebu A large humped animal similar to an ox.

zenith The point in the heavens directly above the observer.

zenithal projection See: azimuthal projection

zephyr 1. In general, a gentle breeze. 2. A warm summer breeze, especially one from the west. 3. (cap.) The west wind personified, or the god of the west wind (Zephyrus).

Zeus The chief Greek god; god of the elements and of moral law and order.

zinjanthropus A man-like individual whose remains were discovered in Tanganyika in 1959. Its age is established at about 1,750,000 years; therefore, it is the oldest skeletal remains of man yet found.

zodiac The belt of twelve constellations through which the sun seems to pass in the course of a year. Each of the constellations has been identified with a sign that corresponds to a particular time of the year.

zoisia Creeping grasses native to Australia and Asia. They are used as lawn grasses in parts of the United States.

Zeus

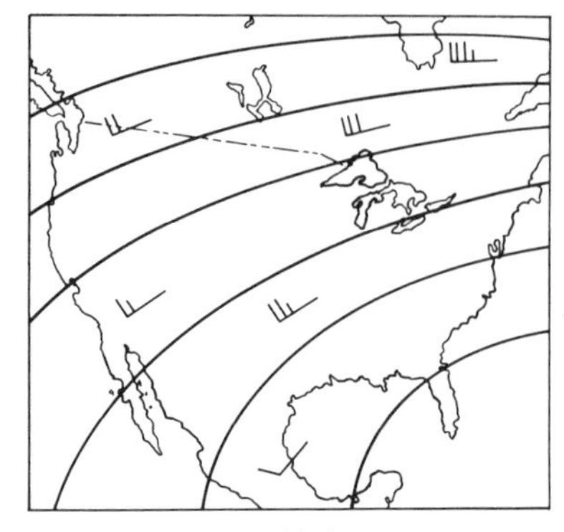

zonal index

zonal index A measure of the intensity of the west-east component of the general atmospheric circulation, usually between 35° and 55°N. latitude, expressed either by the difference in millibars between the mean pressures prevailing along these latitude circles or by the mean geostrophic zonal wind velocity corresponding to this pressure difference. It is taken as positive when the pressure is higher at 35°N. than at 55°N. and the circulation component is directed eastward. When the index amounts to about +15 millibars, indicating a strong eastward flow in middle latitudes, it is called a high index. Conversely, when the pressure difference between the parallels is small, or when the pressure may happen to be lower at 35° than at 55°. this low index shows that the west-east flow is negligible or even reversed. A low index may range from about +5 to –2 millibars.

zonal soils Those soils with mature profiles developed under specific conditions of climate and vegetation.

zonda 1. The sondo: a sultry and enervating north wind in Argentina. 2. A hot, dry foehn-type west wind that occurs in winter in Argentina on the eastern slopes of the Andes.

zone 1. In the old classifications of climate, a latitudinal band encircling the earth, such as the Torrid Zone, which was the entire area between the tropics of Cancer and Capricorn. 2. In present-day usage, any area of the earth's surface or layer of the atmosphere regarded as a meteorological entity. Thus there are rainfall zones, climatic zones, frontal zones, wind zones, zones of convergence, etc.

zone of aeration The zone above the water table. Water in the zone of aeration is under atmospheric pressure and will not flow into a well.

zone of saturation The space below the water table in which all the interstices are filled with water. Water in the zone of saturation is called ground water.

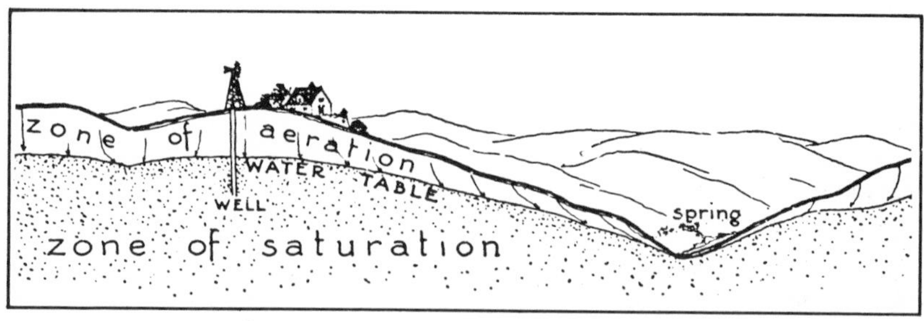

zone of saturation

The water-table, forming the top of the zone of saturation, is highest under the hills and lowest in the valleys where the water seeps out in springs.

zoochore A plant whose structure adapts it for dispersion by animals.

zoogeography The study of the distribution of animals over the earth.

zoonoses Those diseases and infections that are naturally transmitted between vertebrate animals and man, e.g. bubonic plague.

zoophyte An animal that looks like a plant; e.g., coral.

zooplankton The microscopic animal life found floating in the ocean. It is used as food by fishes and other water creatures. It includes larvae and eggs as well as adult animals.

Zuider Zee A portion of an inlet of the North Sea in the Netherlands. It is now reclaimed land.

Zulus A negro people of the moist semitropical region of south-eastern Africa.

Zuni A tribe of American Indians living on the Colorado Plateau.

zygote A fertilized egg or cell that is the result of the fusion of two gametes.

APPENDIX I

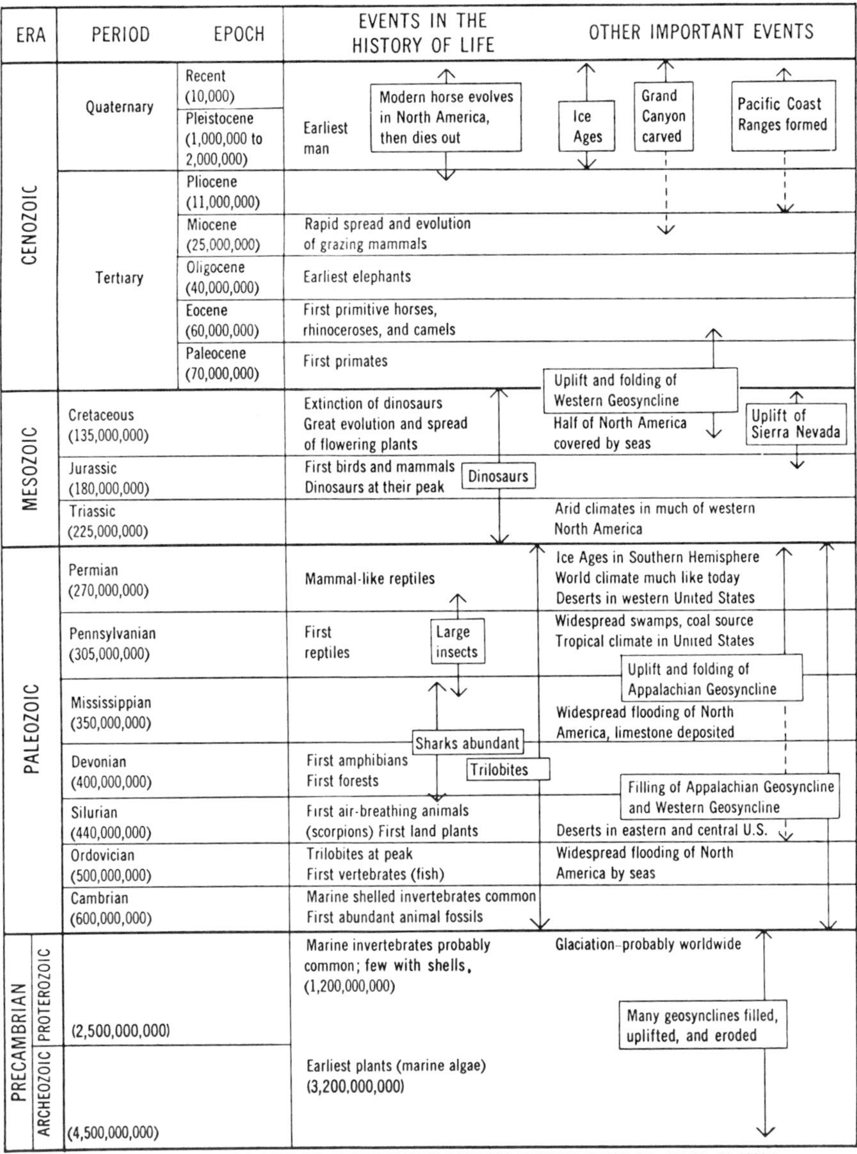

ERA	PERIOD	EPOCH	EVENTS IN THE HISTORY OF LIFE	OTHER IMPORTANT EVENTS
CENOZOIC	Quaternary	Recent (10,000)		
		Pleistocene (1,000,000 to 2,000,000)	Earliest man; Modern horse evolves in North America, then dies out	Ice Ages; Grand Canyon carved; Pacific Coast Ranges formed
	Tertiary	Pliocene (11,000,000)		
		Miocene (25,000,000)	Rapid spread and evolution of grazing mammals	
		Oligocene (40,000,000)	Earliest elephants	
		Eocene (60,000,000)	First primitive horses, rhinoceroses, and camels	
		Paleocene (70,000,000)	First primates	Uplift and folding of Western Geosyncline
MESOZOIC	Cretaceous (135,000,000)		Extinction of dinosaurs; Great evolution and spread of flowering plants	Half of North America covered by seas; Uplift of Sierra Nevada
	Jurassic (180,000,000)		First birds and mammals; Dinosaurs at their peak; Dinosaurs	
	Triassic (225,000,000)			Arid climates in much of western North America
PALEOZOIC	Permian (270,000,000)		Mammal-like reptiles	Ice Ages in Southern Hemisphere; World climate much like today; Deserts in western United States
	Pennsylvanian (305,000,000)		First reptiles; Large insects	Widespread swamps, coal source; Tropical climate in United States; Uplift and folding of Appalachian Geosyncline
	Mississippian (350,000,000)		Sharks abundant	Widespread flooding of North America, limestone deposited
	Devonian (400,000,000)		First amphibians; First forests; Trilobites	Filling of Appalachian Geosyncline and Western Geosyncline
	Silurian (440,000,000)		First air-breathing animals (scorpions) First land plants	Deserts in eastern and central U.S.
	Ordovician (500,000,000)		Trilobites at peak; First vertebrates (fish)	Widespread flooding of North America by seas
	Cambrian (600,000,000)		Marine shelled invertebrates common; First abundant animal fossils	
PRECAMBRIAN — PROTEROZOIC	(2,500,000,000)		Marine invertebrates probably common; few with shells, (1,200,000,000)	Glaciation–probably worldwide; Many geosynclines filled, uplifted, and eroded
PRECAMBRIAN — ARCHEOZOIC	(4,500,000,000)		Earliest plants (marine algae) (3,200,000,000)	

NUMBERS REFER TO TIME IN YEARS B.P. (BEFORE PRESENT) SINCE THE BEGINNING OF THE ERA, PERIOD, OR EPOCH

geologic time scale

APPENDIX II

measures

length

equivalents	*conversions*
1 inch = 25.4 millimeters	inches x 25.4 = millimeters
1 millimeter = 0.039 inch	millimeters x 0.039 = inches
1 inch = 2.54 centimeters	inches x 2.54 = centimeters
1 centimeter = 0.39 inch	centimeters x 0.39 = inches
1 foot = 0.3048 meter	feet x 0.3048 = meters
1 meter = 3.280 feet	meters x 3.280 = feet
1 yard = 0.9144 meter	yards x 0.9144 = meters
1 meter = 1.093 yards	meters x 1.093 = yards
1 statute mile = 1.609 kilometers	miles x 1.609 = kilometers
1 kilometer = 0.621 statute mile	kilometers x 0.621 = miles
1 statute mile = 0.868 nautical mile	statute miles x 0.868 = nautical miles
1 nautical mile = 1.152 statute miles	nautical miles x 1.152 = statute miles

other length equivalents

1 rod = 16½ feet = 5½ yards
1 furlong = 660 feet = 220 yards = 40 rods
1 statute mile = 63,360 inches = 5,280 feet = 1,760 yards = 320 rods = 8 furlongs
1 league = 15,840 feet = 5,280 yards = 3 statute miles
1 nautical mile = 6080.2 feet
1 degree of latitude = 60 nautical miles
1 minute of latitude = 1 nautical mile
1 second of latitude = 88 feet
1 fathom = 6 feet
1 vara = 31 to 33 inches (value varies among Latin American countries)

area

equivalents	*conversions*
1 square inch = 6.45 centimeters	square inches x 6.45 = square centimeters
1 square centimeter = 0.16 square inch	square centimeters x 0.16 = square inches
1 square foot = 0.09 square meter	square feet x 0.09 = square meters
1 square meter = 10.76 square feet	square meters x 10.76 = square feet
1 square yard = 0.836 square meter	square yards x 0.836 = square meters
1 square meter = 1.195 square yards	square meters x 1.195 = square yards
1 hectare = 2.471 acres	hectares x 2.471 = acres
1 acre = 0.4047 hectare	acres x 0.4047 = hectares
1 acre = 43,560 square feet	acres x 43,560 = square feet
1 square foot = 0.000023 acre	square feet x 0.000023 = acres
1 acre = 4,047 square meters	acres x 4,047 = square meters
1 square meter = 0.00025 acre	square meters x 0.00025 = acres
1 acre = 160 square rods	acres x 160 = square rods
1 square rod = 0.00625 acre	square rods x 0.00625 = acres

other equivalents

1 acre = a square with sides of 209 feet
1 centiare = 1 square meter = 1,550 square inches = 172+ square feet
1 are = 100 square meters = 119.6 square yards = 0.247 acres
1 hectare = 10,000 square meters
1 square mile = 640 acres = 2.59 square kilometers
1 township = 36 square miles = 93.24 square kilometers = 9,324.0 hectares

volume

equivalents	*conversions*
1 cubic inch = 16.39 cubic centimeters	cubic inches x 16.39 = cubic centimeters
1 cubic centimeter = 0.06 cubic inch	cubic centimeters x 0.06 = cubic inches
1 cubic foot = 0.028 cubic meter	cubic feet x 0.028 = cubic meters
1 cubic foot = 1,728 cubic inches	cubic feet x 1,728 = cubic inches
1 cubic meter = 35.3 cubic feet	cubic meters x 35.3 = cubic feet
1 cubic yard = 27 cubic feet	cubic yards x 27 = cubic feet
1 cubic yard = 0.765 cubic meter	cubic yards x 0.765 = cubic meters
1 cubic meter = 1.3 cubic yards	cubic meters x 1.3 = cubic yards

other dry measures

1 pint = 33.60 cubic inches = 0.55 liter
2 pints = 1 quart = 67.2 cubic inches = 1.1 liters
8 quarts = 1 peck = 537.6 cubic inches = 8.8 liters
4 pecks = 1 bushel = 2,150.4 cubic inches = 35.2 liters
1 liter = 0.908 quarts
10 liters = 1 decaliter = 0.284 bushels
10 decaliters = 1 hectoliter = 2.84 bushels = 3.53 cubic feet
10 hectoliters = 1 kiloliter = 28.4 bushels = 35.3 cubic feet
1 cord foot = 16 cubic feet = 0.453 cubic meter
1 cord = 8 cord feet = 128 cubic feet = 3.625 cubic meters

liquid measures

1 gill = 4 fluid ounces = 7.219 cubic inches = 0.118 liter
4 gills = 1 pint = 28.875 cubic inches = 0.47 liter
2 pints = 1 quart = 57.75 cubic inches = 0.95 liter
4 quarts = 1 gallon = 231 cubic inches = 3.79 liters = 0.134 cubic feet
1 barrel = 31½ gallons
1 acre foot = 325,851 gallons = 43,560 cubic feet
1 million gallons = 3.07 acre feet
1 liter = 1.057 liquid quarts or 0.908 dry quart
10 liters = 1 decaliter = 2.64 gallons
10 decaliters = 1 hectoliter = 26.418 gallons
10 hectoliters = 1 kiloliter = 264.18 gallons
1 Imperial gallon = 1.2 gallons

weight

avoirdupois weights

equivalents	*conversions*
1 grain = 0.0648 gram	grains x 0.0648 = grams
1 gram = 15.43 grains	grams x 15.43 = grains
1 dram = 27.34 grains	drams x 27.43 = grams
1 grain = 0.037 dram	grains x 0.037 = drams
1 ounce = 16 drams	ounces x 16 = drams
1 dram = 0.063 ounce	drams x 0.063 = ounces
1 pound = 16 ounces	pounds x 16 = ounces
1 ounce = 0.063 pound	ounces x 0.063 = pounds
1 ton = 2,000 pounds	tons x 2,000 = pounds
1 pound = 0.0005 ton	pounds ÷ 2,000 = tons

metric weights

10 milligrams = 1 centigram = 0.1543 grain = 0.000353 ounce
10 centigrams = 1 decigram = 1.5432 grains
10 decigrams = 1 gram = 15,432 grains
10 grams = 1 decagram = 0.3527 ounce
10 decagrams = 1 hectogram = 3.527 ounces
10 hectograms = 1 kilogram = 2.205 pounds
10 kilograms = 1 myriagram = 22.05 pounds
10 myriagrams = 1 quintal = 220.46 pounds
10 quintals = 1 metric ton = 2,204.6 pounds
1 gallon of water = 8.34 pounds
1 cubic foot of fresh water = 62.4 pounds at 4°C.
1 cubic foot of sea water = 64.05 pounds at 4°C.
1 acre foot of water = 2,722,500 pounds

pressure

equivalents	*conversions*
1 atmosphere = 29.92 inches of mercury	atmospheres x 29.92 = inches of mercury
1 inch of mercury = 0.0335 atmosphere	inches of mercury x 0.0335 = atmosphere
1 inch of mercury = 33.86 millibars	inches of mercury x 33.86 = millibars
1 millibar = 0.0295 inches of mercury	millibars x 0.0295 = inches of mercury
1 millibar = 0.75 millimeter of mercury	millibars x 0.75 = millimeter of mercury
1 millimeter of mercury = 1.33 millibars	millimeter of mercury x 1.33 = millibars

flow

equivalents	*conversions*
1 cubic foot per second = 74.8 gallons per second	cubic feet per second x 74.8 = gallons per second
1 gallon per second = 0.013 cubic foot per second	gallons per second x 0.013 = cubic feet per second
1 cubic foot per second = 1.98 acre-feet per day	cubic feet per second x 1.98 = acre-feet per day
1 acre-foot per day = 0.504 cubic foot per second	acre-feet per day x 0.504 = cubic feet per second
1 cubic foot per second = 40 miner's inches*	cubic feet per second x 40 = miner's inches
1 miner's inch = 0.025 cubic foot per second	miner's inches x 0.025 = cubic feet per second
1 cubic foot per second = 0.028 cubic meter per second	cubic feet per second x 0.028 = cubic meters per second
1 cubic meter per second = 35.3 cubic feet per second	cubic meters per second x 35.3 = cubic feet per second
1 cubic foot per minute = 7.48 gallons per minute	cubic feet per minute x 7.48 = gallons per minute
1 gallon per minute = 0.134 cubic foot per minute	gallons per minute x 0.134 = cubic feet per minute
10^6 gallons per day = 3.07 acre-feet per day	10^6 gallons per day x 3.07 = acre-feet per day
1 acre-foot per day = 0.326 10^6 gallon per day	acre-feet per day x 0.326 = 10^6 gallons per day
1 acre-foot per day = 20.17 miner's inches	acre-feet per day x 20.17 = miner's inches
1 miner's inch = 0.0496 acre-foot per day	miner's inches x 0.0496 = acre-feet per day
1 gallon per second = 0.534 miner's inch	gallons per second x 0.534 = miner's inches
1 miner's inch = 1.87 gallons per second	miner's inches x 1.87 = gallons per second

*Value varies among states and countries.